Elements and the Cosmos contains the proceedings of the 31st Herstmonceux Conference of the Royal Greenwich Observatory which was held in honour of Professor B. E. J. Pagel. It echoes his wide interests in the determination and interpretation of the abundances of the chemical elements, and the importance which such studies have for the whole of astrophysics. In these proceedings leading experts review progress in nuclear astrophysics, element synthesis in the early universe and, in particular, the chemical analysis of stars, nebulae and galaxies. There is extensive coverage of the interpretation of relative element abundances in the context of the history of our own and other galaxies.

Elements and the Cosmos

Professor B. E. J. Pagel

Elements and the Cosmos

Proceedings of the 31st Herstmonceux Conference held in Cambridge, England 16–20 July 1990

In honour of
Professor Bernard E. J. Pagel, FRS

Edited by
Mike G. Edmunds
and
Roberto Terlevich

CAMBRIDGE
UNIVERSITY PRESS

Published by the Press Syndicate of the University of Cambridge
The Pitt Building, Trumpington Street, Cambridge CB2 1RP
40 West 20th Street, New York, NY 10011–4211, USA
10 Stamford Road, Oakleigh, Victoria 3166, Australia

First published 1992

Printed in Great Britain at the University Press, Cambridge

A catalogue record for this book is available from the British Library

Library of Congress cataloguing in publication data available

ISBN 0 521 41475 X hardback

CONTENTS

NEBULAE

GALAXIES

COSMOLOGY

Preface

"Elements and the Cosmos' is an appropriate title for a conference in honour of Bernard Pagel. The theme of the determination of cosmic element abundances and their interpretation has never been far away from his work, but no one could suggest this meant a narrow approach. The breadth of his knowledge in the whole of astronomy, indeed of science as a whole, has been a source of strength in his research, and a constant delight and example to his colleagues. It was a great pleasure to honour Bernard at the 31st Herstmonceux Conference, held in July 1990, and for the first time in the Royal Greenwich Observatory's new home in Cambridge.

Bernard took his undergraduate and postgraduate degrees at Cambridge, and was subsequently a research fellow at Sidney Sussex College, and spent time at Michigan. He moved to the RGO soon after it had arrived at Herstmonceux in Sussex, and began to work in earnest on the problems of determining reliable chemical abundances for stars. The names that he and his wife Annabel chose for their successive houses in Sussex - "Red Dwarf", and later "Groombridge" - testify to his stellar interests at that time! Many astronomers have been grateful for the warm hospitality that they have received at the Pagels' home over the years. Bernard became a visiting professor at the University of Sussex, and his interest in abundances had widened to embrace nebulae, the chemical evolution of galaxies and cosmochronology. On reaching the statutory retiring age for RGO, it was inevitable that Bernard - with his insatiable curiosity for astronomy - should seek for pastures new, and he has already established himself as a lively and stimulating professor of astrophysics at the Nordic Institute for Theoretical Physics in Copenhagen. This conference was for his 60th birthday. The topics covered here echo his wide range of interests.

It is inevitable in compiling a conference proceedings that not every talk is followed by a manuscript, but remarkably few escaped this time. Willy Fowler gave a talk on inhomogeneous Big Bang models but it is not included here as it was published in detail in the *Astrophysical Journal* in 1991 (L.H. Kawano, W.A. Fowler, R.W. Kavanagh and R.A. Malaney, *Ap.J.*, **372**, 1). Mike Seaton gave a report on the stellar opacity project - an outline of this important endeavour may be found in *J.Phys.* B, 20, 6363, 1987, and there were talks on quasar absorption lines by Gene Smith and Alec Boksenberg. The only main area of Bernard's work which was neglected at the conference was cosmochronology - it is here that his remarkable and acute critical ability has been particularly felt - and perhaps he has demolished so many myths that there is not a lot left to say!

Bernard has been a most stimulating colleague to work with, and there must be many researchers worldwide who have benefited from his penetrating, but always fair and constructive, criticism. The remarkable accuracy that is now being reached in abundance determinations - and its consequent harvest of new astrophysical results - is in no small way due to Bernard's life work in insisting that things should be done properly. We wish him many more years of active and fruitful research, excellent red wine to drink - to exercise that critical facility - and many years of enjoyable travel to indulge his polyglot linguisic abilities!

Mike Edmunds
Roberto Terlevich

LIST OF PARTICPANTS

Abia, Carlos; Belgium
Alloin, Danielle; France
Aretzaga, Itziar; Spain
Arimoto, Nobuo; UK
Axon, David; UK
Baldwin, J.; Chile
Beer, Hermann; Germany
Beers, Timothy; USA
Blackwell, D.; UK
Bland, Joss; USA
Boffin, H.; Belgium
Boksenberg, Alec; UK
Bravo, Eduard; Spain
Burbidge E M; USA
Canal, Ramon; Spain
Campbell, Alison; USA
Carr, Bernard; UK
Carswell, Bob; UK
Chiara Pardi, M; Italy
Clegg, Robin; UK
Deliyannis, Constatine; USA
Diaz, Angeles; Spain
Drake, J; USA
Edmunds, Michael; UK
Esteban C; Spain
Fall, Michael; USA
Ferland, Gary; USA
Fernandez, Roberto Cid; Brazil
Folgheraiter, Emilio; UK
Fierro, J.; Mexico
Fowler, William; USA
Francois, Patrick; France
Garcia Vargas, Maria Luisa; Spain
Garnett, Don; USA
Gilmore, Gerry; UK
Gonzalez-Delgado, Rosa; Spain
Grevesse, Nicolas; Belgium
Gustafsson, B.; Sweden
Hillebrant, Wolfgang; Germany
van den Hoek, L; Holland
Iben, I.; USA
Irvine, J.; UK
Isern, Jordi; Spain
Kajino,Toshitaka; Japan
Kinman, Tom; USA
Koppen, Joachim; Germany

Lambert, David; USA
Lorente, M M; Spain
Lynden-Bell, Donald; UK
Maciel, Walter; Brazil
Melnick, I; Chile
Moles, Mariano; Spain
Murdin, P.; UK
Nissen, Poul; Denmark
Nomoto, Kenichi; Japan
Nordstrom, Birgitta; Denmark
Novotny, Eva; UK
Oey, Sally; USA
Pagel, Bernard; Denmark
Peimbert, Manuel; Mexico
Penny, Alan; UK
Penston, Michael; UK
Pena, Miriam; Mexico
Perez, Enrique; Spain
Peterson, Ruth; USA
Prantzos, Nikos; France
Prieto, Elmudena; Spain
Rakos, Karl; Austria
Rauch, Micahel; UK
Rebolo, Rafael; Spain
Reeves, Hubert; France
Riley, Simon; UK
Rocca-Volmerange, Brigitte; France
Rubin R; USA
Ryan, Sean; Australia
Sahu, Kailash; Spain
Sanchez, Miguel; Madrid
Schild, Hans; UK
Schmidt, A; UK
Sciama Denis; Italy
Searle, Leonard; USA
Seaton, Mike; UK
Skillman, Evan; USA
Smith, G.; UK
Smith, Linda; UK
Stasinska, G.; France
Terlevich, Elena; UK
Terlevich, Roberto; UK
Thielemann, Friedrick-Karl; USA
Trimble, Virginia; USA
Vilchez J M; Spain
Wayman, P.; Ireland

Cosmological Nucleosynthesis

HUBERT REEVES

Service d'Astrophysique,
C.E.N.S. Saclay,
Institut d'Astrophysique de Paris.

SUMMARY

A review is given of the situation of the Big Bang nucleosynthesis of the nucleides D, ^{3}He, ^{4}He and ^{7}Li, taking into account the latest experimental data (number of neutrino species, lifetime of neutron) and theoretical developments (quark-hadron phase transition).

1. COSMOLOGICAL NUCLEOSYNTHESIS

It is for me a pleasure to paticipate in this symposium in honour of Bernard Pagel. I have always been a faithful reader of his papers. I appreciate in particular his physical intuition. Looking backwards in time, I think that he has been right more often than most of our colleagues......

In this talk I will try to give an up-to-date review of the subject of cosmological nucleosynthesis, a subject in which he has been very active in the past years.

The game of Big Bang Nucleosynthesis (BBN) is to compute the abundances of light nuclides X_i (the "yields") generated around T = 10^9 K and to compare them with the observed abundances. To reach this goal we must face two different tasks: (1) to follow the physics of the Big Bang forward in time until 10^9 K, in order to obtain a set of physical parameters from which the X_i can be computed, (2) to extrapolate the chemical abundances from the observed data, all the way back to 10^9 K. Then, try to match the calculated and extrapolated abundances X_i at the "interface" of BBN , around 10^9 K.

I discuss first the computation of the yields. In recent years we have had important progress, but also some unforseen and unpleasant complications.

Two new important laboratory experiments have resulted in a decrease in the number of "effective" parameters needed for the computations. From the recent LEP results (CERN collaboration 1990) we have obtained the number of neutrino species (3.2±0.2). This number is of importance in relation to the cosmic energy density and hence the rate of expansion of the Universe. This rate, in turn, fixes the value of the decoupling temperature of the weak interaction. The neutron density at the decoupling temperature fixes the calculated abundance of helium-4. In the early computation of BBN, this number was left as a "parameter" to be fitted by comparison with the observed abundances. For years it had been forseen that the number of neutrino species had to be close to the presently mesured value of three (Yang *et al* 1984). It is worthwhile recalling here that in BBN calculations, "fractional" numbers of relativistic species are not necessarily meaningless, since they could correspond to species which would have decoupled at earlier times.

From the Grenoble cold neutron laboratory (Mampe *et al* 1989) we have an improved value of the neutron lifetime. The value: lifetime of 890±4 sec (corresponding to a half-life of 10.3±0.06 minutes) is accurate enough that one does not have to include the influence of its uncertainty on the yields. This fact effectively reduces again the number of parameters of BBN.

The only remaining parameter of BBN is the density of baryonic matter. In the past years, it was generally assumed that this density was spatially homogeneous throughout the Universe at the moment of BBN. In this case (usually called the "standard" BBN), the computed yields are found to match reasonably well the extrapolated observations of light nuclides in the density range from 2 to 4 x 10^{-31} g/cm^3 (or $\eta = 3$ to 6 x 10^{-10} : the ratio of the number of nucleons to the number of photons). (Yang *et al* 1984, Pagel 1989, Beaudet and Reeves 1984).

2. INHOMOGENEITIES

Over recent years we have progressively come to realise that a realistic computation of BBN should take into account the possible effects of the quark-hadron phase transition around 200 MeV (Witten 1984, Satz 1985). In view of this development, it seems that we should not use the word "standard" for the models assuming homogeneity in baryonic density, nor should we use the words "non-standard Big Bang" to describe models taking into account the reality of this phase transition.

One major question here is the order of the quark-hadron phase transition. If it is of second order, it creates no density inhomogeneities and we can simply use the homogeneous density model. If it is of first order, baryonic density inhomogeneities are likely to have been created during the transition, when the Universe was approximately twenty microseconds old (on the conventional cosmic clock). The size of the horizon was then ten kilometers and the mass within the horizon around 10^{26} grams.

This number gives an upper limit to the size of the density inhomogeneities possibly created at this instant.

From this time until about one second, the weak interactions were fast enough to ensure a uniform neutron to proton ratio (given by the Boltzman factor: n/p = exp(M_n-M_p)/kT) throughout the expanding inhomogeneities, despite the fact that the neutrons diffused away from the density condensations while the protons were kept inside by their electromagnetic interactions with the photon gas. Around one second, the weak interactions became too slow to ensure this weak interaction equilibrium any more. The neutrons diffused away faster than the protons and n/p inhomogeneities were created. Around 0.1 MeV (one hundred seconds) primordial nucleosynthesis began, first in the high density regions. The capture of neutrons to form deuterons decreased the neutron density and generated neutron back-diffusion from the low density to the high density regions. The computations have to take into account these phenomena, in a dynamical way, in order to obtain realistic yields. This crucial question of the order of the transition should be answered by high energy collisions of heavy nuclei. Experiments have already started at CERN. The preliminary tests, based on numbers of J/Ψ particles emitted (Satz 1987, Potvin 1989), indicate that the state of the quark-gluon plasma has been reached during the collisions. However, the correct interpretation of the results is a matter of controversy. It is expected that definite conclusions are still a long way in the future.

The question can also be studied through QCD calculations on lattices along a method initiated by Wilson and Polyakoff. The results already published in the past years are not free of difficulties and still involve a number of simplifying assumptions (Ukawa 1989). Year after year, the situation improves, but we are still far from having definite answers. The last "news from the lattice" (Fukujita *et al* 1990) favours a first order phase transition. This result is based, however, on so-called "pure gauge fields", meaning that only the effects of the gluons have been taken into account, but not the effects of the quarks themselves. Pending better QCD results, and forthcoming experimental data, I will assume that the transition is indeed first order. In order to cover all uncertainties, the effects of the phase transition are taken into account by introducing a new set of parameters. As far as the yields are concerned, we have to face spatial variations of the baryonic density: ρ_b becomes $\rho_b(r)$. This space dependence is usually dealt with by the introduction of three new parameters: d, the mean distance between the condensation peaks; R, the density contrast between maxima and minima; and f_v, a measure of the "clumpiness" of the medium.

The value of the mean distance d depends on the surface energy of the "bubbles" of hadrons nucleating in the sea of quarks during the transition. If the surface density (s) is high, overcooling will be extended, only large bubbles will form and there will be few of them: thus the value of d will also be large. If s is low, many small bubbles will form early and d will be low. We expect the value of s to be given by QCD calculations on a lattice (Potvin 1989). In our present ignorance of the value of s, we have to face the possibility that d could extend all the way from zero to the value of the horizon at Q-H phase transition (ten kilometers).

Qualitatively the situation is the following. In the upper part of the range (1 < d < 10km at Q-H transition) the distances between the bubbles are too large for appreciable neutron diffusion before and during BBN. In consequence we have density inhomogeneities but no p/n inhomogeneities. The yields are obtained by summing and averaging the results of constant density yields over the various density regions. At the other extreme, if d is less than 0.1m, both the neutrons and the protons diffuse and the density is homogenized before BBN. The intermediate case (1 km > d > 0.1 m) is the crucial one where the neutrons diffuse effectively, but not the protons.

The contrast R depends, in part, upon the value of the critical temperature T_C which is expected to be between 100 and 250 MeV. In the lower part of the range (T_C < 150 MeV), this computed contrast R tends to be large (several tens). It decreases gradually toward the upper end of the range, as simple phase space argument would predict. The value of T_C should come out from QCD calculations. There are already some indications, from perturbative approaches, that it should lie in the upper part of the range (Glasser and Leutwyler 1987a,b).

Finally, the value of f_v depends in a complicated way on the hydrodynamics of the hadron bubble growth. The problem is that QCD calculations on lattices can only deal with statistical equilibrium situations; they are unable to treat dynamical processes (in these calculational techniques, the parameter "time" is replaced by the "temperature"). The problem is not with the universal expansion (too slow to influence the course of events) but with the rate of growth of bubbles.

Simplified models have been made of these processes, based on the hypothesis of weakly interacting particles (Miller and Pantano 1989, Fuller *et al* 1988). In these models the pressure and energy density are given by the product of the number of particle species (with appropriate multiplicity factors) and the fourth power of the temperature. The transition then corresponds to an abrupt change of this demographic factor (from 37 for the quark-gluon plasma to 3 in the hadronic phase). However, there are reasons, based on QCD calculations, to doubt the validity of the weak interaction hypothesis close to the critical temperature.

3. PARAMETER CHOICE

In view of all these uncertainties in the values of our three parameters, the standard proceedure has been to cover the whole parameter space, computing with appropriate averaging the corresponding yields. These calculations, including fine zoning and neutron back-diffusion, have been made by several groups (Mathews *et al* 1988, Reeves *et al* 1990, Terasawa and Sato 1989, Kuri-Suonio *et al* 1989, Applegate *et al* 1988). The results are in general agreement.

With all these numbers, one is in a position to evaluate the range on mean baryonic density compatible with the observations. More specifically we ask the question: in what fraction of parameter space of d, R, f_v do we find appropriate yields? This question necessarily introduces an element of subjectivity in the decision of the minimum fraction acceptable. However, the expected chaotic hydrodynamic processes accompanying the bubbling and percolation probably imply a rather wide dispersion of these effective parameters around their mean values. Since the yields are often strong functions of these parameters, this dispersion would seem to argue against any scenario based on a very small fraction of the parameter space (i.e. a narrow choice of parameters).

I will discuss the situation with respect to the baryonic density range (Reeves 1990, Reeves *et al* 1990). The lower range (from 1 to 2 x 10^{-31} g/cm³) would require heavy D astration (initial D/H > 3 x 10^{-4}) which is made unlikely by the present upper limit on the value of (D + ^{3}He)/H < 10^{-4}. In the range form 2 to 10 x 10^{-31}, we find acceptable yields in a large parameter space, with a tendancy toward larger R values (R > 10) as we move toward higher densities. In the range from 10 to 20 x 10^{-31}, the acceptable area in the parameter space shrinks rapidly. Values of R $\gg$ 10 and narrow ranges of f_v are required in order to fit the data. Above 20 x 10^{-31} the required values of R reach several thousands. Even at that, the He yields are always larger than 0.25 and the Li/H always larger than 10^{-9}. My subjective feeling is to select a compatible mean baryonic density range between 2 and 10 x 10^{-31}, (3 x 10^{-10} < ρ < 15 x 10^{-10}). Rather similar bounds on the mean baryonic density acceptable have been obtained by Kurki-Suonio *et al* (1990) and Pagel (1989).

As the QCD calculations on lattices proceed, quantitative evaluations of the parameters will improve progressively, and the range of acceptable baryonic density will undoubtedly narrow down.

Some authors (Melaney and Fowler 1989, Mathews *et al* 1989, Applegate, Hogan and Scherrer 1987, Bond and Kajino 1989) have considered the possibility of $\Omega_b =$ 1, corresponding to $\rho_b = 10^{-29}$ for H_o of 75 km/sec/Megaparsec. Besides requiring unrealistically large values of the primordial Li abundance, this scenario corresponds to very narrow choices of the parameters (which also appears somewhat unrealistic). These authors have studied the formation of the light elements Li, Be, B and the r-elements in this scenario. It already appears unlikely that, even with these extreme assumptions, they could produce interesting amounts of these elements. Observational data on old stars will be of prime importance to decide.

References

Applegate,J.H.,Hogan,C. and Sherrer,R.J. 1987, Phys.Rev.,**D35**,1151.

Applegate,J.H.,Hogan,C. and Sherrer,R.J. 1988, Ap.J.,**329**,572.

Beaudet,G. and Reeves,H. 1984, Astron.Astrophys.,**134**, 240.

Boyd,R.N. and Kajino,T. 1989, Ap.J.Lett.,**336**,155.

CERN Collaboration 1989, ALEPH DELPHI, L3 OPAL MARK II, Geneva *preprint*.

Fukujita *et al* 1990, Nucl.Phys.Proc.Suppl.

Fuller,G.M.,Mathews,G.J. and Alcock,C.R. 1988, Phys.Rev.,**D37**,1380.

Gasser,J. and Leutwyler,H. 1987a, Phys.Lett.,**B184**,83.

Gasser,J. and Leutwyler,H. 1987b, Phys.Lett.,**B188**,477.

Kurki-Suonio,H.,Matzner,R.A.,Centrella,J.M.,Rothman,T. and Wilson,J.R. 1988, Phys.Rev.,**D38**,1091.

Kurki-Suino,H.,Matzner,R.A.,Olive,K.A. and Schramm,D.N. 1990, Ap.J. ,**353**,406.

Malaney,R.A. and Fowler,W.A. 1988 *in* Origin and Distribution of the Elements, ed. G.J.Mathews, World Scientific, Singapore.

Malaney,R.A. and Fowler,W.A. 1989, Ap.J.Lett.,**345**,L5.

Mampe,W.,Ageron,P.,Bates,J.,Pendlebury,J. and Steyerl,A. 1989, Phys.Rev.Lett.,**63**,593.

Mathews,G.,Fuller,G.M.,Alcock,C.R. and Kajino,T. 1988, UCRL *preprint* 98943

Mathews,G.,Meyer,B.,Alcock,C.R. and Fuller,G.M. 1989, UCRL *preprint* 102320.

Miller,J.C. and Pantano,O. 1989 SISSA Trieste *preprint*.

Pagel,B.E.J. 1990 *in* Baryonic Dark Matter, eds D.Lynden-Bell, and G.Gilmore, Kluwer, Dordrecht,p237.

Pagel,B.E.J. 1987 *in* A Unified View of the Micro- and Macro- Cosmos, eds A.de Rujula,D.V.Nanopoulos and P.A.Shaver, World Scientific, Singapore,p399.

Potvin,J. 1989, Can.J.Phys.,**67**,1236.

Reeves,H. 1990, Physics Reports,**201**,335.

Reeves,H.,Richer,J.,Sato,K. and Terasawa,N. 1990, Ap.J.,**355**,18.

Satz,H. 1985, Ann.Rev.Nucl.Sci.,**35**,

Satz,H. 1987 *in* Proceedings of the Strasbourg Conference on Quark-Hadron Phase Transition.

Terasawa,N. and Sato,K. 1989, Prog.Theor.Phys.Lett.,**81**,254.

Terasawa,N. and Sato,K. 1989, Phys.Rev.,**D39**,2893.

Ukawa,A. 1988, "Status of lattice QCD at finite temperature", CERN-TH 5266.

Witten,E. 1984, Phys.Rev.,**D30**,272.

Yang,J.,Turner,M.S.,Steigman,D.,Schramm,D.N. and Olive,K. 1984, Ap.J.,**281**,493.

DISCUSSION

D.W.Sciama: 1) LEP now gives $N_\nu = 3.01\pm0.01$. 2) Since particles like photinos would not couple to Z, LEP does not count them. It is therefore necessary to determine N_ν directly from considerations of nucleosynthesis.

H.Reeves: You are right. The Z decay does not probe exactly the same thing as BBN. However, the particles probed by BBN are weighted according to the strength of their coupling constants. A right-handed neutrino, if it exists, would barely count for BBN as it would have decoupled at high temperature, probably before pion and muon annihilation. In this sense, the LEP number remains the "best buy" for BBN.

J.Primack: K.Freeze and others have suggested that the QCD transition could result in a "tangled seaweed" picture of the high density regions. Do other calculations agree? Have the BBN calculations taken this possibility into account?

H.Reeves: The present calculations of the hydrodynamics of bubble formation and evolution are based upon simplified assumptions which are highly questionable. The assumption of states of weakly interacting particles close to the critical temperature does not appear to be valid, according to the QCD calculations on a lattice. The possibility of a "tangled seaweed" picture cannot be excluded at the present time. It would correspond to a wide distribution of the effective parameter d, for instance. As a result, it would probably discourage any narrow choice of these parameters to fit the BBN abundances.

T.Kajino: QCD physics infers very large baryon-number density fluctuations at the end of QCD phase transition, but primordial nucleosynthesis favours small fluctuations. We must understand what happened after the QCD phase transition and before the nucleosynthesis epoch.

H.Reeves: Certainly. This will be a very difficult problem, as the present QCD calculations on a lattice do not handle dynamic processes (they are equilibrium calculations).

B.Pagel: I believe we know $Y_p < 0.24$, with 0.23 as the preferred value. If you want to be more conservative, you have to say 0.21 to 0.25, not 0.23 to 0.25. Although there are uncertainties in galactic chemical evolution of ^{3}He, they all go in the direction of strengthening Schramm's upper limit on primordial $(D + {}^3He)/H$.

BARYON INHOMOGENEITIES IN THE EARLY UNIVERSE AND PRIMORDIAL NUCLEOSYNTHESIS OF Li-Be-B

T. KAJINO

Department of Physics, Tokyo Metropolitan University
Setagaya, Tokyo 158, Japan

1 INTRODUCTION

Although the standard big-bang models[1] have enjoyed a success for primordial light element abundances, they assume homogeneous baryon distribution and infer rather dilute universe in baryons, $\Omega_B \approx 0.04$, at the risk of missing mass problem. In order to explain the observed total Ω value, which has recently proved to be $\Omega = 0.1 \sim 0.4$, the standard models must assume an existence of unknown non-baryonic dark matter. This discrepancy becomes even more serious in modern cosmological theories of inflation which requires marginally closed universe, $\Omega + \Lambda = 1$. Although inflation is now being tested by both theoretical and observational studies of the Ω- and Λ-problems, inhomogeneous nucleosynthesis models[2-10] have shed light on this, suggesting more relaxed Ω_B-value which is not very far from unity. The $\Omega_B = 1$ inhomogeneous cosmologies were unlikely in some nucleosynthesis calculations. However, the late-time homogenization effect[11] has revived a possibility of $\Omega_B = 1$ universe. In the inhomogeneous cosmologies, not only the lightest elements $A \leq 7$ but also the rare Li-Be-B species[7,8] and intermediate-to-heavy mass nuclei[2,4-6] could be produced by way of radioactive nuclei.

All these interesting consequences, however, are subject to unknown fluctuation shapes of inhomogeneous baryon distribution which were attained at the end of cosmic QCD phase transition. The first purpose of this paper is to discuss time evolution of baryon number densities during the cosmic QCD phase transition in order to obtain the possible fluctuation shapes.[12,13] The second purpose is to extend the primordial nucleosynthesis calculations to the production of Be and B isotopes[7,8] and intermediate-mass nuclei[5,6] in order to look for several observational tests for or against the cosmological theories of baryon inhomogeneous cosmologies.

2 QCD PHASE TRANSITION IN THE EARLY UNIVERSE

Quantum chromodynamics (QCD) is presumed to be a perfect theory of strong interaction. There are three fundamental parameters in QCD. They are the coexistence temperature of quark-gluon plasma phase and hadron phase, T_C, surface tension of the phase boundaries, σ, and baryon penetration rate through the phase boundaries, λ.

2.1 Time Evolution of Baryon Number Densites

The evolution in time of baryon number densities at the coexistence epoch $T = T_C$ is described by

$$\frac{dn_B{}^q}{dt} = -\lambda n_B{}^q + \lambda' n_B{}^h - \left\{\frac{\dot{V}}{V} + \frac{\dot{f_V}}{f_V}\right\} n_B{}^q, \tag{1a}$$

$$\frac{dn_B{}^h}{dt} = \frac{f_V}{1-f_V}\left\{-\lambda' n_B{}^h + \lambda n_B{}^q + \frac{\dot{f_V}}{f_V} n_B{}^h\right\} - \frac{\dot{V}}{V} n_B{}^h, \tag{1b}$$

where $n_B{}^q$ and $n_B{}^h$ are the net baryon number densities in the quark-gluon plasma phase and hadron phase, V is the horizon volume and f_V is the volume fraction of quark-gluon plasma. In the first and second terms of eq. (1a), λ and λ' are the baryon number transfer rates through the phase boundaries. Although these values have been taken to be free parameters in the previous studies, they are now calculated in chromoelectric flux tube model[12] which is a phenomenological model of QCD. The calculated rates are three or four orders of magnitude smaller than those assumed in the previous calculations. In the third and fourth red-shift factors, $V(t)$ and $f_V(t)$ are the known functions of time as solutions of Einstein equation. Equation (1b) also has similar source terms. Because of low penetrabilities of baryons, the solutions of eqs. (1a) and (1b) show that the baryon number density in the quark-gluon plasma phase increases progressively as the time goes.[13]

Primordial nucleosynthesis is very different from that in the standard homogeneous cosmologies if the baryon number-density distribution is largely inhomogeneous.[2-4] Distinctive inhomogeneous nucleosynthesis, however, arises only from specific distributions of protons and neutrons at the time of nucleosynthesis. First, in order to get the uniform neutron distribution at the time of nucleosynthesis, the separation length l_B between the fluctuations must be comparable to the neutron diffusion length, $10\text{ m} < l_n < 100\text{ m}$. If this condition is satisfied, then neutrons in the high baryon density zones can diffuse out into the low density zones.[9] The second constraint from the light element abundances is $Rf_V > 20\Omega_B$, where $R = n_B{}^q/n_B{}^h$ and f_V are the values at the end of the phase transition. We adopt the condition $Rf_V > 20\Omega_B$ as the criterion for "interesting" inhomogeneous nucleosynthesis based on numerical simulations.[3,6]

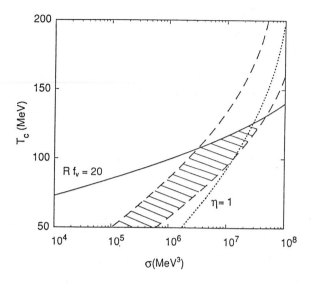

Figure 1 displays two criteria on the parameter plane T_C and σ. The solid curve corresponds to the marginal value of $Rf_V = 20$ in the $\Omega_B = 1$ universe. Below this curve Rf_V is larger than 20 and the fluctuation is large enough to lead to inhomogeneous nucleosynthesis. The other two dashed curves display the separation length, l_B. The upper and lower curves correspond to $l_B = 10$ m and 100 m, respectively. The baryon number density fluctuations for the parameter region between the two dashed curves and below the solid curve (shaded region) lead to typical inhomogeneous nucleosynthesis, whereas in the other region baryon number density fluctuations may make an ordinary standard big-bang nucleosynthesis.[13]

Fig. 1. The T_C - σ plane. $Rf_V > 20\Omega_B$ with $\Omega_B = 1$ below the indicated curve. Severe supercooling occurs below the $\eta = (T_C - T)/T_C = 1$ curve. $l_B = 10$ m (upper dashed curve), $l_B = 100$ m (lower dashed curve).

fluctuations may make an ordinary standard big-bang nucleosynthesis.[13]

Recent lattice QCD simulations including the effects of dynamical quarks suggest 100 MeV $\leq T_C \leq 150$ MeV in very weak first order phase transition, though not definitely confirmed. σ shoud be smaller than 10^7 MeV^3. These constraints make the physics of the early universe very interesting as displayed in Fig. 1.

2.2 Strange Matter Formation

The formation of strange quark matter in the QCD phase transition of the early universe was first suggested by Witten[14] as a possible candidate for baryonic dark matter. However, Witten's equilibrium condition alone cannot produce strange matter. We have recently found[13] by solving eqs. (1a, b) that it could be really formed in strongly non-equilibrium condition if the two QCD parameters have values $\sigma < 5 \times 10^5$ MeV^3 and $T_C = 100 - 150$ MeV. It was also found[15] that they can survive until the present time, satisfying causal relation.

3 PRIMORDIAL NUCLEOSYNTHESIS

3.1 Primordial Li-Be-B and Galactic Chemical Evolution

Inhomogeneous models[2,3] predict larger ^{7}Li abundance by at least an order than the standard model prediction. However, the observed primordial component[16,17] of ^{7}Li is quite controversial for uncertain chemical evolution of this nucleus which is easily destroyed by astration at stellar temperatures as low as $T \approx 10^6$ K. It has recently been found[7] that the new reaction path

$$^7\text{Li}(^3\text{H,n})^9\text{Be}(\text{n},\gamma)^{10}\text{Be}(\text{e}^-\,\nu)^{10}\text{B}, \tag{2}$$

in neutron-rich zones provides extremely enhanced ^{9}Be and ^{10}B abundances. It is known[4,9,10] that the effect of neutron back-diffusion is not very efficient to destroy overproduced ^{7}Li. It is therefore premature to conclude that the neutron back-diffusion alone makes ^{9}Be abundance small.

Since ^{9}Be, ^{10}B and ^{11}B are usually thought of as having been produced by spallation in cosmic rays, it is important to consider the evolution of these abundances after their formation in big-bang. The time evolution of any element after the big-bang nucleosynthesis is given approximately by[8]

$$N_A(t_{\text{NOW}}; t) = \exp\left\{ - \int_t^{t_{\text{NOW}}} \mu(t')\,dt' \right\}$$

$$\times \left[\int_0^t \sum_i \sum_j \lambda_{i+j \to A}(t') N_i(t') N_j(t')\,dt' + N_A(0) \right], \tag{3}$$

where λ and μ are, respectively, the production and destruction rates of $A = ^9$Be, ^{10}B and ^{11}B by spallation, astration and others. t is the time in which the star in question was formed from interstellar materials, and $N_A(0)$ is the initial abundance which is taken to be the theoretical primordial abundance. Since the iron abundance is known to increase with time as

$$t = 10^{10+[\text{Fe/H}]} \qquad \text{yr}, \tag{4}$$

where metallicity [Fe/H] is defined as the logarithm of the abundance ratio of Fe to H in the star divided by that in the sun, comparisons of abundance with [Fe/H] and time are equivalent.

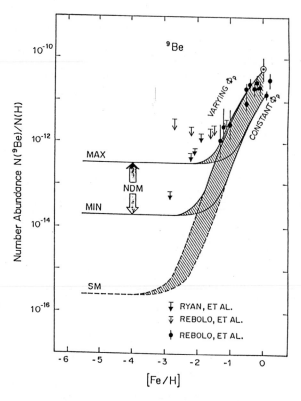

Fig. 2. Predicted relationship between the number abundance ratio $N(^9Be)/N(H)$ and [Fe/H] in the standard model (SM) and the inhomogeneous model (NDM; non-uniform density model). Data points are from refs. 18 and 19.

Figure 2 shows a comparison of the ^9Be abundance with [Fe/H] between theory[8] and observation.[18,19] Theoretical curve consists of two components that come from a primordial origin at [Fe/H] < -3 and from a production in cosmic rays at -3 < [Fe/H], with possibly minor modification by stellar processings. Although the theoretical uncertainties in inhomogeneous density models make the acceptable range rather large, the predicted differences of the ^9Be abundance between the standard model and the inhomogeneous density models become striking as one gets to very metal poor stars. Observed data of Ryan et al.[18] are approaching the sensitivity required to test the two models and one of them is in our predicted range in inhomogeneous models. That constraint has

recently been relaxed[20] by factor of three or so by reanalysing the datum, making almost all inhomogeneous values of the ^9Be abundance acceptable. Most of our theoretical error bars come from unknown QCD parameters. Actual detection of the ^9Be abundance is highly desirable not only for testing the cosmological theories but for constraining the baryon inhomogeneities made from cosmic QCD phase transition.

The only data[21] which exist for B limits $N(^{10}B+^{11}B)/N(H) < 10^{-11}$ in Pop II stars, a value which is much too high to constrain cosmological theories. Note, however, the importance of discriminating between ^{10}B and ^{11}B in metal poor stars; ^{11}B is predicted to be made with

10 to 50 times the abundance of ^{10}B in the inhomogeneous models, in marked contrast to the observed solar system value of ^{11}B/^{10}B = 4.1 and the spallation produced value both in the present and in previous models of 2.0.

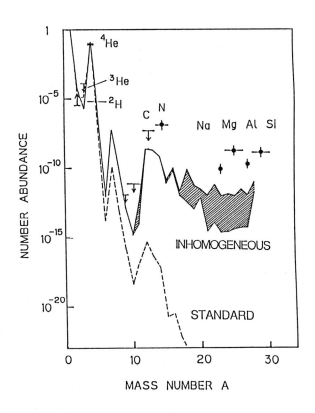

Fig. 3. The calculated primordial abundances of light-to-intermediate mass-nuclei as a function of mass number. Data points are from ref. 22.

3.2 Intermediate-Mass Nuclei

Although the standard homogeneous big-bang models cannot produce any heavy elements, the inhomogeneous models[2,4-6] can do so by way of radioactive nuclear reactions. We demonstrated this in Fig. 3. We adopted here the same inhomogeneous models as in ref. 3, assuming $\Omega_B = 1$. The abundances of intermediate mass nuclei are enhanced very much, and they are approaching the observed values in Population II giants,[22] which should be taken to be an upper limit of primordial abundances because giants have been processed. The calculated C/N abundance ratio, almost constant floor of CNO, and isotopic ratios of C, N, O, Ne, and Mg, etc., are very different from solar system values, providing observational tests for or against inhomogeneous cosmologies.[6]

4   CONCLUSION

Dynamical process of generating baryon inhomogeneities at the epoch of cosmic quark-hadron phase transition is studied in effective model of QCD. It is found that the appearance of baryon number-density fluctuations or the formation of strange quark matter depends strongly on the two QCD parameters T_C and σ. Primordial abundances of rare Li-Be-B species and intermediate-mass nuclei are very different in inhomogeneous nucleosynthesis models from that in the standard model prediction, providing several observational tests for or against the inhomogeneous cosmologies.

This work has been performed under the auspices of Ministry of Education, Science and Culture, contract number 01540253 and also partially supported by JSPS-NSF grant EPAR071-8815999 of the Japan-U.S. Joint Research Project.

REFERENCES

1.   R. V. Wagoner, W. A. Fowler, and F. Hoyle, Ap. J. **148** (1967) 3.
2.   J. H. Applegate and C. J. Hogan, Phys. Rev. **D31**, 3037 (1985).
3.   C. R. Alcock, G. M. Fuller and G. J. Mathews, Ap.J. **320** (1987) 439.
4.   R. A. Malaney, and W. A. Fowler, Astrophys. J. **333** (1988) 14.
5.   T. Kajino, G. J. Mathews, and G. M. Fuller, in Heavy-Ion Physics and Nuclear Astrophysical Problems, S. Kubono ed. (World Scientific, 1988), p.51.
6.   T. Kajino, G. J. Mathews, and G. M. Fuller, Ap. J. **364** (1990), 7.
7.   R. N. Boyd and T. Kajino, Astrophys. J. **336** (1989) L55.
8.   T. Kajino and R. N. Boyd, Ap. J. **359** (1990) 267.
9.   G. J. Mathews, C. R. Alcock, G. M. Fuller and T. Kajino, in Dark Matter, Proc. 8-th Recontre de Moriond, ed. J. Audouze and J. Tran Thanh Van (Gil-sur-Yvette: Editions Frontieres, 1988), p.303.
10.   G. J. Mathews, B. Meyer, C. R. Alcock, and G. M. Fuller, Ap. J. **358** (1990) 36.
11.   C. R. Alcock, D. S. Dearborn, G. M. Fuller, G. J. Mathews, and B. S. Meyer, Phys. Rev. Lett. **64** (1990) 2607.
12.   K. Sumiyoshi, K. Kusaka, T. Kamio and T. Kajino, Phys. Lett. **225B** (1989) 10.
13.   K. Sumiyoshi, T. Kajino, C. Alcock and G. Mathews, Phys. Rev. **D42** (1990), in press.
14.   E. Witten, Phys. Rev. **D30**, 272 (1984).
15.   K. Sumiyoshi, T. Kajino, C. Alcock and G. Mathews, (1990), in preparation.
16.   F. Spite and M. Spite, Astr. Ap. **115** (1982) 357; Nature **297** (1982) 483.
17.   L. M. Hobbs and C. Pilachowski, Ap. J. **334** (1988) 734.
18.   S. G. Ryan, M. S. Bessell, R. S. Sutherland and J. E. Norris, Ap. J. **348** (1990) 57.
19.   R. Rebolo, P. Molaro, C. Abia and J. E. Beckman, Astr. Ap. **193** (1988) 193.
20.   C. P. Deliyannis, (1990), talk presented in this Conference.
21.   P. Molaro, Astr. Ap. **183** (1987) 241.
22.   M. S. Bessell and J. Norris, Ap. J. **285** (1984) 622.

Primordial Nucleosynthesis Revisited

S.P. RILEY AND J.M. IRVINE

Department of Physics,
The University,
Manchester,
M13 9PL, U.K.

1. INTRODUCTION

In a discussion of the early universe, the standard Big Bang cosmology confronts the standard model of particle physics and a substantial body of low-energy nuclear physics data. The initial euphoria which greeted the apparent complementarity of these two fields has given way to a more critical analysis of their compatibility in the face of improved data from both astrophysics and nuclear physics laboratories.

According to the standard Big Bang cosmology, once the universe has evolved for 10^{-2} seconds, thermal equilibrium has been established, hadronisation has taken place, and all unstable particles, with the exception of the neutron, have decayed. The subsequent evolution can be modelled using laboratory assessible, low energy nuclear physics. The principal uncertainty in following the thermal evolution of the universe during this period is the number of neutrino generations and their masses. In the standard model of particle physics, it is assumed that there are three generations of massless neutrinos.

In a period of a few minutes, as the universe cools from 10 MeV to 10 KeV, primordial nucleosynthesis takes place. The light isotopic abundances produced are determined by the network of thermally averaged nuclear reaction rates.

2. OBSERVED ABUNDANCES

Boesgaard and Steigman (1985) provide an excellent review of attempts to extract the primordial light elements from observational data.

2.1 Deuterium

^{2}H is a very fragile nucleus, and its observed abundances are generally regarded as being lower limits to the primordial abundance. Following Yang et al 1984, we shall adopt $(D/H)_p \geq 1(2)$ x 10^{-5} as our "conservative" ("best bet") lower limit to the primordial deuterium-to-hydrogen ratio.

2.2 Helium-3

Yang *et al* have argued that an upper bound on the sum of the primordial ^{2}H and ^{3}He abundances can be derived by ensuring that they were not overabundant at the birth of the solar system. Following Boesgaard and Steigman, we shall adopt [(D + ^{3}He)/H]$_p$ ≤ 10(6) x 10^{-5} as our "conservative" ("best bet") upper limit.

2.3 Helium-4

The most precise method of estimating the primordial ^{4}He is currently believed to involve the observation of hydrogen and helium recombination lines in the emission line spectra of bright ionised HII regions. Following the observations of Pagel and Simonson (1989) and Torres-Peimbert *et al* 1989, we shall adopt Y_p = 0.230 ± 0.010(0.005) as or "conservative" ("best bet") estimate for the primordial ^{4}He mass fraction.

2.4 Lithium-7

Observations of the interstellar medium and Population I stars suggest Li/H $\sim 10^{-9}$. Taking stellar contamination and cosmic ray spallation into account, we shall adopt $(^7$Li/H$)_p$ ≤ 8 x 10^{-10} as our "conservative" upper limit to the primordial abundance. Spite and Spite (1982) first suggested that the lithium abundance Li/H $\sim 10^{-10}$ obseved in metal deficient Population II stars corresponds to the primordial value. Deliyannis *et al* (1989) have recently modelled the evolution of surface abundances of ^{6}Li and ^{7}Li in such old halo stars, and we shall adopt their upper limit of $(^7$Li/H$)_p$ ≤ 2.3 x 10^{-10} as our "best bet" limit to the primordial abundance.

3. NUCLEAR REACTION RATES

As the temperature falls below 10 MeV, primordial nucleosynthesis begins. The number abundances of each nuclear species "i" evolve through the rate equations

$$\frac{dY_i}{dt} = \sum_{j,k,l} N_i \left[\frac{Y_l^{N_l} Y_k^{N_k}}{N_l! \, N_k!} [l,k]_j - \frac{Y_i^{N_i} Y_j^{N_j}}{N_i! \, N_j!} [i,j]_k \right]$$

where N_i are the number of nuclei per unit mass with mass number A_i and charge number Z_i. Baryon conservation ensures that the mass fractions X_i = $A_i Y_i$ satisfy $\Sigma_i X_i$ = 1. $[i,j]_k$ and $[l,k]_j$ are respectively the thermally averaged production and anihilation rates for the reaction i + j → k + l. The nuclear physics input to the nucleosynthesis calculations is in the form of 31 low-energy reaction rates.

In order to study the sensitivity of the predicted primordial abundances to nuclear physics uncertainties, each of the reaction rates was individulally varied by a factor of 2 about the value recomended by Caughlan and Fowler (1988). This allowed us to identify the eleven reactions to which the predicted abundances are most sensitive. Maximal and minimal rates were then derived for each of these reactions using the published experimental uncertainties (see Table 1).

The neutron half-life, τ_n, is extremel difficult to measure with any accuracy. The present recomendation of the Particle Data Group (Yost *et al* 1988) is that $\tau_n = 10.35\pm0.12$ minutes. However, the value of $\tau_n = 10.25\pm0.04$ minutes obtained by Mampe *et al* (1989) using stored ultracold neutrons has significantly smaller uncertainty associated with it compared to all previous results.

We now use these nuclear reaction rate uncertainties to optimise the agreement between the predicted and observed abundances plotted in the N_ν-$\rho_B(0)$ plane, where N_ν is the number of light neutrino flavours, and $\rho_B(0)$ is the present baryon density of the universe. For $A \equiv (D/H)_p$ x 10^5, $B \equiv [(D+^3He)/H]_p$ x 10^5, and $C \equiv (^7Li/H)_p$ x 10^{10}, the predictions must lie within the "conservative" and "best-bet" limits of $0.22 \leq Y_p \leq 0.24$ and $0.225 \leq Y_p \leq 0.235$, respectively. These bounds define an approximately quadrilateral region in the N_ν-$\rho_B(0)$ plane, providing stringent limits for both the present baryon density, and the number of light neutrinos.

Reaction	Uncertainty
n(p,γ)d	30 – 40%
d(p,γ)^{3}He	22%
d(d,p)^{3}H	35 – 45%
d(d,n)^{3}He	30 – 40%
^{3}H(d,n)^{4}He	10%
^{3}He(n,p)^{3}H	20%
^{3}He(d,p)^{4}He	10%
^{3}H(α,γ)^{7}Li	55 – 75%
^{3}He(α,γ)^{7}Be	17%
^{7}Li(p,α)^{4}He	17%
^{7}Be(n,p)^{7}Li	18 – 20%

Table 1: Nuclear reaction rate uncertainties ($0.5 \leq T_9 \leq 1$)

4. RESULTS

We begin with the nuclear reaction rates recomended by Caughlan and Fowler (1988), together with the mean value of the neutron half-life recommended by the Particle

Data Group (Yost *et al* 1988)

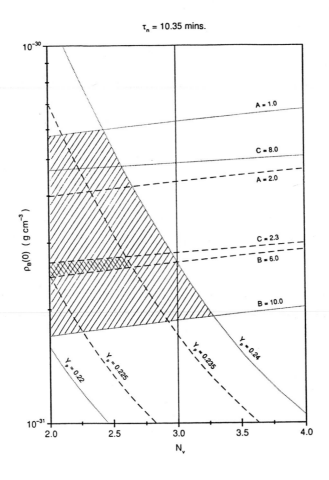

Figure 1.

It will be seen from Figure 1, that although the "best bet" constraints on the primordial abundances are incompatible with three light neutrino flavours, the abundance predictions with $N_\nu = 3$ are compatible with the more "conservative" constraints. However, in no case can a fourth light neutrino species be contemplated.

The cosmological limit on the number of neutrino flavours is determined by the constraints on Y_p and $[(D+{}^3He)/H]_p$ derived from observation. We now examine the effect that nuclear reaction rate uncertainties have in relaxing this limit on N_ν (see Figure 2). Since the primordial deuterium and ${}^3He$ abundances are far more sensitive than ${}^4He$ to changes in the nuclear reaction rates, we select the maximal and minimal rates that minimise the $[(D+{}^3He)/H]_p$ predictions. We find that this has only a minor effect on the cosmological limit on N_ν, with $\Delta N_\nu \approx +0.06$. While this helps slightly, the "best bet" bounds are only consistent with $N_\nu = 3$ if $\tau_n \leq 10.10$ minutes.

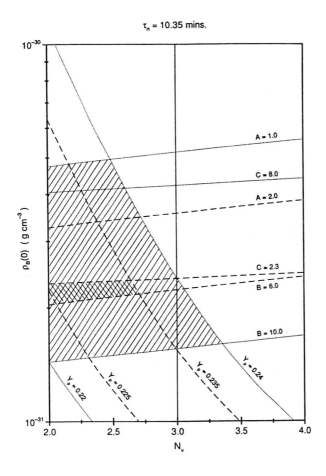

Figure 2.

5. CONCLUSIONS

The initial results from LEP suggest that $N_\nu \approx 3.0\pm0.3$. If we accept the standard model of particle physics with three light neutrinos, and the value of the neutron half-life recommended by the Particle Data Group, the isotopic abundances predicted by the standard model of Big Bang nucleosynthesis are incompatible with our "best bet" analysis of the observed abundances, and are barely consistent with the more generous "conservative" limits. This apparent conflict between these standard models is not significantly eased if we allow for uncertainties in the nuclear reaction rates, and we do not believe that any significant relief will come from changes in the nuclear physics input. We are left with five obvious possibilities:

(1) The primordial abundances derived from observation are incorrect. Unfortunately, real relief only comes if the ^{4}He abundance is in error.

(2) The Galactic chemical evolution model, upon which the upper bound to the sum of the primordial deuterium and ^{3}He abundance is based, is incorrect. It is therefore possible that the conservative upper limit of $[(D+^3He)/H]_p \leq 10^{-4}$ is invalid.

(3) The neutron half-life is lower than the recomended value. The three light neutrino species could be accommodated with the "best-bet" limits if $\tau_n \leq 10.1$ minutes.

(4) The standard, homogeneous model of Big Bang nucleosynthesis is incorrect. Various inhomogeneous, baryon-segregated models have been developed in recent years.

(5) The standard model of particle physics, with three massless neutrino generations, is incorrect. The present limit on the tau-neutrino mass ($m_{\nu_\tau} \leq 35$ MeV) leaves open the question of whether or not they were relativistic during primordial nucleosynthesis.

References

Boesgaard,A.M. and Steigman,G. 1985, Ann.Rev.Astron.Astrophys.,**23**,319.

Caughlan,G.R. and Fowler,W.A. 1988, Atomic Data and Nuclear Data Tables,**40**,283.

Deliyannis,C.P.,Demarque,P.,Kawaler,S.D.,Krauss,L.M. and Romanelli,P. 1989, Phys.Rev.Lett.,**62**,1583.

Mampe,W.,Ageron,P.,Bates,C.,Pendlebury,J.M. and Steyerl,A. 1989, Phys.Rev.Lett.,**63**,593.

Pagel,B.E.J. and Simonson,E.A. 1989, Rev.Mex.Astr.Astrofis.,**18**,153.

Spite,F. and Spite,M. 1982, Astron.Astrophys.,**115**,357.

Torres-Peimbert,S.,Peimbert,M. and Fierro J. 1989, Ap.J.,**345**,186.

Yang,J.,Turner,M.S.,Steigman,G.,Schramm,D.N. and Olive,K.A. 1984, Ap.J.,**281**,493.

Yost,G.P. *et al* 1988, Phys.Lett.,**B204**,1.

The Extremely Low He Abundance of SBS 0335-052

ELENA TERLEVICH [1], ROBERTO TERLEVICH [1], EVAN SKILLMAN [2],
JEVAN STEPANIAN [3] and VALENTIN LIPOVETSKII [4].

[1] *Royal Greenwich Observatory, Madingley Rd., Cambridge, CB3 0EZ, U.K.*
[2] *Astronomy Department, Univ. of Minnesota, USA*
[3] *Byurakan Astrophysics Observatory, Armenian Academy of Sciences*
[4] *Special Astrophysical Observatory, USSR Academy of Sciences*

ABSTRACT

Extremely metal-poor galaxies have been recognized as critical to the understanding of several fundamental problems in astronomy, but nearly impossible to find. In the past three years we have had great success in discovering them. Now, with a significant sample of these galaxies, several observational programmes are possible. Foremost of these is the measurement of the primordial helium abundance, which provides strong constraints on the Big Bang model of the origin of the universe. The helium abundance measurement in SBS 0335-052, the first one from our sample of extremely metal-poor HII galaxies, is reported here. The derived value of He/H=0.21 ± 0.01, is the lowest one known.

1. INTRODUCTION

The Big Bang model of the origin of the Universe is based on three observational findings: 1) the expansion of the Universe; 2) the relic 3 K microwave background radiation; and 3) the relative abundances of the light elements (H, D, ^{3}He, ^{4}He, Li). Accurate measurements of these observables provide better constraints on the details of the model. Compared to the first two cosmological observables, the study of the primordial helium abundance has suffered from a disproportionately low share of observational effort. Over the last two decades, the measurement of the rate of expansion of the universe has been one of the primary goals of the largest ground based telescopes, and this task is now being taken up by the Hubble Space Telescope. Last winter, NASA's Cosmic Background Explorer satellite has produced a dramatic improvement in the 3K radiation measurement. Even though an accurate measurement of the primordial helium abundance is just as critical to our understanding of the origin of the universe it has not been the target of any comparable observational effort. In particular, the relative abundances of the light elements just after the Big Bang is directly related to the nucleon mass density, the number of neutrino species, and the half-life of the neutron. Recently, experimental progress has been made on the number of neutrino species and the half-life of the neutron. Therefore, a measurement of the primordial helium abundance, under the assumption of the standard hot Big Bang model of nuclesynthesis, provides a determination of a fundamental parameter in cosmology: the nucleon density in the early universe. In addition, the agreement of the abundances of the light elements produces a constraint on the standard model.

2. MEASURING THE PRIMORDIAL HELIUM ABUNDANCE.

The abundance of helium in any object today represents the contribution from two components, helium produced in the Big Bang and helium subsequently produced by fusion in the centers of stars. One can gauge the relative contributions of primordial and post-primordial helium by measuring the abundance of a heavier element (e.g. oxygen) that was not produced in the Big Bang. That is, stars with larger fractions of oxygen have larger helium abundances, while stars of low oxygen content should have nearly primordial compositions.

Due to the physics of stellar atmospheres, old stars do not provide accurate measurements of the helium abundance (leaving aside the complication of any mixing of processed material from the core to the surface). The most accurate method of measuring the helium abundance is to observe diffuse gas which has been ionized by the ultraviolet photons of a nearby star. The goal then is to measure the helium abundance in ionized gas of low oxygen abundance. Early studies (Peimbert and Torres-Peimbert 1974, 1976; Lequeux *et al.* 1979) pioneered the method of extrapolating the He/H *vs.* O/H relation to a value of the helium abundance where the oxygen abundance is zero. See, for example, Figure 1 taken from Simonson (1990).

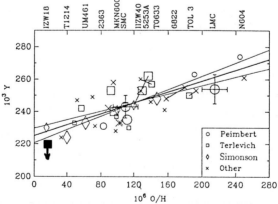

Figure 1 - He/H vs. O/H for well studied extragalactic HII regions (from Simonson, 1990). SBS 0335-052 is the filled square and the arrow points at the corrected value (see text).

It becomes clear how critical it is to find objects of very low oxygen abundance. The smaller the oxygen abundance, the smaller the extrapolation, the more accurate the measure of the primordial helium abundance, and, finally, the more severe the constraint on the Big Bang model.

Unfortunately, several attempts to find regions of very low oxygen abundance failed. Candidates have been obtained amongst the highest excitation objects from surveys capable of detecting intense bursts of star formation. As of 1985, only one object of suitably low abundance was known (IZw 18; an object with an oxygen abundance of only 2 percent of that of the sun), and no one seemed to know how to find more (Shields 1986). It has been known for more than a decade that low mass irregular galaxies are much less evolved chemically than our own galaxy (Lequeux *et al.* 1979, Talent 1980); in the last three years, we have looked therefore for very low metal-abundance regions in extremely low mass galaxies and we have been very successful (Skillman *et al.* 1988; Skillman, Terlevich and Melnick 1989: STM; Skillman, Kennicutt and Hodge 1989). Figure 2 (from STM) emphasizes that success. The histogram compares the number of objects versus their oxygen abundance for three different studies. The lowest panel shows the sample used in a recent determination of the primordial helium abundance (Kunth and Sargent 1983). The middle panel shows the best candidates (Campbell, Terlevich and Melnick 1986), chosen from a survey of over 800 objects (Terlevich *et al.* 1991). The top panel shows the objects we have discovered in the last three years, restricting our search to extremely low mass galaxies. Clearly this represents a superior way to find low abundance objects.

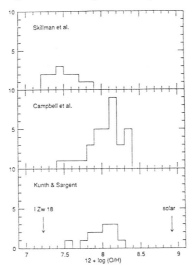

Figure 2 - The distribution in oxygen abundance of HII regions, from studies aimed at finding low abundance regions (from STM). The efficiency of selecting low luminosity galaxies (top plot) is emphasized.

Searching for these metal poor galaxies we have compiled a list of low luminosity candidates from the Second Byurakan Survey (SBS) of ultraviolet excess galaxies, which have been studied spectroscopically by Lipovetskii and Stepanian. We obtained blue spectra of these galaxies with the INT and have selected a subsample of the brightest and lowest oxygen abundance objects. We are conducting follow-up, high precision, high spacial resolution spectroscopy both in the blue and in the red with the WHT in La Palma.

The bulk of the spectroscopic observations has come in the last year, and the data reduction is in progress. While we have discovered objects with oxygen abundances in the range 2% to 5% of the solar value, neither have an oxygen abundance lower than that of IZw 18. The objects from the SBS are most valuable for the study of the primordial helium abundance, because they are, on average, higher excitation, and the derived helium abundance is therefore less vulnerable to the uncertainty of a correction for neutral helium.

Among these objects, we have observed SBS 0335-052, which is particularly appropriate for these studies because of its high surface brightness. The discovery of this object has been reported by Izotov *et al.* (1990). In fact, they claim in their paper that its oxygen abundance is 77 times lower than the solar value, which makes it the object with the lowest heavy-element abundance known. We will discuss our observations in some detail in what follows.

3. OBSERVATIONS AND RESULTS

SBS 0335-052 was observed in 1990 January 31st with the Cassegrain spectrograph of the Isaac Newton Telescope at the Roque de los Muchachos Observatory, La Palma, Canary Islands, using the 235-mm camera and an IPCS detector. The grating used, a blue-blazed 1200 grooves mm^{-1}, gives a dispersion of 35 Å mm^{-1}, and provided a total spectral coverage from λ 3550 Å to λ 5300 Å using two different grating angles. With the slit width used for the observations, (270 μ) the spectral resolution achieved was 0.52 Å/pix.

After wavelength calibration and sky subtraction using standard procedures, we extracted 1-d spectra and calibrated in flux, using a standard spectrophotometric star taken from Oke and Gunn (1983). A flux calibrated spectrum is shown in Figure 3. Both temperature and electron density can be readily determined from our spectra, because the [OII] $\lambda\lambda$ 3726,3729 Å lines are resolved allowing a good estimation of electron density ($n_e \sim 350$ cm^{-3}) and the temperature sensitive [OIII] λ 4363 Å line is very intense. The [ArIV] $\lambda\lambda$ 4711/4741 ratio, also a density indicator (Aller 1984) has to be corrected for HeI λ4713 Å using the HeI λ 4471 Å line; the [ArIV] ratio indicates a higher density, but because of the amount of the correction and also because the ratio falls in a flat range of Aller's diagram, the uncertainty is large and therefore we use the [OII] density value. New data kindly obtained for us by Robin Clegg and reaching the [SII] lines at $\lambda\lambda6717,6731$ Å, gives a value for n_e (both from [SII] and [ArIV]) which is consistent with that estimated from [OII].

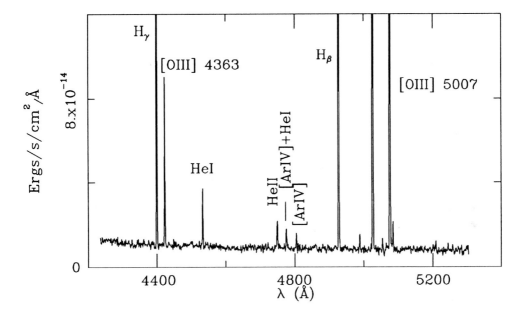

Figure 3 - IPCS spectrum of SBS 0335-052 obtained with the INT. The scale of the plot is such as to see clearly the weaker lines; peak intensity of H_γ is 4×10^{-13}.

The electron temperature t([OIII]) was calculated using the [OIII] line ratio ($\lambda5007+\lambda4959)/\lambda4363$ following the scheme proposed by Seaton (1975) and Aller (1984) and using the most recent atomic data in the literature (Mendoza 1983). The derived value of t([OIII]) is 19500 K; ionic oxygen abundances were calculated assuming a constant temperature throughout the nebula. The total oxygen abundance thus obtained is 12+log[O/H]=7.26 or 0.022 of solar, similar to that of IZw 18.

Assuming that hydrogen and helium lines are produced from recombination, the abundance of He^+ was determined using the strength of λ 4471 Å relative to H_β (Kunth and Sargent 1983). We obtained thus an abundance by number $y^+(4471)=0.071$, which gives a value of helium abundance by mass Y=0.221. From the ratio of (O^+/O^{++}), the maximum correction for neutral helium is 2% . The correction for collisional excitation, for the derived values of T_e and n_e is, at most, 5% (Clegg 1987). The corrected value is thus Y=0.215±0.01, which makes SBS 0335-052 the galaxy with the lowest He abundance known, embarrassingly low for our present understanding of standard Big Bang and nuclear physics (Pagel 1990).

Melnick (private communication) has obtained an image of this HII galaxy in sub-arcsecond seeing conditions, with the NTT at La Silla, and he discovered that it consists at least of two objects, separated by 1."3. Stratification, if present, might mean a higher value for the abundances as in Campbell (1989) for IZw 18. Data with high spacial resolution (as well as high spectral resolution) is therefore of paramount importance.

These results stress the need for further high S/N observations both of this galaxy and of others with very low metal content. Increasing the number of low metallicity objects with reliable helium abundance from 2 (IZw 18 and SBS 0335-052) to 10 (the rest of our sample) should allow a better determination of the "primordial" He value, with a less uncertain extrapolation.

ACKNOWLEDGEMENTS

The INT is operated on the island of La Palma by the RGO at the Observatorio del Roque de los Muchachos of the Instituto de Astrofísica de Canarias. We thank PATT for awarding observing time and acknowledge fruitful discussions with Don Garnett, Jorge Melnick and Manuel Peimbert. Yuri Izotov, promptly understood our panic and sent us by telex the correct coordinates of the object in the evening of the observation.

REFERENCES

Aller, L.H., 1984. *Physics of Thermal Gaseous Nebulae, Reidel, Dordrecht, Holland.*
Campbell, A. W. , 1988. *Astrophys. J. ,* **335**, 644.
Campbell, A. W. , 1989. *Bull.Am.Astr.Soc.,* **20**, 1038.
Campbell, A.W., Terlevich, R. & Melnick, J., 1986. *Mon. Not. R. astr. Soc. ,* **223**, 811.
Clegg, R.E.S., 1987. *Mon. Not. R. astr. Soc. ,* **229**, 31p.
Izotov, Yu.I., Lipovetskii, V.A., Guseva, N.G., Kniazev, A.Yu. & Stepanian, J.A., 1990. *Nature,* **343**, 238.
Kunth, D. and Sargent, W., 1983. *Astrophys. J. ,* **273**, 81.
Lequeux, J., Peimbert, M., Rayo, J.F., Serrano, A. & Torres-Peimbert, S., 1979. *Astr. Astrophys. ,* **80**, 155.
Mendoza, C., 1983. in *"Planetary Nebulae", IAU Symp. # 193, p. 143.,* ed. *Flower, D.R., Reidel, Dordrecht, Holland.*
Oke, J.B. and Gunn, J.E., 1983. *Astrophys. J. ,* **266**, 713.
Pagel, B.E.J., 1990. in *"Baryonic Dark Matter", NATO ASI, Cambridge,* eds. *D. Lynden-Bell and G. Gilmore, p.237, Kluwer, Dordrecht, Holland.*
Peimbert, M. & Torres-Peimbert, S., 1974. *Astrophys. J. ,* **193**, 327.
Peimbert, M. & Torres-Peimbert, S., 1976. *Astrophys. J. ,* **203**, 581.
Searle, L. & Sargent, W.L.W., 1972. *Astrophys. J. ,* **173**, 25.
Seaton, M.J., 1975. *Mon. Not. R. astr. Soc. ,* **170**, 475.
Shields, G.A., 1986. *Publ. astr. Soc. Pacif. ,* **98**, 1072.
Simonson, E.A., 1990. *D.Phil. Thesis, Sussex University.*

Skillman, E.D., Kennicutt, R.C. & Hodge, P.W., 1989. *Astrophys. J.* , **347**, 875.
Skillman, E.D., Melnick, J., Terlevich, R. & Moles, M., 1988. *Astr. Astrophys.* , **196**, 31.
Skillman, E.D., Terlevich, R. & Melnick, J., 1989. *Mon. Not. R. astr. Soc.* , **240**, 563.
Talent, D.L. , 1980. *Ph.D. Thesis, Rice University.*
Terlevich, R., Melnick, J., Masegosa, J., Moles, M. & Copetti, M.V., 1991. *Astr. Astrophys. Suppl.* , **91**, 285.

QUESTIONS

M.V. Penston [AIV] is also used to estimate densities in Seyfert galaxies and there the [NeIV] lines are a problem.

E. Terlevich We have good spectral resolution and we do not see the stronger [NeIV] line at λ 4724 Å; I think therefore that we don't have to worry about [NeIV] in this object.

B. Pagel Why not ignore the high density derived from [ArIV]? It probably refers only to an inner region containing mainly He^{++}.

E. Terlevich In fact, that is what we did. The reported value corresponds to n_e as obtained from the [OII] ratio.

J.M. Vilchez A comment on the correction to He due to collisions: in some normal giant HII regions you can find (with high S/N CCD spectroscopy) differences in density that can be very high; so some care has to be taken when correcting for collisions in more distant objects.

E. Terlevich True, and that is why we need the high spatial resolution observations.

Data for Absorption Line Analysis

D.E.BLACKWELL

Oxford University

1 INTRODUCTION

This talk should properly include such topics as atomic and molecular oscillator strengths, wavelengths, hyperfine structure and collision damping constants. However, I am going to speak almost solely on atomic oscillator strengths, more especially those for the iron group, and make only an occasional reference to damping constants.

A common attitude to laboratory astrophysics is that surely there are enough measured and calculated oscillator strengths of sufficient accuracy now available, and there cannot be a need for more experiments. So, oscillator strengths, at least for the neutral and singly ionised atoms of the commoner elements, are now a bit like the stoats and weasles of zoology – unloved, but necessary for a general understanding of the animal world. But what is meant by 'sufficient accuracy'? Using an FTS spectrometer, the equivalent width of a carefully selected line in a stellar spectrum can be measured with an accuracy approaching one per cent (0.004 dex). So, an ideal data base should contain oscillator strengths of at least this accuracy. In a partial answer to our doubting astrophysicists, I cite the example of NiI. In their valuable critical compilation of atomic transition probabilities, Fuhr et al.(1988) rate most measurements of NiI transition probabilities as having an uncertainty between 25 and 50 per cent. Such uncertainties exceed those needed for a proper interpretation of stellar spectra by factors of up to 100. There are similar results for all other elements. For example, transition probabilities for the higher excitation lines of FeI and all lines of FeII are described as having uncertainties within 50 per cent.

2 NEED FOR RELIABLE PROBABLE ERRORS

I am sure that all are agreed about the need for accuracy, but I want to go further than that and emphasise the need for a reliable probable error of measurement. To illustrate this I use as an example the celebrated controversy in the 1920's about the distances of galaxies: are they at galactic or extragalactic distances? A powerful argument against extragalactic distances was advanced by van Maanen, whose measurements of proper motions in galaxies apparently showed clear evidence of rotation (e.g. van Maanen, 1922). Taken at their face value, these observations appeared to settle the matter decisively, and it was not until 1928 that Hubble identified Cepheid variable stars in nearby galaxies, showing that they are indeed at extragalactic distances. My point now is that this correct conclusion was delayed for several years because it was supposed that van Maanan's measured proper motions had an accuracy many times better than reality. When they apparently showed galaxy rotation, there seemed to be no room for doubt.

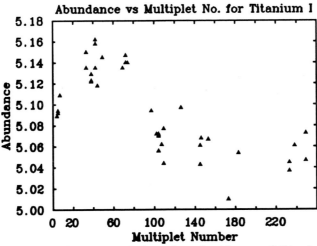

Figure 1: Measures of the solar abundance of titanium.

A similar situation arises with oscillator strengths. Figure 1 plots measures of the solar abundance of titanium determined using Oxford oscillator strengths for several TiI lines covering a range of multiplet number (equivalent to excitation potential). It shows an apparent variation of the solar abundance with excitation. The diagram can be regarded as analogous to the van Maanen diagrams of the rotation of galaxies, in which a vectorial representation of proper motions shows an apparent rotation. Is the apparent variation of abundance in Figure 1 real, or is it only due to inaccuracies in the adopted oscillator strengths, together of course with inaccuracies in the measured equivalent widths? In this example, the spread of abundance is 0.1 dex (about 25 per cent), but as the claimed accuracy of the oscillator strengths is about one per cent (0.004 dex), we suggest that the apparent variation is real and chiefly an effect of non-LTE.

3 IMPORTANCE OF HIGH EXCITATION LINES AND COLLISION DAMPING

Figure 1 shows that the measured abundance of titanium rises to a maximum with excitation potential, and then seems to decrease to a constant plateau. If this interpretation is correct, the true abundance for titanium, and possibly for other elements too, is only found from higher excitation levels. Consequently there is a need for oscillator strengths for the higher excitation lines. Indeed, it might be in this case that we should go to even higher excitation potentials than are shown here. However, this proposition leads to a difficulty because higher excitation is associated with a larger collision damping. The connection is illustrated by the profile of the high excitation (4.99eV) line FeI 7742.72Å in the solar spectrum, shown in Figure 2 (Delbouille *et al.*, 1973). Although this is a weak line, its profile is clearly dominated by wings due to collision broadening. This is also apparent from a comparison between its broadened profile and the sharper profile of the low excitation (1.94eV) line NiI 7788.9Å, which is of similar depth and is plotted at nearly the same wavelength dispersion.

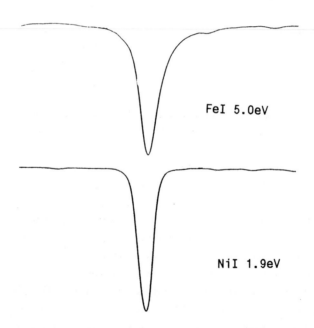

Figure 2: Profiles of the solar lines, FeI 7742.72Å and NiI 7788.72Å.

If this FeI line were to be used to obtain a solar iron abundance, an error in the adopted damping constant of a factor ×2 would lead to an error in abundance of nearly a factor ×2. Knowledge of damping constants for high excitation lines is poor, and such an error is possible. So it is clear that any gain obtained by using a high excitation line, formed more closely in LTE, with a one per cent accurate oscillator strength, could be more than lost by poor knowledge of collision damping. Of course, apart from stellar abundances and applications to stellar atmospheres, collision damping for high excitation lines is of intrinsic interest because the upper level may be very close to the ionisation limit.

4 SINGLY IONISED SPECIES

I want to stress the importance of oscillator strengths for singly ionised species. An abundance determination for iron, for example, in a solar type star, using FeII lines is less sensitive to atmospheric parameters than one using FeI lines because of the high degree of ionisation. Also, oscillator strengths of ionised species are important because they can be used to check the degree of ionisation in a computed model stellar atmosphere. An example of precise modern work on ionised species is that of Biemont and Grevesse and others, for VII (Biemont *et al.*, 1989), who found a solar abundance of 4.04±0.06 compared to the value found from VI lines of 4.01±0.04. In this example the measurements are difficult because of blending, but the final result is satisfactory. Similar work has also been done recently on FeII by Pauls, Grevesse and Huber (1990) and by Holweger, Heise and Kock (1990), who have made new determinations of the solar abundance free from the uncertainties of ionisation.

An important area where very few measured oscillator strengths exist is the far infrared region from about one micron, although there are splendid solar spectra in this region, and the possibility of good stellar spectra. There also remains much work to be done in the far ultraviolet.

5 MEASUREMENTS

5.1 Experiments
Now I propose to speak about methods for obtaining oscillator strengths. Nobody would willingly adopt a cat without first enquiring about its pedigree, and one should not adopt an oscillator strength at random from the literature without a thorough examination of its origin. We begin with experiments. An enormous variety have been used with varying degrees of success over the last seven or so decades. There is not time to even mention all of them, and I propose to speak only of the more relevant ones, with a brief discussion of the results.

5.2 Furnace absorption
I begin with the classical furnace absorption method for measuring relative

oscillator strengths, pioneered by King and King in the 1920's. We have used it at
Oxford (e.g. Blackwell et al., 1982) for some twenty years as part of an on-going
programme to determine accurate stellar parameters, especially mass and
temperature. The method measures relative oscillator strengths, which are placed
on an absolute scale knowing one absolute oscillator strength. In outline, the
element to be investigated is placed in a carbon tube which is heated to a high
temperature, where it forms a column of vapour that is effectively 3m long. Light
is passed through the vapour into two spectrometers which simultaneously
measure the equivalent widths of pairs of absorption lines, using photoelectric
scanners. The temperature of the vapour, which determines the relative
populations of excited levels, is measured photoelectrically. The aim has been to
measure relative oscillator strengths to an accuracy of one per cent or better, and
for this the temperature must be known with an accuracy of order 1K, which is
possible using modern photoelectric techniques.

The prime advantage of the method is that it is simple, so that it is well
understood and the sources of error are easily and reliably assesed. It is capable of
good accuracy, and we have proposed accuracies of between 0.5 and 1 per cent for
many of our measurements of some 700 relative oscillator strengths. However,
there are several disadvantages. As used at Oxford, the method is slow and very
expensive, and because it relies on thermal excitation, it is only useful for
excitation potentials of up to about 3.5eV. Such excitations need a temperature of
up to 2800K, and a power of about 150 KW, and even then only stronger lines
may be measured. Apart from using a higher temperature, which is impractical,
the only way to increase the maximum measurable excitation is to increase the
path length. It is easy to plan such an experiment, but less easy to implement it.
To measure an FeI line of excitation 5eV would require a carbon tube with one
end in Oxford and the other end in Cambridge. Further, to run it at the right
temperature would take a power of 4×10^8MW.

5.3 The hook method
The hook method is a variation of the straightforward furnace method, which
depends on measurement of optical dispersion within an absorption line. The
method is nearly 80 years old, but in spite of its age, it is still used and it is one of
the most accurate available. It is suitable only for low excitation lines.

5.4 The simple emission method
The more accurate application of this method was pioneered by Corliss and
Bozman (1962), who used it for a large range of elements. Their experiments used
a simple high temperature arc with the element under investigation fed into it.
This arrangement could produce high excitation lines, but the accuracy was not

good, sometimes only of order a factor 2. The method has now been developed to a highly sophisticated level and many good oscillator strengths are being measured by its use. We illustrate its applications by referring again to FeI, for which the most accurate and most extensive emission measurements have been made by May, Richter and Wichelmann (1974), and by Bridges and Kornblith (1974). The former work extends to an excitation of 4.9 eV, and the latter to 4.5 eV. Each proposes a probable error of about 0.06 dex or 15 per cent, and this is borne out by a comparison between the two, which shows a standard deviation of the ratio of order 0.07 dex, depending on the range of excitation potential. Further, a plot of differences in the log gf's against upper excitation potential shows little trend with upper excitation, suggesting that the temperatures of the sources have been accurately measured. A comparison of this kind between the results of completely indepndent experiments, although preferably ones that use very different techniques, is one of the best ways of checking measures of oscillator strengths. The results are uniformly good and have been adopted by Fuhr *et al.*(1988) as the best high excitation values for FeI at the time of compilation. However, the accuracy of these experiments is still less than the desirable accuracy discussed earlier by a factor of between 5 and 10.

5.5 Use of transformation formulae
A difficulty with the emission method is that it contains many small imponderable sources of error which cannot easily be identified and for which an allowance cannot readily be made. But these sources of error surely depend on various parameters such as excitation potential, wavelength, oscillator strength. With this assumption a possible way forward is to transform the system in question to another system which does not suffer from these small sources of error, using a multivariate analysis. One choice of reference system might be the Oxford one, although any other would serve. A possible objection to this procedure is that it is wrong to adjust published data, but this is done, properly, in the Fuhr et al. (1988) compilation where each set of data is adjusted to be on the same absolute scale. A typical transformation relation is

$$\log gf(\text{other}) - \log gf(\text{Oxford}) = A + B \times \log gf(\text{Oxford}) + C \times \chi_{upper} + D \times \lambda + E \times \chi_{lower} \times \log gf(\text{Oxford})$$

As an illustration of the method, Figure 3 compares the FeI oscillator strengths measured by Bridges and Kornblith (1974) with the Oxford furnace measurements, whilst Figure 4 shows the effect of transformation to the Oxford system using a relation of the above form. The resulting oscillator strengths combine the accuracy of the Oxford system with the greater range of line strength and excitation potential of the Bridges and Kornblith system. Note that the

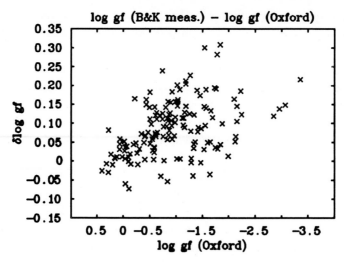

Figure 3: Comparison between FeI oscillator strengths measured by Bridges and Kornblith and by Oxford.

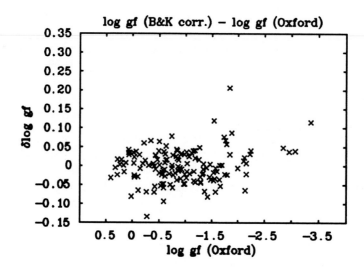

Figure 4: Effect of transforming FeI oscillator strengths measured by Bridges and Kornblith to the Oxford system.

transformation of this large number of oscillator strengths has been accomplished using only five parameters.

5.6 Use of lifetime measurements

A very different kind of experiment involves the measurement of lifetimes of excited levels. If there is only one transition from an excited level, i, at wavelength λ, the lifetime, τ, of the level is given in terms of the oscillator strength, f_{ji}, by the relation

$$\frac{1}{\tau} = \frac{6.67 \times 10^{15}}{\lambda^2} \cdot \frac{g_j f_{ji}}{g_i}$$

where g_i and g_j are the statistical weights of the initial and final levels. In this case, a measurement of the lifetime will give at once the absolute oscillator strength of the transition. If there are several transitions and their relative oscillator strengths are known from a separate experiment, a measurement of the lifetime will place all these on an absolute scale. This is a potentially powerful method, but it was at a disadvantage during its early years because of uncertain lifetime measurements, made using beam foil spectroscopy. In this technique an atomic beam is directed through a carbon target to generate a beam of excited neutral and ionised atoms. These excited atoms radiate, and the lifetimes of the excited levels can be found from the rate of decrease of brightness in a spectrum line along the beam. However, an uncertainty arises because the excited levels may also be fed by transitions from higher excited levels, so giving an erroneous value for the lifetime. Putting this aside for the moment, the second stage, that of measuring the relative strengths of the various transitions from the chosen upper level, is fairly straightforward. As knowledge of the temperature of the source is not required, the only measurement involved is that of the ratios of the line strengths, in absolute units.

The method has now been transformed in accuracy and reliability, partly through the use of a Fourier transform spectrometer to measure the branching ratios, but also by using a pulsed tuneable dye laser for excitation of an upper level. By this means, only one chosen level is excited, and the measured lifetime corresponds only to the downward transitions from it. The method has become very popular, and has been used by, among others, Lawler, Whaling, Brault, Grevesse, and Biemont. The accuracy of the measured lifetimes is good; those of Lawler for TiI, for example, at Wisconsin, have an accuracy of about 5 per cent. A further advantage of the method is that it is fast, and there is no reason why a large number of lines, of any excitation and deriving from ionised atoms as well as neutral atoms, should not be examined by its means. Also, in comparison with the furnace method, its running costs are small. The results so far are of fine accuracy, in some cases better than 5 per cent, and represent an impressive advance; I believe that use of the method is beginning a new era in stellar spectroscopy.

There are many variations on the technique of using lines that originate from one common upper level, and I cannot describe all of them now. However, I would like to mention one that we are using at Oxford. The purpose of it is to extend existing furnace measurements to lines of higher excitation. It is very simple. In it, one line from a set originating from a common upper level is measured in the furnace. This then serves as a calibration for all the other lines, whose relative oscillator strengths have been obtained from their relative intensities. Lines of longer wavelength than the furnace line can be measured in emission, and terminate on levels of higher excitation. The result is that we have proceeded from a low excitation furnace line to a higher excitation line which could not be measured with the furnace. As with all measurements of branching ratios, there is apparently no need for an accurate measurement of temperature. However, this is a fallacy because the spectrometer has to be calibrated to give absolute fluxes. Such a calibration requires a standard flux source, and the only convenient fundamental one at the required wavelengths is a blackbody of known temperature, using the Planck radiation law. Given accurate furnace oscillator strengths, this method should give secondary oscillator strengths of good accuracy – we are striving for an accuracy of 4 per cent, or better.

6 CALCULATION

The most extensive results are those of Kurucz and Peytremann (1975), and later those of Kurucz alone. The second set was published about one year ago and covers some 40 million lines. Its oscillator strengths are more accurate than those in the first publication, and they are on tape only. This is an immensely valuable and impressive work, and is of interest to assess its accuracy, at least for low excitation FeI lines, by comparing with the Oxford results. Figure 5 shows a comparison between the two for all Oxford FeI lines (which includes an additional set of of recently measured high excitation lines). It will be seen that the spread is large, although many of the Kurucz values are within $\delta\log gf = 0.1$, i.e. within 30 per cent.

However, there is some regularity in the spread, for the values of $\delta\log gf$ are closely similar for lines in individual multiplets (Blackwell *et al.*1983); quite often the standard deviation within a multiplet is of order only 0.02 dex (5 per cent).

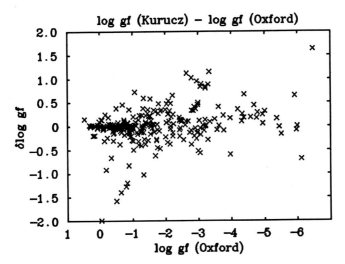

Figure 5: Comparison between calculated oscillator strengths (Kurucz) and measured oscillator strengths (Oxford furnace).

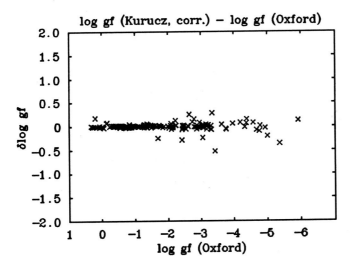

Figure 6: Comparison of corrected oscillator strengths (Kurucz) with measured values (Oxford).

This regularity can be used to normalise the calculated values within a multiplet to Oxford values, or any other system of values, by applying one correction to each multiplet. Figure 6 shows the effect of this normalisation for multiplets that contain two or more Oxford lines. The method offers a means of correcting all the Kurucz FeI oscillator strengths by using one measured line in each multiplet.

Finally, using the sun as a furnace of known properties gives oscillator strengths from measured equivalent widths, a solar abundance and a model solar atmosphere. Several studies of this kind have been made by Gurtovenko and Kostik (e.g. 1981,1982) and others, but of course the values obtained have some uncertainty because of any deficiency in knowledge of solar abundances, solar atmospheric structure, equivalent widths, damping constants, hyperfine structure, microturbulence, blending, non-LTE, etc., and the technique is not suitable for strong lines that are nearly saturated. Fuhr et al. (1988) list 277 solar FeI oscillator strengths with accuracy D (within 50 per cent). These solar values are important because sometimes they are the only available values, especially for weak lines, and are often more reliable than measured values.

7 CONCLUSIONS

What is the present situation? We can see what has happened during the seven years between 1981 and 1988 by looking at the NBS compilations for these two dates, summarised in Table 1 for FeI lines. It is encouraging that the number of lines having an uncertainty within 10% has increased by 81, but the number with uncertainty within 25% has actually decreased by 396. This is because the more recent compilation has been even more critical and has downgraded some measurements. Altogether, the total number of lines has increased only by 354, and these are mostly in the 50% uncertainty class.

Table 1: Number of FeI lines in the critical compilations of Fuhr et al. (1981, 1988): measurements and solar oscillator strengths.

UNCERTAINTY	NUMBER OF LINES	
	1981	1988
Within 10 per cent	130	211
Within 25 per cent	1392	996
Within 50 per cent	43	712

And the future? It will be very interesting to see the compilation for 1995, but I believe it will contain several thousand new lines with accuracies of better than 10 per cent, and the needs of most stellar spectroscopists should be satisfied by the end of the century. However, it is a very different matter for data concerning quantities such as hyperfine structure and collision damping constant, and I cannot see such a rapid advance there.

8 REFERENCES

Biemont, E., Grevesse, N., Faires, L.M., Marsden, G., Lawler, J.E., Whaling, W., 1989. Astron. Astrophys., **209**, 391.

Blackwell, D.E., Petford, A.D., Simmons, G.J., 1982. Mon. Not. R. Astr. Soc., **201**, 595

Blackwell, D.E., Booth, A.J., Menon, S.L.R., Petford, A.D., Smith, G., 1983. Mon. Not. R. Astr. Soc., **204**, 141.

Bridges, J.M., Kornblith, R.L., 1974. Ap. J., **192**, 793.

Corliss, C. and Bozman, W.R., 1962. Nat. Bur. Stand. Monogr., No. 53.

Delbouille, L., Roland, G., Neven, L., 1973. Photometric Atlas of the Solar Spectrum from 3000 A to 10,000 A, Institut d'Astrophysique de l'Universite de Liége, Belgium.

Fuhr, J.R., Martin, G.A., Wiese, W.L., 1988. J. Phys. and Chem. Ref. Data, **17**, Supplement No. 4.

Gurtovenko, E.A., Kostik, R.I., 1981. Astron. Astrophys. Suppl. Ser., **46**, 239.

Holweger, H., Heise, C., Kock, M., 1990. Astron. Astrophys., **232**, 510.

Kurucz, R.L., Peytremann, E., 1975. A table of semi-empirical gf values. Smithson. Inst. Astrophys. Obs., Rep. 362.

May, M., Richter, J., Wichelmann, J., 1974. Astron. Astrophys. Suppl., **18**, 405.

Pauls, U., Grevesse, N., Huber, M.C.E., 1990. Astron. Astrophys., **231**, 536.

van Maanen, A., 1922. Ap. J., **56**, 208.

COMMENT

Professor N. Grevesse: Professor Blackwell has been too modest! You succeeded in convincing solar and stellar spectroscopists about the need for very accurate transition probabilities and showed how to use these data as diagnostic tools in the solar atmosphere (e.g. for accurate abundances, tests of models, possible non-LTE effects, microturbulence etc.)

Capture Cross Sections, s-Process Analysis and the Key Branchings at [85]Kr and [95]Zr

Hermann Beer

Kernforschungszentrum Karlsruhe, Institut für Kernphysik III

1 INTRODUCTION

The quality of s-process calculations depends strongly on the reliability of the input data. Three groups of data are needed: (1) stellar reaction rates of the nuclei on the synthesis path, (2) beta half lives of certain unstable species; perhaps vs. temperature kT and electron density n_e (Takahashi and Yokoi 1987), and (3) s-process abundances, e.g. from the solar abundance distribution (Anders and Grevesse 1989). The most important group of data is the group of stellar reaction rates or Maxwellian averaged capture cross sections.

2 MAXWELLIAN AVERAGED CAPTURE CROSS SECTIONS

Capture cross sections are determined by activation or time–of–flight (TOF) measurements or statistical model calculations. The final goal of these efforts is a Maxwellian averaged capture cross section $\langle\sigma\rangle$ as a function of temperature kT:

$$\langle\sigma\rangle(kT) = (2/\sqrt{\pi})(kT)^{-2} \int_0^\infty \sigma_{n\gamma}(E)\, E \exp(-E/kT)\, dE \qquad (1)$$

where $\sigma_{n\gamma}(E)$ is the excitation function vs. neutron energy E. $\langle\sigma\rangle$ is needed in the interval 5 keV<kT<100 keV. A characteristic of the (n,γ) reaction is the absence of a reaction threshold. Therfore, the computation according to equation (1) required $\sigma_{n\gamma}$ over the whole neutron energy range especially down to the thermal cross section at 0.025 eV to obtain good accuracies in $\langle\sigma\rangle$(kT).

Experimental data of $\sigma_{n\gamma}$, if they are comnplete, consist in a resolved resonance part, where the resonance parameters are specified and in an unresolved resonance part, where the excitation function $\sigma_{n\gamma}$ is given (Mughabghab et al. 1981, McLane et al. 1988). In the unresolved part the integration according to eq. (1) was directly performed. In the resolved resonance part the integration over the resonances was carried out using the Breit Wigner formula. Narrow resonances can also be treated as delta functions (Macklin and Gibbons 1965). Regions, where $\sigma_{n\gamma}$ is unknown have been completed by statistical model calculations which required average level density, radiation width and strength functions. Special treatment required the integral in eq. (1) at the integration limits. The unresolved resonance part of the excitation function was extrapolated, if necessary, to higher neutron energies with an 1/E energy dependence (Kompe 1969). The lower limit demanded an understanding of the capture cross section at thermal energy (E=0.025 eV). The thermal cross section is analyzed in Mughabghab et al. (1981), in general, as the sum of resonance tails from positive and negative resonances (bound states). The two parts were treated separately in the integration. Sometimes, especially for light isotopes, the thermal cross section contains a direct capture contribution instead of a contribution from bound states. This required a different treatment than bound state capture.

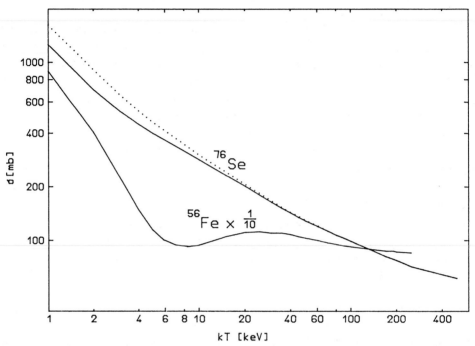

Fig. 1 Maxwellian averaged capture cross sections of ^{56}Fe and ^{76}Se vs. temperature

^{76}Se has a typical shape for a Maxwellian averaged capture cross section: $\langle\sigma\rangle\sim(kT)^{-b}$, where b~0.4–0.8. In the computation of the dotted curve (Fig. 1) direct capture is assumed instead of bound states capture for the interpretation of the thermal cross section. At kT=10 keV the solid and dotted curves are different by 6 %. The Maxwellian averaged cross sections of a number of isotopes is quite different from a temperature dependence $\langle\sigma\rangle\sim(kT)^{-b}$. ^{56}Fe is shown as an example (Fig. 1). Each individual isotope required special consideration and calculation from the existing data.

An additional temperature dependence of $\langle\sigma\rangle$ has been taken into account for nuclei with low lying levels significantly populated at kT=5–100 keV so that excited state capture is non negligible. This effect was large, e.g. for ^{187}Os but also important for some even–even s-only nuclei with a 2^{+} first excited state, e.g. ^{154}Gd, ^{160}Dy, ^{170}Yb etc.

3 s-PROCESS NUCLEOSYNTHESIS

With the improved set of Maxwellian averaged capture cross sections as a function of temperature new calculations have been carried out in order to obtain improved values for the fundamental parameters of the s-process. Fig. 3 shows the result of this analysis. We can distinguish two groups of parameters according to the structures exhibited by the empirical data. There is a gross structure characterized by flat portions and precipices at magic neutron numbers 50, 82, 126 and a fine structure at individual isotopes, e.g. ^{152}Gd, ^{164}Er. The gross structures suggest an s-process with an exponential exposure distribution, an irradiation by a series of discrete neutron bursts (main component), and due to the extra steepness for A<90 a single exposure component (weak component) superimposed. The fine structures are interpreted as s-process branchings. With respect to these features two groups of data can be distinguished. From the gross structures we obtained average exposure, average number of captured neutrons per iron seed and the fraction of iron seed nuclei, from the fine structures neutron density, temperature, electron density and pulse duration of the s-process. The following astrophysical parameters for the main component were found:

Fraction of iron seed F_{Fe}=(0.063 ±0.004) %

Average number of captured neutrons per iron seed:

$AVN^H = \Sigma_A (A-56)[N_S^A - N^A(seed)] / N^{56}(seed) = 9.20 \pm 0.53$

Average time integrated neutron flux τ_0(kT=25 keV)=(0.276 ±0.0098) mbarn^{-1}, τ_0 as a function of kT in the interval kT=10 – 30 keV: τ_0(kT)=τ_0(25) (kT/25)$^{0.518}$, Temperature kT=(25±2) keV, Neutron density n_n=(2.0$^{+1.4}_{-0.8}$) 10^8 cm^{-3}, Pulse width Δt=(16$^{+15.6}_{-9.2}$) yr.

The branchings were described satisfactorily with constant values of these quantities, but one can assume that n_n, kT, and n_e are really time dependent. The values determined represent actually effective values. A simple time dependence would be $n_n(t)=n_0$ for $0 \le t < t_0$ and $n_n(t)=n_0 \exp(-t/\tau_p)$ for $t \ge t_0$. The effective neutron density $n_n^{eff.}$ can be calculated. $n_n^{eff.}$ is dependent on the specific parameters of the branching:

$$n_n^{eff.} = n_0 \frac{t_0(1-f_1^m-f_2^m)+\tau_p\{1-(f_1^m+f_2^m)(1-f_1^m-f_2^m)^{-1} \ln[(f_1^m+f_2^m)^{-1}]\}}{t_0(1-f_1^m-f_2^m)+\tau_p \ln[(f_1^m+f_2^m)^{-1}]} \qquad (2)$$

where we have defined, e.g. $f_1^m=\lambda_{\beta 1}/(\lambda_{\beta 1}+\lambda_{\beta 2}+n_0 \sigma v_T)$ with σ, $\lambda_{\beta 1}$ and $\lambda_{\beta 2}$ the capture cross section, the beta and electron capture decay rate of the branch point isotope, respectively. v_T is the thermal velocity. f_2^m is defined in the same way. Equation (4) is for three exit channels.

In the following I will limit my discussion to the special case t_0=0, $n_n(t)=n_0 \exp(-t/\tau_p)$. In the stellar model of Gallino (1989) such a neutron density profile is considered to be responsible for establishing the abundance patterns in the s-process branchings (Käppeler et al. 1990), although the model itself is much more complex.

The s-process analysis with a neutron density $n_n(t)=n_0 \exp(-t/\tau_p)$ means to carry out the calculations with effective values (integration over the time dependent neutron density) and adjust to the maximum neutron density n_0. A best fit of the branchings was obtained for $n_o=6\ 10^8\ cm^{-3}$, $kT=26\ keV$, and $n_e=1.9\ 10^{27}\ cm^{-3}$. The effective neutron densities of various important branchings are given in table 1. These values are all within the quoted uncertainty of the value $n_n=(2.0^{+1.4}_{-0.8})10^8\ cm^{-3}$ from the calculation with the constant neutron density assumption. The same holds true for the new values of kT and n_e. It should especially pointed out that the neutron density profile $n_n(t)=n_o\exp(-t/\tau_p)$ allows for an estimate of the maximum neutron density, n_o, but not for τ_p, the decay constant ($n_n^{eff.}$ in equation (2) is independent of τ_p).

The above considerations on effective neutron density are valid if the time variation of the neutron density, expressed by τ_p, is much slower than the effective time duration of the branch point isotope $t_{tot.}^{eff.}$. The branching must have time to adjust quickly to the variations of $n_n(t)$. This condition stated as an inequality is given by:

$$\tau_p \gg t_{tot.}^{eff.} = [\ \lambda_n^{eff.}+\lambda_\beta]^{-1} = [\ 0.5\ n_o\sigma v_T + \lambda_\beta]^{-1} \tag{3}$$

If τ_p is chosen greater than about 2 yr the inequality (5) is fulfilled for the studied branchings except for the ^{85}Kr branching, where an effective time $t_{tot.}^{eff.} = 4.6$ yr comparable to τ_p is computed (Table 1). The ^{85}Kr branching alone is, therefore, dependent on τ_p (Beer and Macklin 1989), and was used for its determination. The pulse width $\Delta t=\tau_p$ for the neutron density spike profile was found to be 5 yr (Fig. 2). The effect of the pulses on the other branchings is too small to be detected, they have lost memory of the duration of the pulse but not on maximum neutron density. Similar cases like ^{85}Kr are ^{63}Ni, and ^{79}Se, but the effects are masked by the weak s-process component. The most sensitive branchings next to ^{85}Kr would be the ^{151}Sm branching (Beer and Macklin 1988) and ^{95}Zr.

Table 1 Effective neutron densities and total effective life times of important branch point isotopes for a neutron density $n_n(t)=n_o\exp(-t/\tau_p)$ with $n_o=6\ 10^8\ cm^{-3}$ and $\tau_p=5$ yr, a temperature $kT=26$ keV, and an electron density $n_e=1.9\ 10^{27}\ cm^{-3}$. It should be pointed out that the final abundance pattern of the s-only isotopes is influenced frequently by additional side branchings and that in some cases a radiogenic contribution (e.g. ^{170}Tm) has to be taken into account. This might explain that in the average n_n in the table is higher than $2.0\ 10^8\ cm^{-3}$.

Branch point isotope	s-only isotope in the branching	$n_n^{eff.}\ 10^8$ (cm^{-3})	$t_{tot.}^{eff.}$ (yr)
^{85}Kr	$^{86,87}Sr$	2.3	4.6
^{95}Zr	^{96}Mo	3.0	0.25
^{148}Pm	^{148}Sm	3.5	0.019
^{151}Sm	^{152}Gd	1.5	0.22
^{163}Dy	^{164}Er	2.6	0.17
^{170}Tm	^{170}Yb	2.3	0.18
^{185}W	^{186}Os	2.8	0.18
^{192}Ir	^{192}Pt	2.6	0.12

In the pulsed s-process n_n, Δt, and τ_o are related via $r=\exp[-(\Delta t\ n_n v_T/\tau_o)]$, where r is the fraction of overlap between successive pulses. From the empirically determined astrophysical parameters we find almost complete overlap between pulses (r=0.92±0.06). This is in contradiction to AGB models, where r≈0.6 is predicted. r=0.6 would mean a higher neutron density and/or a still longer pulse width. Both are not compatible with the ^{85}Kr branching which, therefore, turns out to be the key branching in any s-process analysis. The problem of almost complete overlap between pulses cannot be solved with a single pulse model. It requires an extension to a double pulse model (Gallino 1989). For parametric studies it is useful to formulate a double pulse model phenomenologically. The double pulse solution σN^P is constructed from the unpulsed solutions σN_1 and σN_2 (with the same total exposure) using the convenient matrix notation:

$$\sigma N^P = [\, U - r_1 r_2 \exp(M_2 \Delta \tau_2)\, D_1 \exp(M_1 \Delta \tau_1)\, D_2 \,]^{-1} \{\, r_2 \exp(M_2 \Delta \tau_2)\, D_1 [\, U - r_1 \exp(M_1 \Delta \tau_1)\,]\, \sigma N_1 +$$
$$[\, U - r_2 \exp(M_2 \Delta \tau_2)\,]\, \sigma N_2 \} \qquad (4)$$

where the bold symbols are vectors and matrices, respectively. The indices 1 and 2 stand for pulse 1 and 2. **U**, **D**, and **M** are unity, decay and the coefficient matrices, respectively, as defined in Ward and Newman (1978). σN is the vector of the individual σN_A values.
For $\Delta \tau_1$ sufficiently large ($\sigma \Delta \tau_1 \gg 1$) and $\sigma N_1 \approx \sigma N_2$ which means equal neutron density, temperature and electron density for the pulses 1 and 2 we obtain a solution which resembles the single pulse result, where we have neglected quadratic terms. The solution

$$\sigma N^P \approx [\, U - r_2 \exp(M_2 \Delta \tau_2)\, D_1 \,]^{-1} \{\, U - r_2 \exp(M_2 \Delta \tau_2) \}\, \sigma N \qquad (5)$$

of eq. (5) means that our single pulse model results can be interpreted in a new way which allows us to choose for $r = r_1 r_2 \approx r_1 \approx 0.6$ and to fit the ^{85}Kr branching at the same time. According to eq. (5) the overall structure of the σN curve is chiefly due to the first pulse with a pulse width $\Delta t_1 \gg \Delta t_2$, whereas the second pulse provides the necessary fine tuning to adjust to the ^{85}Kr branching.
As in the analysis of s-process abundances from the stellar atmosphere of AGB stars useful data are available presently only for the ^{85}Kr and the ^{95}Zr barnchings, these two branchings turn out to be the key branchings to obtain details of the pulsed structure of the s-process in these stars. The ^{95}Zr branching allows for an estimate of the neutron density which is then prerequisite for the analysis of the ^{85}Kr branching.
The double pulse model considered by Gallino (1989) is characterized by a combination of the ^{13}C(α,n) and the ^{22}Ne(α,n) neutron sources. The two pulses are different in neutron density and temperature, therefore, although the exposures of pulse one and two have properties as we have stated in deriving eq. (5), the model of Gallino (1989) is only in the gross structures directly comparable with our phenomenological model. Parametric studies using eq. (4) are necessary.

REFERENCES

Anders, E., Grevesse, N., 1989, Geochim. Cosmochim. Acta 53, 197
Bao, Z., Y., Käppeler, F., 1987, Atomic Data and Nuclear Data Tables 36, 411

Beer, H., Macklin, R., L., 1988, Ap. J. 331, 1047

Beer, H., Macklin, R., L., 1989, Ap. J. 339, 962

Gallino, R., 1989, in Evolution of Peculiar Red Giant Stars, ed. H. R. Johnson,
    and B. Zuckerman (Cambridge University Press), p. 176

Käppeler, F., Gallino, R., Busso, M., Picchio, G., Raitieri, C., M., 1990, Ap. J., 354, 630

Kompe, D., 1969, Nucl. Phys. A133, 513

Macklin, R., L., Gibbons, J., H., 1965, Rev. Mod. Phys. 37, 166

McLaine, V., Dunfort, C. L., Rose, P., F., 1988, Neutron Cross Sections, Vol. 2,
    Academic Press New York

Mughabghab, S., F., Divadeenam, M., Holden, N., E., 1981, Neutron Capture Cross
    Sections, Vol. 1, Part A and B, Academic Press New York

Takahashi, K., Yokoi, K., 1987, Atomic Data and Nucl. Data Tables 36, 375

Ward, R., A., Newman, M., J., 1978, Ap. J. 219, 195

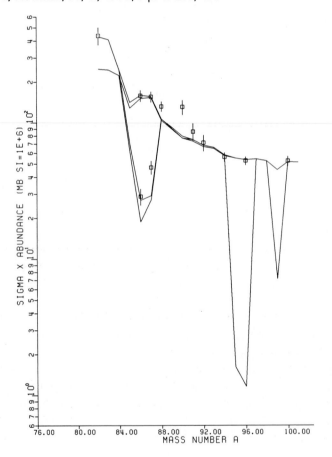

Fig. 2 The mass region of the ^{85}Kr and ^{95}Zr branchings. A pulse width of 5 yr is required to adjust to the empirical values of ^{86,87}Sr and avoid overproduction at ^{86}Kr.

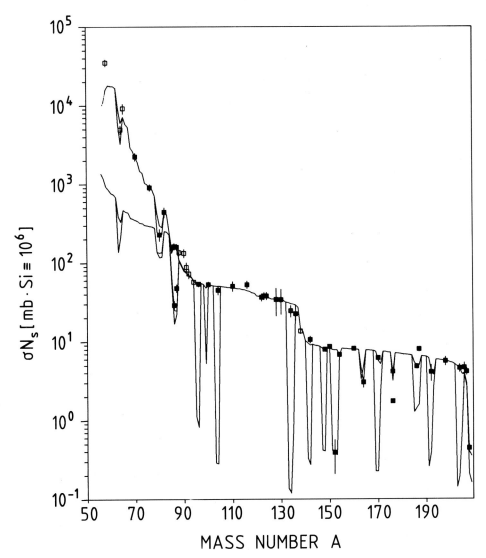

Fig.3 The σN calculation as a function of mass number. The main component s-process curve is plotted from A=56 to 209. In the region A=56-90 the composite curve differs from the main component curve due to the superposition of the weak component. Not all empirical data points shown are used to adjust the theoretical curve. Data points at the magic neutron numbers 50 and 82 (^{88}Sr, the Zr isotopes and ^{138}Ba) as well as the data points for ^{58}Fe, ^{64}Ni and ^{65}Cu are indicated as open symbols and shown only for comparison. Other empirical data too low (^{176}Lu) or too high (^{187}Os, ^{206,207,208}Pb) are affected by long lived radioactive decay and in the case of ^{206,207,208}Pb contain extra r-process contributions from transbismuth nuclei.

B E J Pagel: Would you expect observable differences in abundance ratios resulting from production by progenitors with different properties such as metallicity?

H Beer: This is in principle not a problem to be solved in the frame of a phenomenological model I presented here. The answer can be, of course, found by looking at the adjusted parameters, e.g. the adjusted fraction of iron seed the reverse of which is the overproduction factor.

Updating Solar Abundances

N. GREVESSE

Institut d'Astrophysique
Université de Liège
B-4000 Ougrée-Liège, Belgium

Abstract. Since the recent critical compilation and analysis of the solar and meteoritic abundances by Anders and Grevesse (1989), a few solar photospheric abundances have been revised. We present these new results together with a detailed discussion of the problem of the solar abundance of iron.

1 INTRODUCTION

The recent table of the abundances of the elements in the solar system by Anders and Grevesse (1989) resulted from a critical discussion of the abundance data as obtained from metorites and from the sun.

The accuracy of meteoritic abundance measurements has improved quite a lot and analyses of CI chondrites, which have escaped fractionation processes during their formation, have converged to the point where most elements are known to better than 10 percent.

In Anders and Grevesse (1989; see also the references therein) we discuss the different sources of solar abundance data (photosphere, chromosphere, corona, solar wind, solar energetic particles) and show that separation processes fractionate the gas supplied to the corona. Therefore, the abundances derived from the analysis of the photospheric spectrum provide the most reliable and most accurate set of solar abundances.

There is no doubt that the precision of solar photospheric abundances is a matter of precision in the atomic and molecular data and essentially in the transition probabilities. Professor D.E. Blackwell and his group at Oxford have shown during the last 15 years the crucial role played by atomic transition probabilities in our detailed knowledge of the structure and physical processes of the solar photosphere. Figure 1 illustrates the evolution of the solar abundance of iron and of the dispersion of the results as a function of the quality of the transition probabilities used.

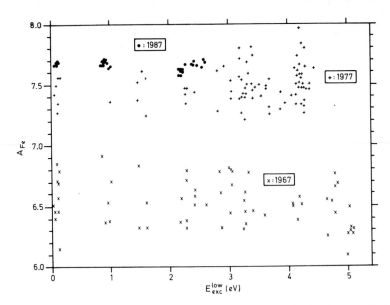

Figure 1 -   Evolution of the solar abundance of Fe, as derived from FeI lines, and
              of the dispersion of the results. The 1987 results are from Blackwell
              et al. (1984) using the very accurate Oxford gf- values of Blackwell
              et al. (1982). Earlier results are from the author, using the same
              solar data but gf-values available in 1967 and 1977 respectively.

Thanks to the improvements in the quality of the transition probabilities, the dispersions in the solar abundance results have decreased to the 10 percent level for many elements and large modifications in the absolute scales of the transition probabilities have changed the solar abundances accordingly.

In each case, past discrepancies between meteoritic and solar results have gone away as the solar abundances have become more accurate. As shown by Anders and Grevesse (1989; see their figure 4) the agreement between meteoritic and solar data is now remarkably good. If we retain only 29 accuratelly known elements in the photosphere whose abudances can be derived using a sufficient number of good quality lines with accurate transition probabilities, then the data agree perfectly : the mean difference is 0.000 $\pm$ 0.036 dex, *i.e.*9 %. There are however a few elements for which there is a rather large disagreement between meteorites and the sun (see Anders and Grevesse for a discussion). The problem of the most important of these disagreements, the case of iron, will be discussed in section 3.

2 NEW SOLAR PHOTOSPHERIC RESULTS

Since our latest compilation of the solar photospheric abundances [Anders and Grevesse (1989); see also Grevesse (1984a and b) and the references therein] a few elements have been revised. The new results are discussed hereafter.

The abundances of carbon and nitrogen have been redetermined from all the best indicators of the solar abundances of these elements including molecular as well as atomic lines. Grevesse *et al.* (1990b) derived the abundance of C from CH vibration-rotation lines, C_2 Swan and Phillips bands, CH electronic transition (A-X), CI and [CI] lines. From a careful discussion of the atomic and molecular data, all these indicators lead to a remarkable agreement, $A_C = 7.60 \pm 0.05$ (where $A_C = \log N_C/N_H$ in the usual scale where $\log N_H = 12.00$). The solar abundance of N has been determined from the vibration-rotation lines of NH in the infrared and from NI lines by Grevesse *et al.* (1990a); both results agree : $A_N = 8.00 \pm 0.05$.

The solar abundances of Cd, Sm and Th have also been revised on the basis of new accurate transition probabilities. The new results are $A_{Cd} = 1.77 \pm 0.11$ (Youssef *et al.*, 1990), $A_{Sm} = 1.01 \pm 0.06$ (Biémont *et al.*, 1989) and $A_{Th} = 0.23 \pm 0.08$ (Lawler et al., 1990). The new results for Cd and Sm agree with the meteoritic values of Anders and Grevesse (1989), $A_{Cd} = 1.76 \pm 0.03$ and $A_{Sm} = 0.97 \pm 0.01$ respectively. The case of Th is more puzzling because the new solar Th abundance is 40 percent larger than the meteoritic result, $A_{Th} = 0.08 \pm 0.02$.

3 THE PROBLEM OF IRON

The solar abundance of iron, as derived from low excitation lines ($E_{exc} \leq 2.6$ eV) of FeI (Blackwell *et al.*, 1984; Blackwell *et al.*, 1986), using the very accurate Oxford transition probabilities (Blackwell *et al.*, 1982), $A_{Fe} = 7.67 \pm 0.03$, is 40 percent higher than the very accurate meteoritic value, $A_{Fe} = 7.51 \pm 0.01$ (Anders and Grevesse, 1989).

The best indicator of the solar abundance of Fe is without any doubt the FeII line spectrum. But unfortunately, very accurate gf-values are still very rare for good quality FeII lines of solar interest (see hereafter).

Very recently O'Brian *et al.* (1990) measured transition probabilities of a large number of high-excitation lines ($E_{exc} \leq 5.00$ eV) of FeI with accuracies ranging from 10 to 25 %. These high-excitation lines should be better indicators of the solar abundance of Fe because they are less sensitive to temperature uncertainties than the low-excitation lines. Furthermore, they are also less sensitive to possible departures from local thermodynamic equilibrium as shown by Steenbock (quoted by Holweger, 1988). These new gf-values by O'Brian *et al.* (1990) agree with the Oxford data for low-excitation lines (Blackwell *et al.*, 1982) : $\Delta = \log gf(\text{Oxf.})$ $- \log gf(\text{O'Brian}) = -0.022 \pm 0.036$ for the lines in common measured by the technique of lifetime together with branching ratios and $\Delta = -0.046 \pm 0.040$ dex for lines measured by interpolating line intensities (see O'Brian *et al.*).

Preliminary results obtained by Grevesse *et al.* (1990c) using about 70 FeI lines with E_{exc} from 0 to 5 eV do not show any dependance against E_{exc}. The dispersion of the results is however larger than that obtained by Blackwell *et al.* (1984) when using the very high accuracy Oxford gf-values and the dip in the abundance observed around 2.2 eV by Blackwell *et al.* (1984) has almost disappeared. The mean iron abundance is $A_{Fe} = 7.61 \pm 0.07$, still larger by 25 % than the meteoritic value, $A_{Fe} = 7.51$.

Milford *et al.* (1989) also measured gf-values for FeI lines of high excitation with a claimed accuracy of 0.01 to 0.04 dex. Using most of their 30 lines, leads to A_{Fe} $= 7.61 \pm 0.14$. This very large dispersion is certainly largely due to uncertainties in the oscillator strengths, much larger than the ones quoted.

Very few accurate transition probabilities exist for good solar FeII lines, the best indicator of the solar abundance of iron. Recent works by Kurucz (1981), Moity (1983), Whaling (1985), Kroll and Kock (1987), Heise and Kock (1990) and Pauls *et al.* (1990), show large disagreements between lines in common. The highest accuracies are reported by Heise and Kock (1990) and Pauls *et al.* (1990), about 15 to 20 %, for a small number of good solar FeII lines.

Whaling (1985) from his gf-values for ultraviolet lines showed that Kurucz (1981) and Moity (19823)'s results were too large. When corrected, these gf-values, applied to derive the solar Fe abundance, lead to large dispersions of the results : A_{Fe} $= 7.73 \pm 0.08$ (gf-Moity corrected) and $A_{Fe} = 7.66 \pm 0.14$ (gf-Kurucz corrected).

Very recently, Pauls *et al.* (1990) used 3 very good near infrared FeII lines for which they measured the gf-values with accuracies of 15 to 20 %. They found $A_{Fe} = 7.66$ ± 0.06. On the other hand, Holweger *et al.* (1990) obtained a much lower value, $A_{Fe} = 7.48 \pm 0.09$, from 13 FeII lines (faint and medium strong lines) together with the gf-values of Heise and Kock (1990).

The explanation of this difference is not entirely clear although part of the difference can be explained as follows. When comparing the gf-value scales of Pauls *et al.* (1990) and of Heise and Kock (1990), if the difference is small, Δ (Pauls-Heise) $= -$

0.013 dex, the dispersion is large, $\pm$ 0.12 dex. And for two lines used by Holweger *et al.* (1990) the differences are rather large, $\Delta = -0.03$ dex (λ 4576.3 Å) and $\Delta -$ 0.12 dex (λ 7515.8 Å). The abundances derived from these two lines are $A_{Fe} = 7.63$ for both lines using the gf-values of Pauls *et al.* (1990) but $A_{Fe} = 7.60$ and $A_{Fe} = 7.51$ respectively using the gf-values of Heise and Kock (1990). Furthermore, part of the difference may come from too large damping factor enhancements used by Holweger *et al.* (1990). Their result might be increased by 0.10 dex if this damping factor enhancement is decreased from 2.5 (their value) to 1.5. To summarize, these few FeII lines could still lead to an iron abundance which is larger by 25 % or more than the meteoritic value.

We also revised the solar abundance of iron as derived from the very faint low excitation forbidden lines of FeII. The interest and the story of the discovery of these lines is described in detail by Pagel (1969; pp. 194-200). These lines should be very good indicators of the solar abundance of Fe because they are nearly model-independent and they should be formed in LTE. Furthermore, their gf-values should be better known then the corresponding theoretical data for permitted lines. Actually, Hansen and Biémont (private communication) recomputed the gf-values for a few [Fe II] lines and found their new data to be 10 % smaller than Garstang (1962)'s values. Using these new gf-values, we now find $A_{Fe} = 7.68 \pm 0.05$ from 10 [FeII] lines.

In summary we believe the solar iron abundance to be higher than the meteoritic value by 25 to 40 %. We nevertheless urgently need very accurate gf-values for the numerous very good FeII lines that are present in the solar spectrum in order to settle definitely the question.

References

Anders, E., Grevesse, N., 1989, Geochim. Cosmochim. Acta **53**, 197.

Biémont, E., Grevesse, N., Hannaford, P., Lowe, R.M., 1989, Astron. Astrophys. **222**, 307.

Blackwell, D.E., Booth, A.J., Petford, A.D., 1984, Astron. Astrophys. **132**, 236.

Blackwell, D.E., Booth, A.J., Haddock, D.J., Petford, A.D., Leggett, S.K., 1986, M.N.R.A.S. **220**, 549.

Blackwell, D.E., Petford, A.D., Simmons, G.J., 1982, M.N.R.A.S., **201**, 595 and references therein.

Garstang, R.H., 1962, M.N.R.A.S. **124**, 321.

Grevesse, N., 1984a, in *Frontiers of Astronomy and Astrophysics*, ed. R. Pallavicini, Florence, p. 71.

Grevesse, N., 1984b, Physica Scr. **T8**, 49.

Grevesse, N., Neuforge, C., Noels, A., 1990c, to be published.

Grevesse, N., Lambert, D.L., Sauval, A.J., van Dishoeck, E.F., Farmer, C.B., Norton, R.H., 1990a, Astron. Astrophys. **232**, 225.

Grevesse, N., Lambert, D.L., Sauval, A.J., van Dishoeck, E.F., Farmer, C.B., Norton, R.H., 1990b, Astron. Astrophys. (in press).

Heise, C., Kock, M., 1990, Astron. Astrophys. **230**, 244.

Holweger, H., 1988, in *The impact of Very High S/N Spectroscopy and Stellar Physics*, eds. G. Cayrel and M. Spite, Kluwer, Dordrecht, p. 411.

Holweger, H., Heise, C., Kock, M., 1990, Astron. Astrophys. **232**, 510.

Kroll, S., Kock, M., 1987, Astron. Astrophys. Suppl. **67**, 225.

Kurucz, R.L., 1981, Semiempirical calculations of gf-values, IV : FeII, SAO Special Report 390.

Lawler, J.E., Whaling, W., Grevesse, N., 1990, Nature **346** , 635.

Milford, P.N., O'Mara, B.J., Ross, J.E., 1989, J.Q.S.R.T. **41**, 433.

Moity, J., 1983, Astron. Astrophys. Suppl. **52**, 37.

O'Brian, T.R., Wickliffe, M.E., Lawler, J.E., Whaling, W., Brault, J.W., 1990, preprint.

Pagel, B.E.J., 1969, *Les Transitions Interdites dans les Spectres des Astres*, Institut d'Astrophysique, Liège, p. 189.

Pauls, U., Grevesse, N., Huber, M.C.E., 1990, Astron. Astrophys. **231**, 536.

Whaling, W., 1985, Kellog Rad. Lab., Caltech, Technical Report 84 A.

Youssef, N., Dönszelmann, A., Grevesse, N., 1990, Astron. Astrophys. (in press).

Discussion

Pagel - Bearing in mind that 20 years ago people were finding ways to explain an order of magnitude discrepancy between solar and meteoritic iron in the opposite sense, I cannot take the present discrepancy of 0.2 dex very seriously.

Grevesse - Things have changed since 1969. Most of the discrepancies between solar and meteoritic abundances have gone away thanks to improved transition probabilities for a large number of elements. The present remarkable agreement between the sun and meteorites indicates that the iron discrepancy must be taken seriously.

NUCLEOSYNTHESIS IN TYPE IB/IC SUPERNOVAE AND CHEMICAL EVOLUTION OF GALAXIES

K. NOMOTO, T. TSUJIMOTO, H. YAMAOKA, S. KUMAGAI, T. SHIGEYAMA
Department of Astronomy, Faculty of Science, University of Tokyo, Tokyo

ABSTRACT

Theoretical models of supernova explosions of various types are summarized to obtain heavy element yields from supernovae, in particular, for new models of Type Ib and Ic supernovae. Maximum brightness and decline rate of their light curves suggest that 12–18 $M_\odot$ stars produce larger amount of ^{56}Ni than more massive stars. We discuss relative roles of various types of supernovae in the chemical evolution of galaxies.

1. SUPERNOVA TYPES

Supernovae have been spectroscopically classified into Type I (SNe I) and Type II (SNe II) according to the absence and presence of hydrogen in their optical spectra. SNe I are further subclassified into Ia, Ib, and Ic. Unlike SNe Ia, SNe Ib exhibit strong absorption lines of [He I] at early times, and prominent emission lines of [O I], [Ca II], and Ca II at late times. SNe Ic spectroscopically resemble SNe Ib, except that the He I lines are absent at early times (Harkness and Wheeler 1990 for a review). SNe II are subclassified into SNe II-P (*plateau*), SNe II-L (*linear*), and SNe II-BL (*bright linear* like SN 1979C) according to the light curve shape (Branch *et al.* 1990).

However, the supernova classification has become more complicated since the discovery of SN 1987K whose spectral classification changed from Type II to Type Ib/Ic as it aged, thereby being called as SNe IIb (Filippenko 1988). Hydrogen feature has also been identified in the early time spectrum of SN Ic 1987M (Jeffery *et al.* 1990). Interestingly the early time spectra of SN 1987K are very similar to SNe Ic (1983V and 87M) (Filippenko 1988; Wheeler and Harkness 1990).

Among these types of supernovae, we mostly discuss nucleosynthesis in SNe Ib and Ic, and briefly mention other types of supernovae. We then discuss their roles in the chemical evolution of galaxies.

2. TYPE IB AND IC SUPERNOVAE

Wolf-Rayet stars with a wide range of masses have been proposed for the progenitors of SNe Ib and Ic, since most of SNe Ib/Ic are associated with star-forming regions

(Wheeler and Harkness 1990 for a review).

Recently Shigeyama *et al.* (1990), Hachisu *et al.* (1991), Nomoto *et al.* (1990a), Kumagai *et al.* (1990), and Yamaoka and Nomoto (1990) have calculated the progenitor's evolution, nucleosynthesis, Rayleigh- Taylor instabilities, and optical, X-ray, γ-ray light curves of exploding helium stars. They have suggested that the helium stars of 3–5 $M_\odot$ (which form from stars with initial masses $M_i \sim$ 12–18 $M_\odot$ in binary systems) are the most likely progenitors of typical SNe Ib/Ic and that SNe Ic progenitors are slightly less massive than those of SNe Ib. Their arguments are as follows.

(1) The light curves of SNe Ic decline faster than SNe Ib.

(2) The early time spectra of SNe Ic show the presence of hydrogen (Jeffery *et al.* 1990), while hydrogen is absent in SNe Ib.

It remains an open question whether the presence of hydrogen causes the difference between SNe Ib and Ic in their early time spectra.

2.1. Binary Evolution

The difference in the spectral feature between SNe Ic/IIb and Ib may be due to the presence of a thin envelope of hydrogen in SNe Ic/IIb immediately prior to the explosion. By evolving massive stars in close binary systems, we examine under what conditions, hydrogen can be left on the helium stars after mass exchange and wind-type mass loss.

We present some preliminary results for two cases 13A and 18A, where the initial masses of the primary stars are M_i = 13 $M_\odot$ (13A) and 18 $M_\odot$ (18A), and their Roche lobe radii are $R_{Roche} = 50\ R_\odot$. These stars evolve as follows.

(1) **Hydrogen-Helium Burning**: After hydrogen exhaustion, the star starts to expand. When the star fills its Roche lobe, the helium core of $\sim$ 3.1 $M_\odot$ (13A) and $\sim$ 4.2 $M_\odot$ (18A) has been formed. Figures 1 – 2 (upper) show the composition structure at the onset of mass transfer during core helium burning.

(2) **Roche Lobe Overflow**: Mass loss is calculated so as to keep the stellar radius equal to R_{Roche}. When the stellar mass has been decreased down to 3.4 $M_\odot$ (13A) and 5 $M_\odot$ (18A), the star starts to shrink toward the size of helium main-sequence stars. Figures 1–2 (middle) show that significant amount of hydrogen remains in a relatively thick layer below the surface (0.5 $M_\odot$ for 13A and 1 $M_\odot$ for 18A).

(3) **Wind Mass Loss**: During helium burning, the star loses its masses in a wind. We assume *mass-dependent* mass loss rate (Langer 1989): $\dot{M} = 2 \times 10^{-8}\ (M/M_\odot)^{2.5}$ $M_\odot$ yr^{-1}, and follow the evolution through carbon burning. Here the coefficient of $\dot{M}$ is a factor of 3 – 5 smaller than that adopted for Wolf-Rayet stars by Langer (1989). This mass loss rate leads to a difference of the surface abundance between 13A and 18A (Fig. 1–2, lower). For case 13A, hydrogen still remains in the layer

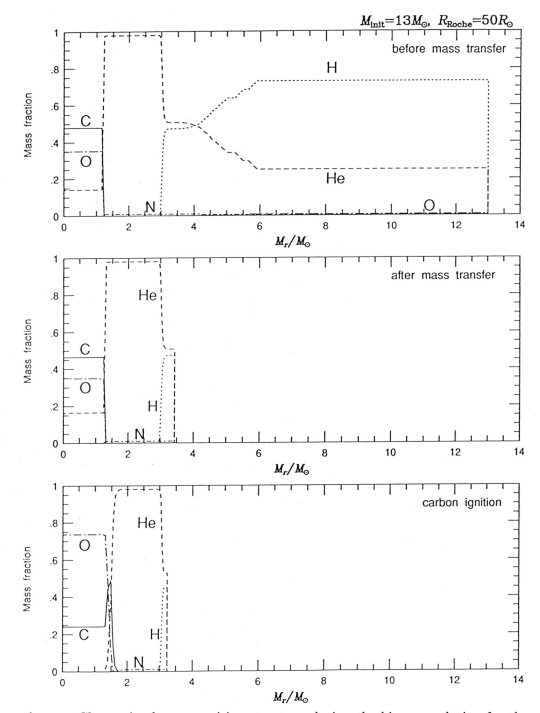

Fig. 1: Change in the composition structure during the binary evolution for the star with $M_i = 13\ M_\odot$ (Case 13A). After carbon ignition, evolution time scale is too short to change the stellar mass and the composition structure.

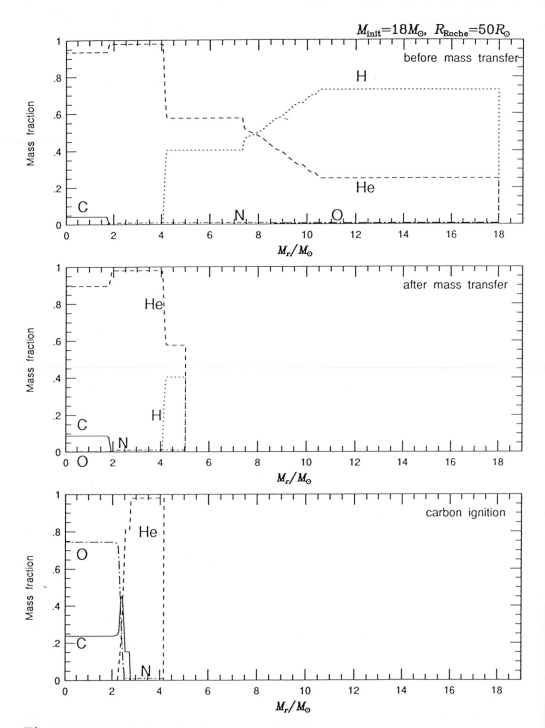

Fig. 2: Same as Fig. 1 but for $M_i = 18\ M_\odot$ (Case 18A).

down to $\sim 0.2\ M_\odot$ below the surface. For case 18A, by contrast, all hydrogen is lost in a wind. The final masses are 3.2 $M_\odot$ (13A) and 4.2 $M_\odot$ (18A).

Although more parameter study is needed, the present results already suggest that more hydrogen can remain on the stellar surface if the helium star had initially smaller main-sequence mass, while all hydrogen is lost in a wind for stars having larger initial masses. In other words, there may exist a critical mass that discriminates between the presence and absence of hydrogen, and that critical mass is larger if larger mass-dependent mass rate is adopted.

2.2. Nucleosynthesis and the Mass of ^{56}Ni

Assuming that the helium star progenitors of SNe Ib/Ic are formed in close binary systems as described above, Shigeyama *et al.* (1990) performed hydrodynamical calculations of the explosion of helium stars with masses $M_\alpha = 3.3, 4$, and 6 $M_\odot$. These are presumed to form from the main-sequence stars of masses $M_i \sim 13, 15$, and 20 $M_\odot$, respectively. These stars eventually undergo iron core collapse as in SNe II. A shock wave is then formed at the mass cut that divides the neutron star and the ejecta, propagating outward to explosively synthesize ^{56}Ni and other heavy elements.

Since the mechanism that transforms collapse into explosion is unclear, the mass cut and explosion energy are not known. The adopted presupernova models (Nomoto and Hashimoto 1988) have the following important difference from the previous models, i.e., the iron core masses are as small as 1.18 $M_\odot$ and 1.28 $M_\odot$ for $M_\alpha = 3.3\ M_\odot$ and 4 $M_\odot$, respectively, significantly smaller than 1.4 $M_\odot$ in the 6 $M_\odot$ star, due to the larger effect of Coulomb interactions during the progenitor's evolution. Because of steep density gradient at the outer edge of the iron core, it is reasonable to assume that the neutron star mass M_{NS} is close to the iron core mass. The final kinetic energy of explosion is assumed to be $E = 1 \times 10^{51}$ erg.

Behind the shock wave, materials are processed into nuclear statistical equilibrium (NSE) composition, mostly ^{56}Ni, if the maximum temperature exceeds 5×10^9 K. As derived from the approximate relation $E = 4\pi r^3/3\ aT^4$ (e.g., Woosley 1988), such a high temperature is realized in a sphere of radius $\sim 3700\ (E/10^{51}\ \mathrm{erg})^{1/3}$ km, which contains a mass M_{NSE} ($\sim 1.44 - 1.46\ M_\odot$ for $M_\alpha = 3.3$ and 4 $M_\odot$). Thus the mass of ^{56}Ni is approximately given by $M_{NSE} - M_{NS}$, which is larger for smaller M_{NS}. The calculated ^{56}Ni masses are 0.26 and 0.15 $M_\odot$ for $M_\alpha = 3.3$ and 4 $M_\odot$, respectively, which are large enough to account for maximum brightness of SNe Ib/Ic. Distribution of the nucleosynthesis products for $M_\alpha = 4\ M_\odot$ is shown in Fig. 3.

The oxygen masses are 0.21 and 0.43 $M_\odot$ for $M_\alpha = 3.3$ and 4 $M_\odot$, respectively, and could be consistent with those inferred from the late time spectra of SNe Ib/Ic in view of the strong dependence of the oxygen mass on the temperature of the ejecta

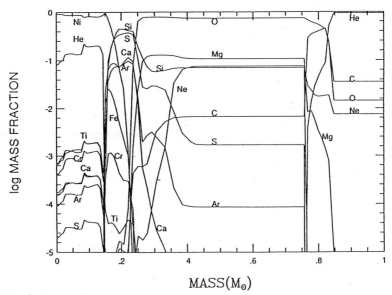

Fig. 3: Explosive nucleosynthesis in the 4 $M_\odot$ helium star (Shigeyama *et al.* 1990). Composition of the innermost 1 $M_\odot$ of the ejecta is shown. (The outermost 1.72 $M_\odot$ helium layer and the 1.28 $M_\odot$ neutron star are not included in the figure.) About 0.15 $M_\odot$ ^{56}Ni and 0.43 $M_\odot$ oxygen are produced.

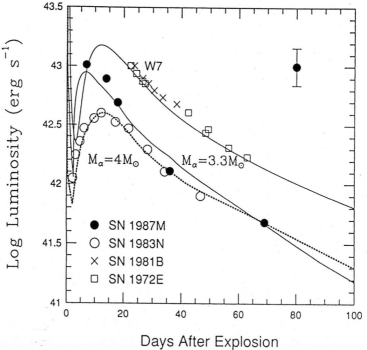

Fig. 4: Approximate bolometric light curve of SN Ic 1987M, and the bolometric light curves of SNe Ia 1972E and 1981B and of SN Ib 1983N. The predicted curves of the 3.3 $M_\odot$ model for SN 1987M, the 4 $M_\odot$ model for SN Ib, and the W7 model for SN Ia are indicated by solid and dotted lines. The error bar illustrates the 2σ photometric uncertainty in the SN 1987M points.

(e.g., Uomoto 1986).

2.3. Light Curve Models

Figure 4 shows the observed bolometric curves of SNe Ia 1972E and 1981B (Graham 1987), SN Ib 1983N (Panagia 1987), and the approximate bolometric light curve of SN Ic 1987M constructed from flux-calibrated spectra (Filippenko *et al.* 1990; Nomoto *et al.* 1990a). In each case, the observed light curve has been shifted along the abscissa to match the corresponding theoretical curve. The peak bolometric luminosities assume $H_0 = 60$ km s^{-1} Mpc^{-1}.

The previous Wolf-Rayet star models have some difficulties (1) in reproducing the light curves of typical SNe Ib which decline as fast as SNe Ia (Panagia 1987; Leibundgut 1988), and (2) in producing enough ^{56}Ni to attain the maximum luminosities of SNe Ib in relatively low mass helium star models (Ensman and Woosley 1988). In particular, Figure 4 demonstrates two important features of SN Ic 1987M: (1) its brightness fell somewhat more rapidly than that of SNe Ia and SN Ib 1983N (also SN Ic 1983I reported by Tsvetkov 1985), and (2) maximum brightness of 1987M is only 0.5 mag dimmer than in SNe Ia, and significantly brighter than in SN Ib 1983N which was ~ 1.5 mag fainter than SNe Ia at maximum (Uomoto and Kirshner 1985; Branch 1986; Panagia 1987). These features are very difficult to understand with the previous massive helium star models.

Figure 4 shows the calculated bolometric light curves of the exploding helium star models with $M_\alpha = 3.3\ M_\odot$ for SN Ic and 4 $M_\odot$ for SN Ib as well as the white dwarf model W7 for SNe Ia (Nomoto *et al.* 1990a). The amount of ^{56}Ni is 0.58 $M_\odot$ (W7), 0.26 $M_\odot$ (SN Ic), and 0.15 $M_\odot$ (SN Ib). The helium star models assume uniformly mixed distribution of elements from the center through the layer at 0.2 $M_\odot$ beneath the surface for both cases. Such a mixing may be due to the Rayleigh-Taylor instability during the explosion (Hachisu *et al.* 1991).

The calculated bolometric light curves of helium stars are powered by the radioactive decays of ^{56}Ni and ^{56}Co. Peak luminosity is reached when the time scale of heat diffusion from the radioactive source becomes comparable to the expansion time scale. Maximum brightness is higher if the ^{56}Ni mass is larger and the date of maximum earlier. After the peak, the optical light curve declines at a rate that depends on how fast γ-rays from the radioactive decays escape from the star without being thermalized. Since the escape probability of γ-rays is determined by the column depth to the Ni–Co layer, the optical light curve declines faster if the ejected mass is smaller and if ^{56}Ni is mixed closer to the surface.

Figure 4 shows that the bolometric light curve of $M_\alpha = 4\ M_\odot$ is in excellent agreement with SN Ib 1983N from the pre-maximum through day 50. Other well-observed SNe Ib generally have early-time visual and bolometric light curves whose shapes are

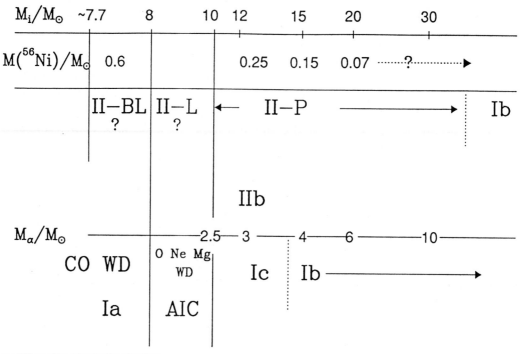

Fig. 5: Hypothetical connection between supernova types and their progenitors for single stars (upper) and close binary stars (lower). M_i and M_α are the initial mass and the helium star mass, respectively. AIC stands for accretion-induced collapse of white dwarfs.

Fig. 6: (a: left) The expanding supernova matter is excited by the decay of ^{56}Co into ^{56}Fe. (b: right) Collapse of an O+Ne+Mg core is induced by electron capture (H. Nomoto 1989).

nearly identical to those of SNe Ia (Porter and Filippenko 1987; Leibundgut 1988).

Compared with the 4 $M_\odot$ model for SNe Ib, the 3.3 $M_\odot$ model has several attractive features for SN Ic 1987M (Fig. 4). First, maximum brightness is higher by more than a factor of two because of the larger ^{56}Ni mass and the earlier date of maximum brightness. Second, decline in the tail is noticeably faster due to the more extensive mixing and smaller ejected mass. Compared with the W7 model for SNe Ia, the 3.3 $M_\odot$ model gives a lower maximum brightness (by about 0.6 mag) and a faster decline, just as observed in SN Ic 1987M (Fig. 4).

3. OTHER TYPES OF SUPERNOVAE

Figure 5 summarizes the initial masses M_i of the progenitors for the various types of supernovae currently proposed by several groups (e.g., Branch *et al.* 1990). The upper and lower rows respectively show the cases of single stars and helium stars of masses M_α (or white dwarfs) in close binary stars. The produced masses of ^{56}Ni are inferred from light curves based on the radioactive decay model (Fig. 6a: H. Nomoto 1989).

For SNe Ia, we adopt the carbon deflagration model W7 (Nomoto *et al.* 1984; Thiele-mann *et al.* 1986) as a standard model which can account for many of the observed features of SNe Ia such as early and late time spectra and light curves (see Woosley and Weaver 1986 and Wheeler and Harkness 1990 for reviews and references therein). In model W7, the amount of ^{56}Ni produced is $M_{Ni} = 0.58\ M_\odot$, and the explosion energy is $E = 1.3 \times 10^{51}$ erg.

In Fig. 5, SNe II-BL and II-L are tentatively assumed to be the explosions of AGB stars having degenerate C+O cores (carbon deflagration) and O+Ne+Mg cores (elec-tron capture collapse; Fig. 6b), respectively (Swartz *et al.* 1990). SNe II-L and II-BL do not show a plateau, which suggests that their progenitors have hydrogen-rich en-velope less than $\sim 1\ M_\odot$. This is consistent with the AGB star models since they are likely to have lost most of their hydrogen-rich envelopes at the explosion.

The above models imply that the yields of SNe II-BL are almost identical to those of SN Ia, while SNe II-L may not produce much heavy elements.

4. CHEMICAL EVOLUTION OF GALAXIES

Supernovae produce heavy elements in galaxies. SNe II and SNe Ib/Ic contribute in the early chemical evolution of galaxies. SNe Ia, on the other hand, enhances mostly iron in the late phase galactic evolution. To quantitatively demonstrate the relative nucleosynthesis roles of SNe Ia and SNe II, Nomoto *et al.* (1990b; also Yanagida *et al.* 1990) sum up the nucleosynthesis products as $[r\, X_{i,Ia} + (1-r)\, X_{i,II}]$ and determine r by minimizing the deviation from the solar abundances (Anders and Grevesse 1989).

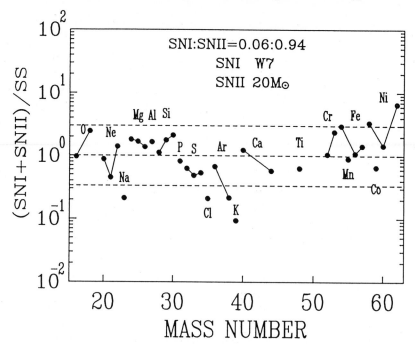

Fig. 7: The combined nucleosynthesis products from SN Ia (W7) and SN II (20 $M_\odot$ star). The SN Ia/SN II ratio of 0.17/0.83 is obtained by minimizing the deviation from the solar abundances.

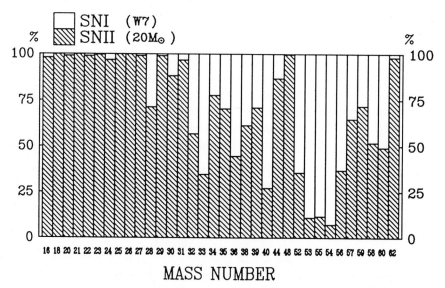

Decomposition into two components
for each isotope in the Solar System.

Fig.   8: The relative contributions of SN Ia and SN II for each species in best reproducing the solar abundances.

For SNe II, we simply assume that SNe II yield is represented by the 20 $M_\odot$ product (Hashimoto *et al.* 1989; Thielemann *et al.* 1990). For SN Ia, the W7 model is adopted. The best fit to the solar abundances is obtained for $r = 0.17$ as shown in Fig. 7, where the deviation from the solar is at most a factor of 3. The contribution of SNe Ia and SNe II for each species is shown in Fig. 8; for example, ~ 63 percent of Fe originate from SN Ia. The obtained ratio may be consistent with that observed estimate by Evans *et al.* (1989) though a large uncertainty is involved.

Nucleosynthesis in SNe II can be compared with the abundance of metal-poor stars. For this purpose, we use the relation between the supernova types and progenitor's masses in Figure 5 as a working hypothesis, and examine the relative contributions of different types of supernovae. Stars more massive than initially $\sim 12\ M_\odot$ end up as SNe II-P if single stars and as SNe Ib/Ic in close binary systems. Figure 5 predicts that, for $12\ M_\odot \lesssim M_i \lesssim 20\ M_\odot$, lower mass stars produce larger amounts of ^{56}Ni than do higher mass stars. On the other hand, the oxygen mass produced from these massive stars is larger for more massive stars, if no mass loss is assumed. However, if the very massive stars undergo significant mass loss to become a relatively small mass Wolf-Rayet stars, their heavy element production is saturated or even smaller than more massive stars. Therefore oxygen mass integrated over initial mass function (IMF) depends mainly on the upper mass limit of stars that do not evolve into Wolf-Rayet stars.

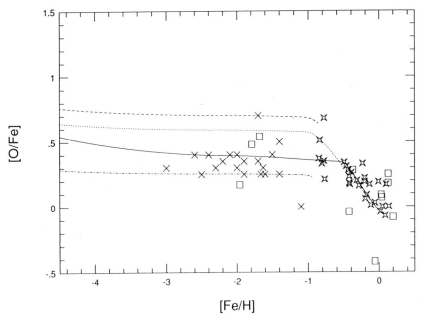

Fig. 9: The [O/Fe] ratio against metallicity [Fe/H] observed in low metallicity stars and calculated with four different theoretical models (see text).

Figure 9 shows the evolutionary change in [O/Fe] = $\log_{10}$ [(O/Fe)/(O/Fe)$_\odot$] as a function of metallicity [Fe/H]. (Observed points are taken from Andersen *et al.* 1988, Barbuy 1988, and Gratton and Ortolani 1986.) The decrease in [O/Fe] at [Fe/H] $\gtrsim$ -1 is due to the iron production from SNe Ia whose lifetime is assumed to be ~ 1 Gyr. For massive stars, integration of the yield over 12 – 100 $M_\odot$ (case A) and 12 – 35 $M_\odot$ (case B) is made with Salpeter's IMF, where the Fe mass produced from stars more massive than 20 $M_\odot$ is assumed to be 0.07 $M_\odot$. The resulting integrated yields of SNe II give [O/Fe] = 0.70 (case A: dashed line) and 0.26 (case B: dash-dotted line) at metallicity of [Fe/H] $\lesssim -1$ (solid line). The dotted line (case C) is the integration over 12 – 100 $M_\odot$ but oxygen yield is saturated at constant value for 40 – 100 $M_\odot$; the resulting [O/Fe] is marginally consistent with the observed [O/Fe] = 0.41 $\pm$ 0.15 averaged over low metallicity stars of [Fe/H] $\lesssim -1$ in the solar neighborhood (Gratton 1991). These cases demonstrate that oxygen production is largely contributed by 25 – 40 $M_\odot$ stars.

The solid line in Fig. 11 includes the production of 0.58 $M_\odot$ Fe from SNe II-BL assuming 7.6 $M_\odot \leq M_i \leq 8\ M_\odot$ for their mass range. To be consistent with [O/Fe] $\gtrsim 0.26$ for metal-poor stars, the mass range for SNe II-BL must be as narrow as 7.2 $M_\odot \lesssim M_i \lesssim 8\ M_\odot$. This is consistent with the fact that only one (or two) example of SN II-BL (i.e., SN 1979C and possibly 1980K) have been observed despite their brightness. If we take 7.6 $M_\odot$ – 8 $M_\odot$ for M_i of SNe II-BL, the Fe mass from SNe II-L is limited to $\lesssim 0.1\ M_\odot$.

We would like to thank M. Hashimoto, F.-K. Thielemann, A.V. Filippenko, D. Branch, S. Yanagida, S. Hayakawa, H. Saio, and Y. Yoshii for collaborative work on the subjects discussed in this paper. This work has been supported in part by the grant-in-Aid for Scientific Research (01540216, 01790169, 02234202, 02302024) of the Ministry of Education, Science, and Culture in Japan, and by the Japan-U.S. Cooperative Science Program (EPAR-071/88-15999) operated by the JSPS and the NSF.

REFERENCES

Anders, E., and Grevesse, N. 1989, *Geochim. Cosmochim. Acta*, **53**, 197.

Andersen, J., Edvardsson, B., Gustafsson, B., and Nissen, P.E. 1988, in *IAU Symp. 132, The Impact of Very High S/N Spectroscopy on Stellar Physics*, ed. G. Cayrel, de Strobel, M. Spite (Dordrecht: Kluwer), p.441.

Barbuy, B. 1988, *Astr. Ap.*, **191**, 121.

Branch, D. 1986, *Ap. J. (Letters)*, **300**, L51.

Branch, D., Nomoto, K., and Filippenko, A.V. 1990, preprint for *Comments on Astrophysics*.

Evans, R., van den Bergh, S., McClure, R.D. 1989, *Ap. J.* **345**, 752

Ensman, L., and Woosley, S.E. 1988, *Ap. J.*, **333**, 754.

Filippenko, A.V. 1988, *A. J.*, **96**, 1941.

Filippenko, A.V., Porter, A.C., and Sargent, W.L.W. 1990, *A. J.*, **100**, 1575.

Graham, J.R. 1987, *Ap. J.*, **315**, 588.

Gratton, R.G. 1991, in *IAU Symposium 145, Evolution of Stars: Photospheric Abundance Connection*, ed. G. Michaud.

Gratton, R.G., and Ortolani, S. 1986, *Astr. Ap.*, **169**, 201.

Hachisu, I., Matsuda, T., Nomoto, K., and Shigeyama, T. 1991, *Ap. J. (Letters)*, in press.

Harkness, R.P., and Wheeler, J.C. 1990, in *Supernovae*, ed. A. Petschek (Springer-Verlag), p. 1.

Hashimoto, M., Nomoto, K., and Shigeyama, T. 1988, *Astr. Ap.*, **210**, L5.

Jeffery, D., Branch, D., Filippenko, A.V., and Nomoto, K. 1990, preprint.

Kumagai, S., Shigeyama, and T., Nomoto 1990, in *SN 1987A and Other Supernovae*, ed. I.J. Danziger (Garching: ESO), in press.

Langer, N. 1989, *Astr. Ap.*, **220**, 135.

Leibundgut, B. 1988, Ph.D. thesis, Universität Basel.

Nomoto, H. 1989, *Exploring SN 1987A* (in Japanese), (Tokyo: Kodansha).

Nomoto, K., Filippenko, A.V., and Shigeyama, T. 1990a, *Astr. Ap.*, in press.

Nomoto, K., and Hashimoto, M. 1988, *Physics Reports*, **163**, 13.

Nomoto, K., Shigeyama, T., Yanagita, S., Hayakawa, S., and Yasuda, K. 1990b, in *Chemical and Dynamical Evolution of Galaxies*, eds. F. Ferrini, J. Franco, and F. Matteucci (Pisa: Giondini), in press.

Nomoto, K., Thielemann, F.-K., and Yokoi, K. 1984, *Ap. J.*, **286**, 644.

Panagia, N. 1987, in *High Energy Phenomena Around Collapsed Stars*, ed. F. Pacini (D. Reidel), p. 33.

Shigeyama, T., Nomoto, K., Tsujimoto, T., and Hashimoto, M. 1990, *Ap. J. (Letters)*, **361**, L23.

Swartz, D.A., Wheeler, J.C., and Harkness, R.P., 1990, *Ap. J.*, submitted.

Thielemann, F.-K., Hashimoto, M., and Nomoto, K. 1990, *Ap. J.*, **349**, 222.

Thielemann, F.-K., Nomoto, K., and Yokoi, K. 1986, *Astr. Ap.*, **158**, 17.

Tsvetkov, D.Yu. 1985, *Sov. Astr.*, **29**, 211.

Uomoto, A. 1986, *Ap. J. (Letters)*, **310**, L35.

Wheeler, J.C., and Harkness, R. 1990, *Phys. Rep.*, in press.

Woosley, S.E. 1988, *Ap. J.*, **330**, 218.

Woosley, S.E., and Weaver, T.A. 1986, *Ann. Rev. Astr. Ap.*, **24**, 205.

Yamaoka, H., and Nomoto, K. 1990, in *SN 1987A and Other Supernovae*, ed. I.J. Danziger (Garching: ESO), in press.

Yanagida, S., Nomoto, K., and Hayakawa, S. 1990, in *Proceedings of the 21st International Cosmic Ray Conference (Adelaide)*, **4**, 44.

TYPE II SUPERNOVA NUCLEOSYNTHESIS
AS A FUNCTION OF PROGENITOR MASSES

Friedrich-Karl Thielemann[1], Ken'ichi Nomoto[2], Takeshi Shigeyama[2], Takuji Tsujimoto[2], Masa-aki Hashimoto[3]

[1]Harvard-Smithsonian Center for Astrophysics, 60 Garden Street, Cambridge, MA 02138, USA

[2]Department of Astronomy, Faculty of Sciences, University of Tokyo, Japan

[3]Department of Physics, College of Arts and Sciences, Kyushu University, Fukuoka, Japan

ABSTRACT

We discuss the nucleosynthesis expected from type II supernovae (SNe II) over the mass range of progenitor stars. Results are presented for progenitor massses of 13, 15, 20, and $25 M_\odot$. SN1987A verified the results for the $20 M_\odot$ star within observational uncertainties. Predictions of detailed nucleosynthesis calculations for the other cases and an analytical model, based on the analysis of the $20 M_\odot$ star, are compared. We address the gradual change in composition as a function of stellar mass and the average contribution of SNe II to galactic chemical evolution. A comparison is made with abundances found in low metallicity stars, which formed from interstellar matter, supposed to have only experienced SN II enrichment. The amount of ejected Fe-group elements is still unclear, as the mass cut between neutron star and ejected envelope depends on the details of the explosion mechanism and cannot be addressed in a simple model with an initiated explosion. However, indications from SN1987A and SN Ib/c light curves are given and applied.

1. INTRODUCTION

Supernovae of type I (SNe I) and type II (SNe II) represent the sites of heavy element production (oxygen and beyond) in galaxies. SNe II, due to their origin from fast evolving massive stars, dominate the heavy element production in the early phase of galactic evolution. Thus, we can derive their nucleosynthesis from abundance observations in low metallicity stars. The aim of this article is to explain these abundance patterns by utilizing our present understanding of SN II nucleosynthesis. In the following we want to discuss briefly the general physics of SN II events. Section 2 will deal with explosive burning in SNe II and presents the predicted yields for the whole mass range of SNe II. In section 3 we predict abundance yields of SNe II, averaged over an initial mass function (IMF) and compare these with abundances in low metallicity stars.

The most desirable way to perform explosive nucleosynthesis calculations in SNe II would come from a hydrodynamical calculation, following the Fe-core collapse, the bounce at nuclear densities, and the propagating shock wave through the envelope which will be ejected. However, there exist still open problems with the supernova mechanism of massive stars (see e.g. Bruenn 1989; Cooperstein and Baron 1989,1990; Myra and Bludman 1989; or Wilson and Mayle 1988; and Mayle and Wilson 1988, 1990). Here we want to make use of the fact that typical energies of 10^{51} erg are observed and light curve as well as explosive nucleosynthesis calculations can be performed with an artificially induced shock wave of the appropriate energy (see Woosley and Weaver 1986; Shigeyama, Nomoto, Hashimoto 1988; Woosley 1988; Thielemann, Hashimoto, Nomoto 1990). There will, of course, be a variation of this value with progenitor mass, but only by small factors of 2-3. As we will show later, the temperatures obtained during the propagation of the shock wave are only proportional to $E^{1/3}$. Thus, we expect that this uncertainty introduces negligible errors. Larger uncertainties are expected from the missing knowledge of the exact structure of the core at the point in time when the shock wave starts propagating outward. The lack of knowledge of the shock structure introduces an additional uncertainty, because different ways of initiating the explosion (piston, thermal bomb, kinetic bomb) can lead to different results. Aufderheide, Baron, and Thielemann (1990) have studied these questions in detail and come to the conclusion that errors in the composition of up to 30% are introduced. While this is not negligible it is small enough in order to proceed with these studies and obtain a reasonable understanding of SN II nucleosynthesis before the SN II mechanism is understood completely.

One remaining problem, when performing calculations with initiated shock waves, cannot be avoided. The location of the mass cut between neutron star and ejected envelope is unclear and can only be deduced from observational constraints. In case of SN1987A the light curve, powered by decaying ^{56}Ni and ^{56}Co, could be utilized to determine the mass of ^{56}Ni produced, which is located in the innermost region of the ejecta and therefore provides information about the mass cut. For progenitors of different mass, where such information is not available, we have to search for other sources of information.

2. EXPLOSIVE NUCLEOSYNTHESIS AND THE COMPOSITION OF SN II EJECTA

Explosive nucleosynthesis has been discussed in many articles, in most detail in early parametrized calculations (Woosley, Arnett, Clayton 1973). The most important parameters are the maximum temperature T_{max} and the initial composition, characterized by $Y_e = \sum Z_i Y_i$ or the neutron excess $\eta = \sum (N_i - Z_i) Y_i = 1 - 2Y_e$, where the individual abundances are related to the number densities via $Y_i = n_i/(\rho N_A)$. Explosive burning stages can be categorized according to T_{max}: (i) explosive Si-burning for $T_{max} > 4 \times 10^9$K, with complete Si exhaustion occurring only beyond

5×10^9K, (ii) explosive O-burning for $T_{max} > 3.3 \times 10^9$K, and (iii) explosive Ne and C-burning for $T_{max} > 2.1 \times 10^9$K. For a more detailed account and a discussion in the context of SN1987A, a massive SN II of 20M$_\odot$, see e.g. Thielemann, Hashimoto, Nomoto (1990).

As discussed above, the most significant parameter in explosive nucleosynthesis is the temperature, and a good prediction for the composition can already be made by only knowing T_{max} without having to perform complex nucleosynthesis calculations. Weaver and Woosley (1980) already recognized that matter behind the shock front is strongly radiation dominated. Assuming an almost homogeneous density and temperature distribution behind the shock (which is approximately correct, see Fig.3 in Shigeyama, Nomoto, and Hashimoto 1988), one can equate the supernove energy with the radiation energy inside the radius R of the shock front

$$E_{SN} = \frac{4\pi}{3} R^3 a T^4. \tag{1}$$

This equation can be solved for R and with $T = 5 \times 10^9$K, the lower bound for explosive Si-burning with complete Si-exhaustion, and $E_{SN} = 10^{51}$erg, the result is $R \approx 3696$km (see Woosley 1988). For the evolutionary model by Nomoto and Hashimoto (1988) of a 20M$_\odot$ star, utilized in this calculation, this radius corresponds to 1.7M$_\odot$, in excellent agreement with the exact hydrodynamic calculation. Temperatures which characterize the edge of the other explosive burning zones correspond to the following radii: incomplete Si-burning (T_9=4, R=4977km), explosive O-burning (3.3, 6432), and explosive Ne/C-burning (2.1, 11750). This relates to masses of 1.75, 1.81, and 2.05M$_\odot$ in case of the 20M$_\odot$ star, which is also in quite good agreement with detailed hydrodynamic calculations. The agreement worsens slightly for lower temperatures because the radiation domination decreases, but on the other hand the exact boundary of explosive Ne-burning is not that important as the difference between abundances in explosive Ne-burning and the convective Ne/C-core is not that large. The radii mentioned are model independent and vary only with the supernova energy.

When applying the same procedure to other SN II progenitor models by Nomoto and Hashimoto (1988), and assuming an average supernova energy of 10^{51}erg, the masses in Table 1 result. Table 1 can be understood in the following way. Matter between the mass cut and the mass indicated with ex Si-c undergoes explosive Si-burning with Si-exhaustion. Then follows the zone of incomplete Si-burning until ex Si-i, explosive O-burning until ex O, explosive Ne/C-burning until ex Ne, and unprocessed matter from the C/Ne-core is ejected until C-core. The masses ΔM involved are displayed in the intermediate line which lists also the most abundant elements in these zones. As discussed before, we do not know a priori where the mass cut is located, and therefore cannot predict the total amount of ejected matter which experienced complete Si-burning (Si-c). The zones beyond explosive Ne/C-

TABLE 1
MASSES IN EXPLOSIVE AND HYDROSTATIC BURNING

BURNING SITE	MAIN PRODUCTS	13M$_\odot$	15M$_\odot$	20M$_\odot$	25M$_\odot$
Fe-core		1.18	1.28	1.40	1.61
mass cut		?	?	1.60	?
	"Fe", He	?	?	0.10	?
ex Si-c		1.42	1.46	1.70	1.79
	Si, S, Fe, Ar, Ca	0.06	0.06	0.05	0.06
ex Si-i		1.48	1.52	1.75	1.85
	O, Si, S, Ar, Ca	0.06	0.05	0.06	0.07
ex O		1.54	1.57	1.81	1.92
	O, Mg, Si, Ne	0.12	0.16	0.24	0.34
ex Ne		1.66	1.73	2.05	2.26
	O, Ne, Mg, Si	0.09	0.29	1.65	3.49
C-core		1.75	2.02	3.70	5.75

TABLE 2
MASS FRACTION OF ELEMENTS IN BURNING SITES

Element	C-core	ex Ne	ex O	ex Si-i	ex Si-c
O	0.72	0.80	0.45		
Ne	0.13	0.04			
Mg	0.09	0.08	0.005		
Si	0.02	0.08	0.30	0.40	
S			0.20	0.25	
Ar			0.025	0.06	
Ca			0.02	0.05	
Fe				0.20	0.70

burning ($T_{max} > 2.1 \times 10^9 \mathrm{K}$) are essentially unaltered and the composition is almost identical to the pre-explosive one.

Thielemann, Nomoto and Hashimoto (1990) gave the average composition for major elements in individual explosive burning stages, based on their calculations for SN1987A. These are displayed in Table 2. When taking the masses ΔM from Table 1 for the individual burning zones and multiplying them with the mass fractions given in Table 2, one can predict abundance yields for this whole mass range of SN II progenitors. The explosive yields should be quite insensitive to this procedure and quite accurate, because they depend only on the peak temperatures, densities, and the neutron excess of matter. The latter should be very similar outside the Fe-core for different stellar masses. An exception might be material which undergoes complete Si-burning, because the mass cut and with it the neutron excess will depend on the individual star and the details of the explosion mechanism. This also underlines the major uncertainty hidden in these yields. With the explosion mechanism not completely understood yet, one has to assume a position of the mass cut, and dependent on that position and the compression occurring during the collapse prior to the ejection of these mass zones, matter will have a variety of neutron excess values and might even contain r-process nuclei. Another uncertainty can be found in the zones containing only hydrostatically processed material. Here we expect a changing C to O ratio, reflecting the increasing temperatures in core He-burning of more massive stars. Ne and Mg as products of C-burning could change as well.

We also performed detailed explosive nucleosynthesis calculations for all four progenitor masses in the same way as in Thielemann, Hashimoto, Nomoto (1990). The results are presented in Table 3 together with the analytic predictions following from the utilization of Tables 1 and 2 (second column). A detailed account for the complete isotopic composition, resulting from the explosion calculations will be given in Thielemann et al. (1991ab). From the comparison we can see that the agreement between analytical predictions and detailed calculations is quite reasonable. One finds differences up to 30%. For the lighter elements O, Ne, Mg, which result from hydrostatic burning stages, the variation of the hydrostatic yields with the progenitor mass is the reason and the composition of a 20M$_\odot$ star is not necessarily representative. For the heavier elements the inaccuracy is due to uncertainties related to the (artificial) initiation of the explosion. Aufderheide, Baron and Thielemann (1990) concluded that all initiated SN II calculations will have such an intrinsic error and one cannot expect the calculations to be more accurate in any case.

TABLE 3
MAJOR NUCLEOSYNTHESIS YIELDS

Element	13M$_\odot$		15M$_\odot$		20M$_\odot$	25M$_\odot$	
C	0.060	0.062	0.083	0.084	0.115	0.148	0.152
O	0.218	0.188	0.433	0.359	1.480	3.000	2.816
Ne	0.028	0.017	0.039	0.044	0.257	0.631	0.467
Mg	0.012	0.018	0.046	0.039	0.182	0.219	0.342
Si	0.047	0.053	0.071	0.058	0.095	0.116	0.142
S	0.026	0.027	0.023	0.025	0.025	0.040	0.029
Ar	5.5-3	5.1-3	4.0-3	4.8-3	4.5-3	7.2-3	5.4-3
Ca	5.3-3	4.2-3	3.3-3	4.0-3	3.7-3	6.2-3	4.4-3
Fe	?0.240?		?0.150?		0.075	?0.050?	

The content of Table 3 indicates an interesting behavior. While the heavier intermediate mass nuclei originate only from explosive O and Si-burning, which contributes similar amounts for all progenitor masses (see also Table 1), the lighter elements, C through Si , have dominant or essential contributions from hydrostatic burning (C/Ne-core) or explosive Ne-burning. For both latter cases we see a tremendous mass dependence and a strong reduction of the involved mass zones for less massive stars. The mass cut between ejecta and neutron star was taken in such a way that the 0.075M$_\odot$ of ejected ^{56}Ni for SN1987A (a 20M$_\odot$ star) could be reproduced and the lightcurves of SNe Ib/Ic by 15 and 13M$_\odot$ stars, respectively, which lost (almost all) their H-envelope by mass transfer in a binary system (for details see Shigeyama et al. 1990, Nomoto et al. 1990, Hachisu et al. 1990).

3. AVERAGED ABUNDANCE YIELDS

Galactic chemical evolution calculations take into account the continuous enrichment of the interstellar medium by SNe I and SNe II, stellar winds (planetary nebulae) etc.. In the very early evolution of the Galaxy only the most massive stars contribute because of their short lifetime. At time t only those stars with $\tau_{MS}(M) < t$ can be considered (using the main sequence lifetime as an approximate measure for the lifetime until the onset of a supernova event in massive stars). If we have varying nucleosynthesis yields with stellar mass, this will lead to varying abundance ratios [x/Fe] in the ISM as a function of time or metallicity [Fe/H]=log$_{10}$[(Fe/H)/(Fe/H)$_\odot$], which can also be taken as a time indicator.

Matteuchi (1987), Matteuchi and Francois (1989) and Mathews, Bazan and Cowan (1990) find that for a typical star formation rate in the solar neighborhood 30M$_\odot$ stars will contribute for the first time at [Fe/H]$\approx$-3.9, 12M$_\odot$ stars at -3 and the least massive SNe II somewhere between -3 and -2. Intermediate mass stars will enrich the interstellar medium for [Fe/H]$\geq$-2 via planetary nebula ejection. SNe I, which come from binary systems of intermediate mass stars, are further delayed in time and appear at [Fe/H]$\approx$-1. The elements listed in Tables 2 and 3 can only derive from supernovae (with the exception of C) and therefore can be contributed solely to SNe II for [Fe/H]<-1. In the range -2.5$\leq$[Fe/H]$\leq$-1 we will expect averaged values for [x/Fe], because SNe II of the whole progenitor mass range contribute. Only below -3 do we expect deviations due to the selection effects, singling out more and more massive SNe II with decreasing [Fe/H]. Therefore it is not surprising that observations show a constant [x/Fe], x being O, Mg, Si, S, Ca, Ti, Cr, Ni between -2.5 and -1. The integrated yields of SNe II should therefore result in an abundance pattern as found in low metallicity stars.

In a first attempt we have tried a crude method to test whether these considerations are consistent with the results from our explosion calculations. We utilized the coarse grid given by a set conisisting of 13, 15, 20, and 25M$_\odot$ stars – neglecting the contribution from less massive stars – and performed an integration over the abundance yields of the individual elements $M_{ej,el}(M)$, weighted with a Salpeter initial mass function (IMF)

$$M_{el} \propto \int_{13}^{100} M_{ej,el}(M) M^{-2.35} dM. \qquad (2)$$

We extrapolated the yields smoothly up to M=40M$_\odot$, from where they were kept constant. This relates to the fact that more massive stars will loose large amounts of envelope mass, in form of stellar winds, during their early evolution, so that the later evolution resembles that of less massive stars. The integration over an IMF gives the same abundance ratios as full scale galactic chemical evolution calculations, provided that the IMF and the star formation rate (SFR) are constant in time. The results are given in Table 4 and compared to the values found in low metallicity stars. The observational values are taken from the reviews by Gehren (1988), Wheeler, Sneden, and Truran (1989), and Lambert (1989). The observational data show a scatter of 0.1–0.2.

TABLE 4

INTEGRATED ABUNDANCE YIELDS

Ratio	detailed model	analytical estimate	observed for [Fe/H]$<$-1
[C/Fe]	-0.13	-0.11	0.0
[O/Fe]	0.68	0.64	0.5
[Ne/Fe]	0.76	0.58	-
[Mg/Fe]	0.59	0.90	0.4
[Si/Fe]	0.41	0.50	0.4
[S/Fe]	0.22	0.06	0.5
[Ar/Fe]	0.13	0.01	-
[Ca/Fe]	0.24	0.10	0.3

The results are encouringing, considering the crude method and the fact that the observations probably have errors of 0.1-0.2 dex. They agree within the observational errors for all elements, with the exception of S. The analytical model gets a wrong Ne/Mg ratio, due to the utilization of the the hydrostatic composition of the 20$M_\odot$ star, and lower ratios for S, Ar, and Ca, because the detailed calculations show an increase for these yields from 20 to 25$M_\odot$, an effect not observed for the analytical yields. In order to obtain these results it was, however, necessary to assume mass cuts for the 13, 15, and 25$M_\odot$ stars. These were taken from papers discussing the explanation of SNIb/c light curves with stars of initial main sequence masses of 13 and 15$M_\odot$ (Shigeyama et al. 1990, Nomoto et al. 1990, Hachisu et al. 1990), and require the ejection of 0.24 and 0.15$M_\odot$ of ^{56}Ni for the 13 and 15$M_\odot$ stars, respectively. For the 25$M_\odot$ and more massive stars we assumed the ejection of 0.05$M_\odot$ of ^{56}Ni. This reflects the fact that the exponential tails of SNe II do not seem to get much dimmer than in case of SN1987A with 0.075$M_\odot$. This result depends, of course, strongly on the correct reproduction of SN Ib/c light curves and spectra, interpreting these events as explosions of massive stars in binary systems which lost the complete H-envelope to the binary companion (Ib) or retained a minor amount of the H-envelope (Ic), which is sufficient to suppress the spectral He-lines and leeds to only minor Hα-features. The need for the relatively small He-cores (3.3 and 4$M_\odot$, respectively) is given by the steep slope of the light curve (Ic similar to Ia) which requires an early escape of x-rays and gamma-rays from ^{56}Co-decay, thus steepening the light curves in comparison to the pure ^{56}Co exponential decay slope.

We also neglected SNe II in the mass range 8-10$M_\odot$ which will undergo collapse to neutron star densities as well as C- through Si-burning in one continuous burning stage initiated by e-capture in a strongly degenerate core. 10-13$M_\odot$ stars are also strongly affected by core degeneracy and have a very steep density gradient at the edge of the Fe-core. In both cases minor amounts of explosive nucleosynthesis ejecta are expected (see e.g. the Crab nebula), although it is not completely clear whether these are negligible. But the C/O ratio in the hydrostatic C-cores will be larger than for more massive stars and with the strong statistical weight of the lower mass stars (IMF), the [C/Fe] will increase and the [O/Fe] ratio decrease towards the observed value. We see that, when doing so, one obtains a picture which is consistent with observations. This is not a direct proof for our correct choice. Ni-ejecta increasing with mass of the progenitor star but giving the same IMF-integrated yield, would produce the same result. Observational evidence can only come from [x/Fe]-ratios for stars with [Fe/H]$\leq$-2.5, where (in time) the lower-mass core-collapse supernovae could not have exploded yet and only more and more massive stars are contributing with decreasing [Fe/H]. At [Fe/H]$\approx$-4 one would expect [x/Fe]-ratios which are only affected by stars with masses M>30$M_\odot$ and dominated by the least massive ones, because of the steep slope of the IMF. Our list in Table 3 does not yet include such massive stars yet, so we take the yields of the 25$M_\odot$ star as a close estimate. Some values would be [Mg/Fe]$\approx$1.11 and [Ca/Fe] $\approx$0.37. This seems to coincide with the very recent observations by Molaro and Bonifacio (1990). This is a very encouraging result and a complete survey over such low metallicity objects (hopefully forthcoming, Beers et al. 1990) can be a final proof for the correct choice of the neutron star mass cut.

This research was supported in part by NSF grant AST 89-13799, NASA grant NGR 22-007-272, the Grants-in-Aid for Scientific Research of the Ministry of Education, Science, and Culture in Japan (01540216, 01790169, 02234202, 02302024), and the U.S.-Japan Cooperative Science Program (INT 88-15999). The computations were performed at the National Center for Supercomputer Applications at the University of Illinois (AST 890009N).

References
Aufderheide, M., Baron, E., Thielemann, F.-K. 1990, *Ap. J.*, in press
Beers, T., Preston, G.W., Shectman, S.A. 1990, this volume
Bruenn, S.W. 1989, *Ap. J.* **340**, 955
Cooperstein, J., Baron, E. 1989, in *Supernovae*, ed A. Petschek, (Springer-Verlag, New York), p.xxx

Cooperstein, J., Baron, E. 1990, in *Supernovae*, ed S.E. Woosley, (Springer-Verlag, New York), in press

Danziger, I.J., Bouchet, P., Gouiffes, C., Lucy, L. 1990, in *Supernovae*, ed S.E. Woosley, (Springer-Verlag, New York), in press

Gehren. T. 1988, in *Rev. in Mod. Astronomy* 1, p.52

Hachisu, I., Matsuda, T., Shigeyama, T., Nomoto, K., 1990, *Ap. J. Lett.*, in press

Lambert, D.L. 1989, in *Cosmic Abundances of Matter*, ed. C.J. Waddington, AIP Conf. Proc. 183, p.168

Mathews, G.J., Bazan, G., Cowan, J.J. 1990, in preparation

Matteucci, F. 1987, in *Stellar Evolution and Dynamics of the Outer Halo of the Galaxy*, eds. M. Azzopardi, F. Matteucci, *ESO Conf. Proc.* 27, p.609

Matteucci, F., Francois, P. 1989, *M.N.R.A.S.*, 239, 885

Mayle, R.W., Wilson, J.R. 1988, *Ap. J.* 334, 909

Mayle, R.W., Wilson, J.R. 1990, in *Supernovae*, ed S.E. Woosley, (Springer-Verlag, New York), in press

Molaro, P., Bonifacio, P. 1990, *Astron. Astrophys.*, submitted

Myra, E.S., Bludman, S. 1989, *Ap. J.* 340, 384

Nomoto, K., Filippenko, A.V., Shigeyama, T. 1990, *Astron. Astrophys.*, in press

Nomoto, K., Hashimoto, M. 1988, *Phys. Rep.* 163, 13

Shigeyama, T., Nomoto, K., Hashimoto, M. 1988, *Astron. Astrophys.* 196, 141

Shigeyama, T., Nomoto, K., Tsujimoto. T.,Hashimoto, M. 1990, *Ap. J. Lett.* 361, L23

Thielemann, F.-K., Hashimoto, M., Nomoto, K. 1990, *Ap. J.* 349, 222

Thielemann, F.-K., Nomoto, K., Hashimoto, M. 1990, in *Chemical and Dynamical Evolution of Galaxies*, eds. F. Ferrini, J. Franco, F. Matteucci (Giardini, Pisa-Lugano), in press

Weaver, T.A., Woosley, S.E. 1980, *Ann. N.Y. Acad. Sci.* 366, 335

Wheeler, J.C., Sneden, C., Truran, J.W. 1989, *Ann. Rev. Astron. Astrophys.*, 27, 279

Wilson, J.R., Mayle, R.W. 1988, *Phys. Rep.* 163, 63

Woosley, S.E. 1988, *Ap. J.* 330, 218

Woosley, S.E., Arnett, W.D., Clayton, D.D. 1973, *Ap. J. Suppl.* 26, 231

Woosley, S.E., Weaver, T.A. 1986, *Ann. Rev. Astron. Astrophys.* 24, 205

Varani, G.-F., Meikle, W.P.S., Spyromilio, J., Allen, D.A. 1990, *MNRAS*, in press

W. Hillebrandt: The abundance compilation for SN1987A of Danziger et al. gives an iron abundance twice the amount of initial cobalt, inferred from the light curve. In your calculations always almost all the Fe comes from ^{56}Ni(Co). Can you explain this apparent discrepancy? If real, could it tell us something about the explosion mechanism?

F. Thielemann: I cannot explain the discrepancy because it is also in disagreement with observations by Varani et al. (1990) which give a strong constraint for the amount of Fe ejected as Fe rather than Ni or Co (already $0.01M_\odot$ of stable Fe would lead to a result outside their error bars). That would not allow for a factor of 2 ($=0.075M_\odot$) and I would rather expect the discrepancy to come from uncertainties in the abundance determinations (which Danziger et al. quote to be a factor of 2-3).

M. Peimbert: Is there a limiting mass above which a SN explosion would produce a black hole?

F. Thielemann: There is probably such a mass. The main question is the value of the maximum neutron star mass which depends on the nuclear equation of state. In our models a $25M_\odot$ star would produce a central remnant of $M=1.55M_\odot$. The maximum neutron star mass lies somewhere between 1.6 and $2M_\odot$.

Stellar Abundances – Recent Progress and Expectations

BENGT GUSTAFSSON

Uppsala University Astronomical Observatory

1 INTRODUCTION

Ten years ago an ESO workshop was arranged on "Methods of Abundance Determination for Stars". It is obvious from a re-reading of the proceedings (Nissen and Kjär, 1980) that a number of contemporary improvements in the observational and theoretical study of stellar spectra contributed an optimistic atmosphere to the conference. To which extent have these expectations been fulfilled? It seems fair to answer that we have seen a revolution in stellar spectroscopic analysis - the field has been cultivated far beyond a state when the question "Can astrophysical abundances be taken seriously?" was seriously asked. A more qualified answer is, however, depending on the context. In applied work we see very great advances - in particular the limiting magnitudes of spectroscopic analyses with a given reasonable accuracy are now raised by at least 5^m. As regards our basic understanding of the physics of stellar atmospheres, there is also very significant recent progress – however, not always yet applied in stellar abundance analyses.

At the ESO workshop Bernard Pagel gave an excellent summary, where he introduced two categories of stellar spectroscopists, "broad sweepers" and "ultimate refiners". The members of the first category are mainly interested in obtaining a vast amount of data for many, often faint, stars in order to study problems of nucleosynthesis, stellar or galactic evolution. They are satisfied with abundance determinations with some uncertainty (say, with typical errors of about 0.2 dex), at least if the systematic errors are under reasonable control. The "ultimate refiners", on the other hand, are worried by the fact that significant errors still exist between accurately observed and computed stellar spectra. They try to use these discrepancies to learn more about the stellar atmospheres as such, with the ultimate hope to improve the abundance determinations to errors less than, e.g., 0.05 dex. In the following review, reflecting my own interests, I will mainly comment on the advance made and the problems met by "ultimate refiners" in recent years, although some reference will also be made to work of immediate significance from a "broad sweeping" perspective. One of my points will be that these two perspectives are more linked together than one would think.

Various aspects of this active field of research have been summarized in a number of recent review papers, e.g., by Gehren (1988), Gustafsson (1987, 1988, 1989), Lambert (1988, 1989, 1990), Nissen (1990), Smith (1989), Spite and Spite (1985), by various authors in the book by Wehrse (1990) and by Wheeler, Sneden and Truran (1989).

2 GREAT IMPROVEMENTS - AN OVERVIEW

The expectations concerning the future of stellar spectroscopy expressed at the 1980 workshop have been fulfilled. The development of effective spectrometers and array detectors with linear response has led to far reaching improvements in abundance analyses, not yet fully exploited systematically. E.g., with a resolution of 2×10^4 and a S/N of 50 one can now make detailed analyses of B stars close to the Main Sequence in the Large Magellanic Cloud ($V \sim 14^m$). With future 10 m or 16 m telescopes this limiting magnitude will increase to 16^m - 17^m, which may enable analyses of A stars or K giants in the LMC or, with some future spectrometer development, of B stars near the main sequence in M31 (cf. Baschek's article in Wehrse 1990). With a S/N degraded to 5 one may even be able to reach F dwarfs in the LMC.

Of great significance, at least for the analysis of stars at the ends of the spectral sequence, is the widened spectral range towards ultraviolet (IUE, HST) and infrared (Reticons, Fourier-Transform Spectrometers and cryogenic echelles under construction). This widening of the accessible spectral range has made it possible to find more suitable abundance diagnostics in many cases.

The impressive improvements in laboratory measurements and calculations of atomic and molecular oscillator strengths (gf values), as well as in other basic physics data, are similarly important for the model atmospheres and for the calculation of detailed spectra. This development, as regards the atomic gf values, has been reviewed at this conference by Donald Blackwell (see also Gustafsson, 1990) and will not be discussed further here.

Since an abundance derived from an unsaturated spectral line is proportional to the observed equivalent width, divided by the gf value, high S/N spectra at high resolution and free of serious systematic errors in combination with accurate gf values should lead to precise abundance estimates, provided that the conversion to abundances via model atmospheres works properly. It was expected at the ESO workshop that the most significant remaining source of error in many analyses published in the eighties would be in the model atmospheres.

The improvement in the modelling of stellar atmospheres and spectra has been truly impressive during the last decade. This is not only due to the new generation of computers, but also due to new numerical schemes and deeper physical understanding. In most grids of classical model atmospheres constructed during the seventies (cf. e.g. Bell *et al.* 1976 and Kurucz 1979) the line blanketing data were still not complete, local thermodynaic equilibrium, hydrostatic equilibrium and plane-parallel stratification were assumed to prevail, and the convective fluxes were calculated using the inadequate "mixing-length theory". In a number of recent studies these "model approximations" have been more or less relaxed.

These improvements in stellar spectroscopy have led to very significant improvements in stellar abundance analysis for the hot stars (in particular for the O- and early B-type stars), as well as for M giants and supergiants, and for carbon stars. Also for Wolf-Rayet stars the progress is striking, although much work remains to be done before accurate quantitative abundance results are obtained for them. For solar-type stars, including G- and K-type giants, the understanding of physical processes is now considerably deepened; moreover stars at much greater distances may be analysed which is of particular significance for these stars, since the contamination of their atmospheres by mixing or diffusion is believed to be rather limited which makes them useful for studies of galactic chemical evolution. Finally, for late K- and M-type dwarfs modern techniques have yet not been applied in systematic analyses, mostly because these stars are fairly faint.

3 EARLY-TYPE STARS

For the stars of early spectral types very impressive methodological advances in modelling were made already in the seventies, in particular by Mihalas and collaborators (cf. Mihalas 1978 for references) and later by Kudritzki's group in Munich (for a review see Kudritzki 1988). The basic assumption of LTE was abandoned in calculations of spectra and, for hydrogen and helium, of model atmospheres. This development has continued. Here, the invention of the Lambda operator perturbation method by Scharmer (1981), partly based on early work by Rybicki (1971) and Cannon (1973), has led to a great revolution in radiative transfer calculations. The flexible and elegant method has been applied to problems with moving media (Scharmer, 1984), to multi-level problems (Scharmer and Carlsson, 1985, Werner and Husfeld, 1985) in stationary as well as expanding atmospheres (Pauldrach and Herrero, 1988, Hamann and Wessolowski, 1990 and papers referenced therein, cf. also Hillier, 1990) and to the calculation of non-LTE blanketed model atmospheres (Werner 1986, 1989). This development (see also Kalkofen, 1987) is of significance for the modelling of atmospheres and spectra of all sorts of stars, as well as of other astrophysical objects; in particular, however, the method has mostly been applied in studies of early-type stars.

One example of the application of the Lambda operator perturbation method is the work by Werner (1988) where non-LTE blanketed model atmospheres for subluminous O stars and central stars of planetary nebulae were calculated. Hundreds of atomic levels and radiative transitions can be taken into consideration with this method. Severe non-LTE effects were found on the temperature structure of the outer layers in the O-star models. However, the effects on abundance determinations when LTE models were replaced by the more detailed non-LTE models for these stars were found to be rather small.

A more statistical method for treating blanketing by a vast number of lines in non-LTE has been developed by Anderson (1985, 1987). The spectrum is represented by about 200 blocks, each containing photons from different wavelength bands but with similar histories. The calculation of the detailed distribution of photons within each block lags one interation behind, which limits the number of unknowns in the primary problem to a few hundred. (The method is somewhat reminiscent of that of the wellknown ODF-method for LTE blanketing and the Variable Eddington-factor technique for the angular dependence of the radiative field.) Anderson has published a couple of blanketed early-type models. For a late O-type model Anderson (1985) finds a electron temperature structure not very much different from LTE models, but with shifted ionization balances towards higher ion states and with a much stronger ultraviolet radiation field for $h\nu > 24$ eV. The hydrogen lines of this model are close to those of less blanketed non-LTE models.

Another major recent improvement for early-type stars is the development of self-consistent models of winds of hot stars, in particular by the Munich group. The effects of the back-scattering of the winds on the outer photospheric layers, proposed by Remie and Lamers (1982) for explaining the helium ionization balance, and later investigated by Abbott and Lucy (1985) and Abbott and Hummer (1985), have been modelled in considerable detail (Pauldrach, Puls and Kudritzki, 1986, Kudritzki, Pauldrach and Puls, 1987) and selfconsistent non-LTE models for the spherically symmetric photosphere and wind have been constructed (Gabler *et al.* 1989). These efforts have led to a number of interesting results. Pauldrach *et al.* (1990) have demonstrated that the spectral wind signatures are depending on stellar luminosity, which should enable an independent primary determination of distances from stellar spectrocopy, a result of great cosmological significance. However, quantitative non-LTE analyses of chemical abundances seem to be only marginally affected by sphericity effects and winds for O-type stars.

The first model-atmospere abundance analyses of some credibility of Wolf-Rayet stars have recently been published. Thus Hamann and Schmutz (1987) and Wessolowski, Schmutz and Hamann (1988) and, independently, Hillier (1987, 1989) have published non-LTE transfer calculations with the approximate Lambda operator technique, or similar methods, for dense expanding shells, schematically assumed to represent Wolf-Rayet star atmospheres. The hydrogen abundances recently reported by Hamann *et al.* (1990), agree astonishingly well with those estimated earlier by Conti *et al.* (1983) with more primitive methods. Hillier (1988, 1989) has analysed one WN and one WC star and confirmed them to be very nitrogen and carbon rich, respectively.

A great number of abundance analyses have been published during the last decade for early-type stars where LTE model atmospheres were used but the calculation of model spectra was made with due consideration of departures from LTE. These departures are often found to be important for O-type stars. For B-type stars the blue spectrum of He shows rather small non-LTE effects for the rather weak lines. The abundance effects are also rather small for O II and N II since many spectral lines of different strengths are available and an artificially enhanced microturbulence parameter is compensating for the strong line effects (cf. Becker and Butler, 1988, 1989). For C II and Mg II, and several other elements represented by few rather strong lines, the effects are, however, very important.

For the early A-type dwarf Vega Gigas (1986, 1988) has determined the abundances of magnesium, iron and barium in a detailed non-LTE analysis. The non-LTE effect on the Fe/H determination was calculated to be < 0.3 dex; the effects on the Mg abundance seem negligible while Ba II is significantly affected (by about 0.3 dex). Gigas found the star to be iron and magnesium poor relative to the Sun by a factor of three, while the barium abundance was less reduced. Also the helium abundance is low. Similarly, Lemke (1989) analysed 16 A-type dwarfs and found one of them (HR 5959) to be quite metal poor ($[Fe/H] < -1.0$), while the helium abundance of this star is 3 times greater than normal. It seems reasonable to ascribe these abnormal abundances to diffusion processes. The most important sources of error in the analysis are the temperature uncertainties (leading to typically 0.1 dex in $[Fe/H]$) and the errors in photoionization cross sections and electron collision cross sections, both contributing uncertainties of the same order of magnitude in the analysis of the Fe I spectrum, but very little for the $[Fe/H]$ determination from Fe II lines.

4 SOLAR-TYPE STARS

A most impressive improvement in the modelling of solar-type atmospheres is the construction of 3D hydrodynamic inhomogeneous models with radiative transfer by Nordlund (1982) and Nordlund and Dravins (1990). In the latter papers models for 4 stars on, or close to, the main sequence, ranging from Procyon (F5 IV-V) to Alpha Cen B (K1 V), are described and compared with spectroscopic observations. The general behaviour of wavelength shifts and line asymmetries, different for the different stars, is found to be reproduced remarkably well, as was earlier found for the Sun by Dravins, Lindegren and Nordlund (1981, cf. also Dravins 1982). Major limitations in these *ab initio* simulations are the relatively sparse grid of points that one can afford in space (32x32x32; in the horizontal directions, however, representing Fourier components) and the assumption of anelastic convection, filtering out sound waves. These limitations are more relaxed in Nordlund's recent work with R. Stein, where compressible convection with sound waves and shocks, treated with adaptive grid methods, is allowed (cf. Stein *et al.* 1989). Steffen (1989) and Steffen, Ludwig and Krüss (1989) also allow compression and a greater spatial resolution, however in a model confined to two space dimensions. They obtain smaller temperature fluctuations and more stationary flows, which is ascribed by Nordlund and Dravins (1990) to be due to their confinement to 2D.

The effects of convection and inhomogeneities on equivalent widths are unfortunately not yet known. Holweger, Heise and Kock (1990), using the 2D simulations by Steffen and collaborators, argue that the effects should be rather small on abundances (on the order of 0.1 dex or less for the iron abundance) with minor differential effects for lines with different excitation. One would guess that Nordlund's 3D models should show greater effects, but this remains to be proven.

The extended new semiempirical atomic-line data by Kurucz (1986) have been used for constructing grids of models within the traditional plane-parallel LTE mixing-length framework (Kurucz 1990, Edvardsson *et al.* 1990). These models are found to reproduce the ultraviolet flux observed for the Sun and similar stars much better than earlier models - the long-standing problem of the "unknown ultraviolet opacity" in the Sun is thus essentially explained by the veil of thousands of very weak metal lines, as was early suggested by Holweger (1970, see also Gustafsson and Bell, 1979). Anderson (1989), using the scheme described in Sec. 3, has calculated a blanketed solar model with departures from LTE allowed for eight elements, including iron. He finds a more shallow temperature gradient in the non-LTE model, compared to a corresponding LTE model, at visual optical depths around 10^{-3}. This agrees well with an empirical solar model based on CO line strengths (Grevesse, private communication). In the outer layers Anderson's non-LTE model is much more cooled through the CO lines. This is basically because the metal lines in the surface layers are decoupled from the thermal gas as they are formed more in scattering processes. The effects of non-LTE blanketing on abundance results can be estimated to be typically 0.1 dex, mostly because of the changed temperature gradient.

Departures from LTE in solar-type stars for individual elements may lead to serious systematic errors in abundances. The tendency for high-excitation metal lines to give higher abundances than lines from lower atomic levels, found for Pollux (K0 III) by Ruland *et al.* (1980) and later approximately reproduced as a over-ionization effect in statistical-equilibrium calculations by Steenbock (1985), has been found also in an interesting study by Magain and Zhao (1990) of 13 metal-poor dwarfs. These authors find the effect to depend on effective temperature and gravity in the expected way, but the size of the effect (0.2 dex in abundance) is astonishingly great and difficult to reproduce in statistical-equilibrium calculations, although considerable over-ionization is found (Lemke, 1990). A similar tendency traced for metal-poor disk dwarfs by Edvardsson *et al.* (1990) may possibly reflect systematic errors in the temperature scale. It should be noted here that the theoretical calculation of non-LTE effects in late-type stars is severely hampered by the uncertainty in cross-sections for inelastic collisions of atoms with hydrogen atoms (cf. Steenbock and Holweger, 1984).

Oxygen abundances for stars of different overall metal abundance are important tracers of nucleosynthesis in the early Galaxy. The conflicting $[O/Fe]$ abundances (by up to 1 dex!) for metal-poor stars derived from the forbidden lines by Barbuy (1988) and others for red giants and by Abia and Rebolo (1989) from the oxygen triplet at 777 nm for dwarfs have been discussed by the latter authors. Recently Kiselman (1990) has carried out detailed non-LTE calculations and found considerable corrections to the LTE abundances from the triplet lines; these corrections make the two sets of abundance data agree fairly well, but some problems remain like a significant discrepancy between observed and calculated solar triplet line strengths.

Empirical consistency tests of abundance estimates are important, not the least for solar-type stars where contemporary standard model atmospheres still are far from the physical reality. Often these tests lead to reassuring results but with some remaining discrepancy which may contain further information about the stellar atmosphere for the "ultimate refiner". Some recent examples of such tests will be given:

A recent discussion of the solar iron abundance (also commented on by Grevesse at the present conference) has a starting point in the problem that the solar Fe/Si abundance ratio, with Fe abundances derived from Fe I lines and oscillator strengths from the measurements of the Blackwell group at Oxford, is about 0.15 dex greater than the corresponding ratio in carbonaceous chondrites, thought to be representative of the Solar System heavy-element composition. Holweger, Heise and Kock (1990) have instead used Fe II lines with recent laboratory gf values. The analysis of these lines should be less model dependent. A lower Fe abundance, in accordance with the chondrite results, is obtained, and the departures from the Fe I result is ascribed to problems with convective inhomogeneities or non-LTE in the solar atmosphere or in the Oxford furnace. Pauls, Grevesse and Huber (1990), however, have measured gf values for three other weak infrared Fe II lines and find from these an iron/silicon ratio consistent with that obtained from Fe I lines. Holweger, Heise and Kock (1990) claim that this may be due to difficulties in estimating proper oscillator strengths in the approach of Pauls *et al.* The situation needs clarification.

Another example of a consistency test is the abundance analysis of the three components of the 36 Oph system by Cayrel de Strobel *et al.* (1989), leading to very similar results for the two hottest components while the coldest one (K5 V) has marginally lower abundances of Si, Fe and Ni (by 0.15, 0.06 and 0.05 dex, respectively). The authors tentatively ascribe this effect to departures from radiation equilibrium in the upper layers of the latter star, caused by a chromosphere which is traced in the profile of the IR Ca II triplet.

In their detailed analysis, entitled "an exercise in caution", of the Hyades giant Gamma Tau (K0 III) Griffin and Holweger (1989) find an Fe abundance significantly lower (by 0.16 dex) than that derived by, e.g., Cayrel *et al.* (1985) for the Hyades dwarfs. This difference could also be the result of enchanced chromospheric activity, modifying the temperature structure.

Boesgaard and Friel (1989, 1990) have measured the scatter in iron abundance and in carbon/iron ratio for stars in a number of galactic open clusters and moving groups in a high-precision study. They find a homogeneous composition for the cluster stars, but report a significant scatter among group members.

5 COOL STARS

The progress in cool star abundance analysis has been very considerble in recent years (cf. the review by Gustafsson 1989), due to the use of new infrared spectrometers. Another reason for this development is the construction of grids of model atmospheres with reasonably adequate data for molecular blanketing. These models are, however, still in LTE and in most cases in hydrostatic equilibrium. Typical examples are the grids published by the Indiana group (e.g., Brown *et al.* 1989) for M giants and the carbon-star models from Uppsala, used e.g. by Lambert *et al.* (1986) in their study of CNO abundances in N-type stars. In the Brown *et al.* grid the opacity data for TiO and H_2O are still not fully satisfactory. In the Uppsala C star models, absorption from diatomic and polyatomic molecules like HCN and H_2C_2 were included (later also C_3, cf. Jørgensen *et al.* 1989), but the polyatomic bands, of vital importance for the atmospheric structure (cf. Eriksson *et al.* 1984, Jørgensen 1990) tend to come out too strong. The reason for this may be partly the temperature scale (cf. Jørgensen 1989) and partly non-thermal heating of the outer atmosphere which has to be inhomogeneous (Jørgensen and Johnson 1990). There is also a possibility that further polyatomic molecular absorption or absorption from dust, formed at unexpectedly high temperatures (Frenklach *et al.* 1989), significantly affects the atmospheres.

The uncertainties, due to remaining model errors, in the abundances of M and C stars are hard to establish; they are thought to exceed 0.2 dex in several cases (cf., e.g., Lambert *et al.* 1986) but few systematic studies of this has been made. One exception is the investigation of the errors made when adopting plane-parallel models instead of spherically symmetric ones, recently carried out by Plez (1990). He finds these errors to be rather small, with some interesting exceptions. E.g., the CO V-R bands behave differently when the atmospheric extension increases; their relative strengths may be used for estimating the extension and thus for estimating both stellar radius and mass from stellar spectra.

Models for pulsating stellar atmospheres like Miras have been published in recent years by Bowen (1988) and by Bessell *et al.* (1989). The hydrodynamic models by Bowen contain dust formation but radiative transfer is treated quite schematically. The Bessell *et al.* models are more detailed in the latter respect but dust formation is not allowed for, and the molecular opacities are rather schematic. They have been used by Brett (1990) for studying the CN and ZrO bands in Miras in the Magellanic Clouds. Bounds on the C/O ratios are obtained but only rough abundance estimates may be obtained as yet with these first-generation models.

As examples of checks of the internal consistency of abundances of cool stars we may mention the use of vibration-rotation lines of NH and CH by Lambert *et al.* (1986) to check their N-type star nitrogen and carbon abundances, primarily derived from electronic transitions of CN and C_2 molecules, respectively. The NH test comes out satisfactorily, while the calculated CH lines are far too strong, compared with the observed ones. This discrepancy is still not understood. Another consistency test for the same set af abundances is the one by Olofsson *et al.* (1990), based on observations of CO and HCN lines from the circumstellar shells of these carbon stars. Within the uncertainties the circumstellar line strengths are consistent with the photospheric abundances; some indication that the photospheric nitrogen abundances may be underestimated by a factor of 2 or so is, however, present.

6 SUMMARY AND COMMENTS ON THE ART OF SWEEPING
We have found above that the introduction of more physically correct models of stellar

atmospheres and spectra seem to lead to important effects on detailed abundance analyses in certain cases, while the effects are smaller or negligible in other cases. Very schematically one may sum up the conclusions as follows:

Non-LTE blanketing: Rather small effects for abundance determinations for hot stars, some effects for solar-type stars.

Non-LTE spectra: Quite important for hot stars, of significance for solar-type stars (possibly more for metal-poor ones), no clear evidence as yet for cool stars.

Winds: Small effects for hot stars (except for Wolf-Rayet stars), possibly important if dust forms for cool giants.

Sphericity: Not so serious for most detailed analyses except for very extended stars, departing from hydrostatic equilibrium.

Convection, inhomogeneities: Of some (not well known) significance for solar-type stars and possibly cool stars. May well be more severe for metal-poor (and thus more transparent) dwarfs.

Pulsations: Of fundamental significance for the analysis of Miras, while the effects for irregular cool variables are not known.

Chromospheres: May possibly change the upper atmospheric structure enough to affect abundance analyses for cool stars and active solar-type stars.

Many of these uncertainties are both greater and more difficult to estimate for abundance criteria used at lower resolution, e.g., at greater distances. Therefore, "broad sweepers" should have serious doubts concerning such criteria as long as these have not been scrutinized very carefully. A calibration of such diagnostics, using high-resolution high S/N observations of weak and well understood spectral lines in a wide set of calibration stars, may be preferred to a purely theoretical calibration based on calculated spectra. The latter are, however, necessary complements to limit systematic errors, e.g., caused by extrapolation of low resolution criteria to stars with abundance patterns that are not represented within the set of calibration stars.

An alternative for abundance analyses of apparently faint stars is to use a high spectral resolution but a low S/N. One example of such a method is that of Carney *et al.* (1987) and Laird *et al.* (1988). The spectrum between 515 and 525 nm was observed for a great number of solar-type stars at a reciprocal resolution R around 30,000 and a S/N of 4. The spectra were calibrated with synthetic spectra (with some empirical correction) and iron abundances were derived with an estimated accuracy around 0.1 dex (Schuster and Nissen, 1989, found independently a good agreement between metal abundances obtained from Strömgren photometry, calibrated with high-resolution spectroscopy, and those of Laird *et al.*) One may, in fact show that, as long as the spectrometer is effective, the read-out noise significantly smaller than the signal and relatively extended spectral ranges are covered (so that many spectral lines can be observed), one should always resolve the spectra, even if the S/N gets disgustingly small. *The art of sweeping broadly is to use a brush with dense bristles.*

7 ENDING REMARKS

Since the pioneering studies, among which was Bernard Pagel's important work on abundances in metal poor stars, there has been an impressive development in stellar abundance analysis. This development has been stimulated and fostered by contributions, keen interest, challenging demands and clear criticism from the pioneers. Still much work remains before we can exploit the high signal/noise that modern observational techniques give or promise also for faint stars, and the high accuracy that laboratory physicists and theorists already achieve for the gf values and other basic data. Maybe, in a decade from now we shall obtain abundances for stars in the Nuclear Bulge, or even in nearby external galaxies, that are good to 0.05 dex or even better, and where this error is safely known. Maybe, we shall then be able to discuss real abundance differences, smaller than 0.03 dex, between more nearby stars. Nobody knows what may come out from such work, but most probably it will be worth the effort. The way there will be hard but interesting, in the spirit of those pioneers that once made this field an intricate but graceful science.

References

Abbott, D. C., and Lucy, L. B. 1985, Astrophys. J. **288**, 679.
Abbott, D. C., and Hummer, D. G. 1985, Astrophys. J. **294**, 286.
Abia, C., and Rebolo, R. 1989, Astrophys. J. **347**, 186.
Anderson, L. S. 1985, Astrophys. J. **298**, 848.
Anderson, L. S. 1987, in Kalkofen (1987), p. 163.
Anderson, L. S. 1989, Astrophys. J. **339**, 558.
Barbuy, B. 1988, Astron. Astrophys. **191**, 121.
Becker, S. R., and Butler, K. 1988, Astron. Astrophys. **201**, 232.
Becker, S. R., and Butler, K. 1989, Astron. Astrophys. **209**, 244.
Bell, R. A., Eriksson, K., Gustafsson, B., and Nordlund, Å. 1976, Astron. Astrophys. Suppl. **23**, 37.
Bessell, M. S., Brett, J. M., Scholz, M., and Wood, P. R. 1989, Astron. Astrophys. **213**, 209.
Boesgaard, A. M. 1989, Astrophys. J. **336**, 798.
Boesgaard, A. M., and Friel, E. D. 1990, Astrophys. J. **351**, 467 and 480.
Bowen, G. H. 1988, Astrophys. J. **329**, 299.
Brett, J. 1990, Monthly Notices Roy. Astron. Soc., in press.
Brown, J. A., Johnson, H. R., Alexander, D. R., Cutright, L. C., and Sharp, C. M. 1989, Astrophys. J. Suppl. **71**, 632.
Cannon, C. J. 1973, Astrophys. J. **185**, 621.
Carney, B. W., Laird, J. B., Latham, D. W., and Kurucz, R. L. 1987, Astron. J. **94**, 1066.
Cayrel, R., Cayrel de Strobel, G., and Campbell, B. 1985, Astron. Astrophys. **146**, 249.
Cayrel de Strobel, G., Perrin, M.-N., Caryrel, R., and Lebreton, Y. 1989, Astron. Astrophys. **225**, 369.
Conti, P. S., Leep, E. M., and Perry, D. N. 1983, Astrophys. J. **268**, 228.
Dravins, D. 1982 , Ann. Rev. Astron. Astrophys. **20**, 61.

Dravins, D., Lindegren, L., and Nordlund, Å. 1981, Astron. Astrophys. **96**, 345.

Edvardsson. B., Gustafsson. B., Lambert. D. L., Nissen, P. E., Andersen, J., and Tomkin, J. 1990, Astron. Astrophys., to be submitted.

Eriksson, K., Gustafsson, B., Jørgensen, U. G., and Nordlund, Å. 1984, Astron. Astrophys. **132**, 37.

Frenklach, M., Carmer, C. S., and Feigelson, E. D. 1989, Nature **339**, 196.

Gabler, R., Gabler, A., Kudritzki, R. P., Puls, J., and Pauldrach, A. W. A. 1989, Astron. Astrophys. **226**, 162.

Gehren, T. 1988, in *Reviews in Modern Astronomy* 1, Cosmic Chemistry, Springer-Verlag, Berlin, p. 52.

Gigas, D. 1986, Astron. Astrophys. **165**, 170.

Gigas, D. 1988, Astron. Astrophys. **192**, 264.

Griffin, R. E. M., and Holweger, H. 1989, Astron. Astrophys. **214**, 249.

Gustafsson, B. 1987, in *ESO Workshop on Stellar Evolution and Dynamics in the Outer Halo of the Galaxy* (ed. M. Azzopardi and F. Matteucci), ESO, Garching, p. 33.

Gustafsson, B. 1988, in *The Impact of Very High S/N Spectroscopy on Stellar Physics* (ed. G. Cayrel de Strobel and M. Spite), Kluwer Academic Publ., Dordrecht, p. 333.

Gustafsson, B. 1989, Ann. Rev. Astron. Astrophys. **27**, 701.

Gustafsson, B. 1990, in 22nd EGAS Conf., Physica Scripta, in press.

Gustafsson, B., and Bell, R.A. 1979, Astron. Astrophys. **74**, 313.

Hamann, W.-R., and Schmutz, W. 1987, Astron. Astrophys. **174**, 173.

Hamann, W.-R., and Wessolowski, U. 1990, Astron. Astrophys. **227**, 171.

Hillier, D. J. 1987, Astrophys. J. Suppl. **63**, 947, 965.

Hillier, D. J. 1988, Astrophys. J. **327**, 822.

Hillier, D. J. 1989, Astrophys. J. **347**, 392.

Hillier, D. J. 1990, Astron. Astrophys. **231**, 116.

Holweger, H. 1970, Astron. Astrophys. **4**, 11.

Holweger, H., Heise, C., and Kock, M. 1990, Astron. Astrophys. **232**, 510.

Jørgensen, U. G. 1989, Astrophys. J. **344**, 901.

Jørgensen, U. G. 1990, Astron. Astrophys. **232**, 420.

Jørgensen, U. G., and Johnson, H. R. 1990, Astron. Astrophys., in press.

Jørgensen, U. G., Almlöf, J., and Siegbahn, P. E. M. 1989, Astrophys. J. **343**, 554.

Kalkofen, W. 1987, (ed.), *Numerical Radiative Transfer*, Cambridge University Press, Cambridge.

Kiselman, D. 1990, Astron. Astophys. Letters, to be submitted.

Kudritzki, R. 1988, in 18th Advanced Course of the Swiss Society of Astron. Astrophys.

Kudritzki, R. P., Pauldrach, A., and Puls, J. 1987, Astron. Astrophys. **173**, 293.

Kurucz, R. L. 1979, Astrophys. J. Suppl. **40**, 1.

Kurucz, R. L. 1986, Highlights of Astron. **7**, 827.

Kurucz, R. L. 1990, private communication.

Laird, J. B., Carney, B. W., and Latham, D. W. 1988, Astron. J. **95**, 1843.

Lambert, D. L. 1988, J. Astron. Astrophys. **8**, 103.

Lambert, D. L. 1989, in *Cosmic Abundances of Matter*, AIP Conf. Proc. **183**, AIP, New York, p. 168.

Lambert, D. L. 1990, in *Astrophysics: Recent Progress and Future Possibilities* (ed. B. Gustafsson and P. E. Nissen), Matematisk Fysiske Meddelelser **42**: 4, p. 75, Royal Danish Acad. Sci. Letters, Copenhagen.

Lambert, D. L., Gustafsson, B., Eriksson, K., and Hinkle, K. 1986, Astrophys. J. Suppl. **62**, 373.

Lemke, M. 1989, Astron. Astrophys. **225**, 125.

Lemke, M. 1990, private communication.

Magain, P., and Zhao, G. 1990, Astron. Astrophys., in press.

Mihalas, D. 1978, *Stellar Atmospheres* (2nd ed.), W. H. Freeman and Co., San Francisco.

Nissen, P. E. 1990, in Proc. of IAU XI Regional Meeting, in press.

Nissen, P.E., and Kjär, K. 1980, *ESO Workshop on Methods of Abundance Determination for Stars*, ESO, Geneva.

Nordlund, Å. 1982, Astron. Astrophys. **107**, 1.

Nordlund, Å., and Dravins, D. 1990, Astron. Astrophys. **228**, 155, 184, 203.

Olofsson, H., Eriksson, K., and Gustafsson, B. 1990, Astron. Astrophys. **230**, 405.

Pauldrach, A., and Herrero, A. 1988, Astron. Astrophys. **199**, 262.

Pauldrach, A., Puls, J., and Kudritzki, R. P. 1986, Astron. Astrophys. **164**, 86.

Pauldrach, A. W. A., Kudritzki, R. P., Puls, J., and Butler, K. 1990, Astron. Astrophys. **228**, 125.

Pauls, U., Grevesse, N., and Huber, M. C. E. 1990, Astron. Astrophys. **231**, 536.

Plez, B. 1990, Mem. Soc. Astron. Italiana, in press.

Remie, H,. and Lamers H. J. G. L. M. 1982, Astron. Astrophys. **105**, 85.

Ruland, F., Holweger, H., Griffin, R. and R., and Biehl, D. 1980, Astron. Astrophys. **92**, 70.

Rybicki, G. 1971, in *Line Formation in Magnetic Fields*, NCAR, Boulder, p. 146.

Scharmer, G. B. 1981, Astrophys. J. **249**, 720.

Scharmer, G. B. 1984, in *Methods in Rad. Transfer* (ed. W. Kalkofen), Cambridge University Press, Cambridge, p. 173.

Scharmer, G. B., and Carlsson M. 1985, J. Comp. Phys. **59**, 56.

Schuster, W. J., and Nissen, P. E. 1989, Astron. Astrophys. **221**, 65.

Smith, V. V. 1989, in *Cosmic Abundances of Matter*, AIP Conf. Proc. **183**, AIP, New York, p. 200.

Spite, M., and Spite, F. 1985, Ann. Rev. Astron. Astrophys. **23**, 225.

Steenbock, M. 1985, in *Cool Stars with Excess of Heavy Elements*, (ed. M. Jaschek and P. C. Keenan), D. Reidel Publ. Co., Dordrecht, p. 231.

Steenbock, W., and Holweger, H. 1984, Astron. Astrophys. **130**, 319.

Steffen, M. 1989, in *Solar and Stellar Granulation* (ed. R. J. Rutten and G. Severino), Kluwer Academic Publ., Dordrecht, p. 425.

Steffen, M., Ludwig, H.-G., and Krüss, A. 1989, Astron. Astrophys. **213**, 371.

Stein, R. F., Nordlund, Å., and Kuhn, J. R. 1989, in *Solar and Stellar Granulation* (ed. R. J. Rutten and G. Severino), Kluwer Academic Publ., Dordrecht, p. 381.

Wehrse, R. (ed.) 1990, *Accuracy of Element Abundances for Stellar Atmospheres*, Lecture Notes in Physics **356**, Springer-Verlag, Berlin.

Werner, K. 1986, Astron. Astrophys. **161**, 177.

Werner, K. 1988, Astron. Astrophys. **204**, 159.

Werner, K. 1989, Astron. Astrophys. **226**, 225.

Werner, K., and Husfeld, D. 1985, Astron. Astrophys. **148**, 417.

Wessolowski, U., Schmutz, W., and Hamann, W.-R. 1988, Astron. Astrophys. **194**, 160.

Wheeler, J. C., Sneden, C., and Truran Jr., J. W. 1989, Ann. Rev. Astron. Astrophys. **27**, 279.

R. E. M. Griffin: Despite the immense improvements in the appreciation and understanding of the details which are now enabling us to generate models which are able to approach asymptotically to reality, nevertheless those who determine metal abundances from photometric low-resolution observations of only a few stellar features claim higher precision in their work than currently can be estimated by those who attempt to take into consideration all the features of the real star. Could you please comment?

B. Gustafsson: I think high spectral resolution is quite indispensible and will, also at rather low S/N, turn out to be superior to photometry at many future applications. HR spectroscopy will also continue to be the most important primary method for stellar abundance determination, by means of which photometric abundances will be calibrated, except for cases where no proper calibration stars can be reached. For the latter cases, and for many other reasons including the proper analysis of HR spectra, synthetic spectroscopy will also be of outmost importance.

V. Trimble: Can you give any sort of rule of thumb about what trends or what kinds of abundance anomalies one ought to believe?

B. Gustafsson: No! I think our experience has shown that we should be humble in this respect. Each case must be scrutinized carefully.

Observational Effects of Nucleosynthesis in Evolved Stars

D. L. LAMBERT

Department of Astronomy
University of Texas
Austin, Texas

1 ABSTRACT

This essay reviews aspects of the chemical composition of three groups of stars whose atmospheres are contaminated with products of internal nucleosynthesis:
- Nitrogen and related elements affected by CN-cycling in B stars;
- Isotopic ratios of C and O in red giants after the first dredge-up;
- The s-process, fluorine, and lithium in asymptotic giant branch stars.

2 INTRODUCTION

A Herstmonceux Conference in 1961 (I think) was my introduction to the world of conferences, symposia, colloquiua and workshops. The report in *The Observatory* magazine on that conference is delightfully short with an italicized preface declaring that it is published by permission of the Astronomer Royal. After a long lapse, I am pleased to return to this Herstmonceux conference in honour of Bernard Pagel.

The topic of this essay, and the focus of my research for the last 20 years, has links to Cambridge, Herstmonceux, and Bernard. *Cambridge*: The initial impetus came when I selected 'Frontiers of Astronomy' by Fred Hoyle as a school prize in 1956. Later at Oxford, I read the classic 'Synthesis of the Elements in Stars' by Burbidge, Burbidge, Fowler and Hoyle (1957). I would suppose that some of the original ideas in that marvellous paper were conceived and developed just a mile or so from the new home of the Royal Greenwich Observatory. *Herstmonceux*: As an undergraduate, I was fortunate to spend three summers at Herstmonceux as a 'Vacation Student'. The second of those long ago summers was spent as an assistant to Sir Richard van der Riet Woolley, the Astronomer Royal. He certainly taught me that vigorous pursuit of research need not cease on the assumption of major administrative responsibilities; he appears at times in my dreams (nightmares?). *Bernard Pagel*: My initial contact with Bernard occurred in those summers at Herstmonceux when he lectured on several weekday afternoons to the audience of the dozen or so undergraduates. The rigorous discussions of stellar structure certainly helped my eventual conversion to stellar astrophysics. Later, Bernard was invited to Oxford as the external examiner for my D.Phil. Until he declared "and I believe we are allowed to ask some general questions", I like to think that the contest between the examiners and the candidate would have been declared a draw by impartial judges. As a result of the rout of the candidate in the final round, I have always tried to delve beyond the narrow confines of the current spectroscopic problem and consider 'the general questions'.

Stellar nucleosynthesis occurs deep within the star and its consequences are not immediately observable by the spectroscopist. (Nuclear reactions at the surface are proposed from time to time to account for specific abundance anomalies.) How may the atmospheric composition betray the effects of the nucleosynthesis? For a stable evolved star, the possible answers include:

- *Deep mixing in red giants.* Today, the terms 'first, second and third dredge-up' are widely understood. I comment later on some recent work on the first dredge-up experienced by stars ascending the giant branch for the first time and on the third dredge-up occurring in giants on the asymptotic giant branch (AGB).

- *Severe mass loss.* Through a strong wind or a by a more explosive event, the stellar envelope may be shed to reveal the former core and expose the ejecta. As examples of this method of gaining spectroscopic access to layers previously exposed to nuclear reactions I cite the Wolf-Rayet stars, the planetary nebulae and their central stars, supernovae remnants, and novae.

- *Mass exchange between the members of a binary system.* This mass exchange may expose the core of the mass-losing star and the nuclear-processed material dumped on the mass-gaining star may remain largely unmixed. Examples of recent lively interest include the Algols and most notably the Barium stars.

The topic assigned to me is now so richly explored that full justice to it demands a book. A ration of 12 pages demands that I be highly selective. I discuss three topics that illustrate, I hope, the breadth and good health of observational studies of nucleosynthesis in evolved stars. Stellar nucleosynthesis and evolution are inextricably linked. On occasions, the atmospheric abundance of an element in a particular type of star may be directly related to the primary site for nucleosynthesis of that element: the cases of lithium, fluorine, and the s-process in AGB stars, are discussed later. On other occasions, the abundances reflect changes of surface chemical composition arising from evolutionary episodes that appear to have little to do with the galactic production (or destruction) of the elements in question. The fascinating case of He, C, and N in the atmospheres of B stars may fall in this latter category.

3   THE PUZZLING CASE OF THE B STARS

There are among the B stars on or near the main sequence several classes of obviously peculiar stars: He-weak, OBC, and OBN, amongst others. In this section, I discuss the little known case of normal B stars whose atmospheres appear unexpectedly contaminated with CN-cycle products; *i.e.*, atmospheres having He and N overabundances and C underabundances. Evidence for such contamination has been assembled by L. S. Lyubimkov of the Crimean Astrophysical Observatory. My attention was drawn to this delightful work by his 1984 paper on the N abundance in B stars where abundances for 36 stars were obtained from the following ingredients:

  • Equivalent widths of the N II 3995 and 4360 Å lines as measured by Kane, McKeith and Dufton (1980) and Dufton, Kane and McKeith (1981);

  • Non-LTE predictions of these equivalent widths as given by Dufton and Hibbert (1981) using non-LTE model atmospheres (Mihalas 1972);

  • The appropriate model atmosphere and corresponding set of predicted equivalent widths were selected using the observed $[c_1]$ and β indices and a calibration based on the non-LTE atmospheres - note: all calculations by Dufton and Hibbert refer to $\log g = 4.0$ models;

  • Zero microturbulence: $\xi = 0$ km s^{-1};

  • Evolutionary tracks (Becker 1981, Brunish and Truran 1982) and the parameters T_{eff} and g provided masses and evolutionary ages of the individual stars.

Lyubimkov's striking discovery that the N abundance increases markedly with evolutionary age was, he points out, overlooked in the original papers reporting the abundances; those papers gave only the mean N abundance of the sample and, perhaps, because that abundance was close to the solar value, the incentive to look in more detail at the results was dulled. For the most massive stars (13-20 M$_\odot$) in the sample, the N abundance increases from $\log \varepsilon(N) \sim 7.6$ to 8.6 in less than 10 million years. The increase in N is less severe at lower masses: $d \log \varepsilon(N)/dt \simeq 0.15$ for 13-20 M$_\odot$, 0.06 for 9.9-12.3 M$_\odot$ and 0.024 for 5.7-8.5 M$_\odot$ where the time is given in units of 10^6 yrs. The initial abundance, $\log \varepsilon(N) \sim 7.6$, is close to the observed value for local H II regions. The extreme abundance, $\log \varepsilon(N) \sim 8.6$, seen in the old B stars was noted by Lyubimkov to be very similar to that of the more evolved F-K supergiants (Luck 1978; Luck and Lambert 1981)(Abundances are given on the usual scale: $\varepsilon(X) = n(X)/n(H)$ with $\log \varepsilon(H) = 12.0$).

The great majority of the F-K supergiants are not expected to have evolved directly from the main sequence, but from red supergiants. The deep convective envelope of a red supergiant is predicted to enrich the atmosphere in CN-cycled products from layers that underwent mild H-burning on the main sequence (the first dredge-up). Standard stellar models do *not* predict the appearance of CN-cycled products at the stellar surface prior to the first dredge-up. Lyubimkov's discovery apparently points to a gross deficiency of standard models. The matching of the N abundances of old B stars to those of the F-K supergiants suggests that the outer layers of the stars are so well mixed prior to the first dredge-up that the red supergiant's convective envelope induces no further marked change in the surface N abundance. (All of the cooler G and K supergiants show a ^{13}C overabundance attributable to CN-cycling; many stars show a lower $^{12}C/^{13}C$ ratio than predicted by standard models.)

If the N overabundances of the B stars are due to CN-cycling, C must be underabundant and He possibly overabundant. Early work on He had hinted at an overabundance correlated with age. Through a reconsideration of the extensive analyses by Wolff and Heasley (1985) and Nissen (1976), Lyubimkov (1988) presents evidence for He enrichment: the He surface abundance increases with evolutionary age at a rate dependent on the stellar mass and the total enrichment over the main sequence lifetime is $\Delta\epsilon(He) \simeq$ 0.04-0.05 for $M \simeq (8\text{-}15)\ M_\odot$ to $\lesssim 0.005$ for $M \lesssim 5\ M_\odot$.

A third paper in the same spirit examines the evidence for C depletion in B stars (Lyubimkov 1989). The evidence must be judged as no more than 'interesting'. Lyubimkov finds in the literature C abundances for 9 stars based on 2 or more C II lines other than the strong 4267 Å feature for which non-LTE effects are large. The mean C abundance of the nine is $\log \epsilon(C) = 8.38 \pm 0.03$, a value less than the solar abundance of 8.60 ± 0.05 (Grevesse *et al.* 1991). (The thorough non-LTE calculations by Eber and Butler [1988] and the equivalent widths of the 4267 Å feature measured by Kane *et al.* [1980] for 14 stars give a mean abundance $\log \epsilon(C) = 8.37 \pm 0.06$ [Lyubimkov 1989]). Such a small (0.2 dex) difference between the stellar and solar abundnces is fragile evidence for a C underabundance attributable to CN-cycling. Lyubimkov shows that the C and N abundances of the 9 stars with a C abundance based on weak C II lines are roughly correlated such that the sum of the C and N abundances is constant and approximately equal to the solar value. The same sum is provided by the (on average) more N-rich and more C-poor F-K supergiants. Conservation of the total number of C and N nuclei is a signature of CN-cycling.

Thanks primarily to Lyubimkov's detective work, it appears that normal B stars are progressively enriched in CN-cycling products as the stars evolve off the main sequence. With today's telescopes and spectrometers and analytical tools (*e.g.*, line blanketed model atmospheres, non-LTE calculations), the claim for early contamination of the atmospheres by material exposed to H-burning may be given a full examination. I close this discussion with a comment on a study nearing completion (Gies and Lambert 1990). High S/N spectra of several wavelength intervals in 39 OB stars have provided C and N abundances from C II and N II lines. If stars of luminosity classes III and II are exluded, there is no evidence for N overabundances and C underabundances. Our selected N II lines are insensitive to non-LTE effects. Sixteen of our stars were in Lyubimkov's sample of 36 stars for which he gave N abundances. In particular, we observed the three stars for which he reported the highest N abundances. We do not confirm the overabundances:

Star	log ε(N)	
	Lyubimkov	Gies and Lambert
HD 30836	8.77	7.71
HD 31237	8.59	7.60
HD 35468	8.65	8.26

A comparison of the N abundances for the two samples is offered in Figure 1 where I plot abundance versus surface gravity. In Lyubimkov's sample, the upper envelope slopes steeply down from left to right, and this trend is roughly equivalent to the reported N enrichment with age. Our sample shows a much weaker trend, and possibly no trend at all.

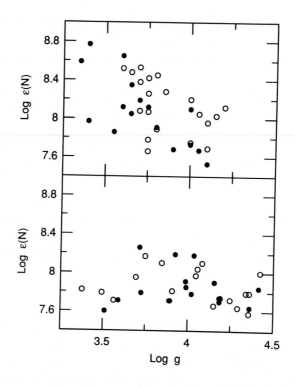

Figure 1. N abundances and surface gravities (g) for B stars: top panel - Lyubimkov (1984) and bottom panel - Gies and Lambert (1990). Stars in common to the two samples are represented by filled circles.

An extensive observing program covering field and cluster stars is warranted. A parallel study of a sample of B-K supergiants should be made, if only because the cooler stars can provide information on nucleosynthesis unobtainable from spectra of hot stars, *e.g.*, the C and O isotopic ratios and the Na abundance which may be enhanced by $^{22}Ne(p,\gamma)^{23}Na$ in the hot interior. The unexpected appearance of CN-cycling products in a B star's atmosphere may be due to rotationally-induced mixing with mass loss by a wind as a contributing factor. It is widely supposed that a N overabundance in a blue supergiant is the signature of a star that has evolved from a red supergiant (see the arguments about the progenitor of SN 1987A). If Lyubimkov is correct, the appearance of CN-cycling products in the stellar atmosphere does not require the star to have been a red supergiant.

4 THE FIRST DREDGE-UP

On exhaustion of H in the core, a main sequence star evolves to the red giant branch (RGB). At the base of the RGB, the star develops a deep convective envelope that extends down into layers that, on the main sequence, experienced some H-burning. As a result of this 'first dredge-up', the RG's atmosphere is predicted to contain signatures of mild H-burning. For red giants with mass $M \lesssim 2.2 \, M_\odot$, the ascent of the RGB ends with the He-core flash which is followed by He-core burning at a luminosity ($\log L/L_\odot \simeq 1.6$) below that of the RGB's tip ($\log L/L_\odot \simeq 3.3$). At the conclusion of He-core burning in stars of 4 to 8 $M_\odot$, the convective envelope is predicted to mix with additional CNO-cycled material (the 'second dredge-up'). In standard models, the first and second dredge-ups are the only processes responsible for changes to the surface chemical composition prior to the star's ascent of the AGB. Of course, the stars in the sky do not recognize that changes to the surface composition are to be limited to those of our standard calculations.

In this essay, I discuss the observed and predicted isotopic ratios $^{12}C/^{13}C$ and $^{16}O/^{17}O/^{18}O$ in stars that have experienced the first dredge-up. A full comparison of the observed and predicted effects of the first and second dredge-up would call for review of the Li, Be, C, N, O, and Na abundances. Predicted changes of the ^3He and ^4He abundances are untestable in cool stars.

The $^{12}C/^{13}C$ Ratio: Standard models of RGs show that the surface $^{12}C/^{13}C$ ratio is reduced by the first dredge-up, *e.g.*, $^{12}C/^{13}C = 90$ is cut to 25 for $M \sim 2 \, M_\odot$ where the final ratio is weakly dependent on the mass (Dearborn, Eggleton, and Schramm 1976). Surveys of the $^{12}C/^{13}C$ in bright field giants showed that many stars had a ratio far below the standard predictions: $^{12}C/^{13}C \sim 5\text{-}10$ was found for many stars. It was suspected that these ^{13}C-rich giants were low mass stars. (Low $^{12}C/^{13}C$ ratios are not the exclusive right of low mass RGs: supergiants such as ε Peg and α Ori with $^{12}C/^{13}C = 4\text{-}7$ are indubitably massive stars.) Gilroy's (1989) study of RGs in open clusters revealed clearly for the first time that ^{13}C-rich RGs are low mass stars. Giants from 19 clusters of near solar metallicity and with turn-off masses from 1 to 6 $M_\odot$ were analysed for the $^{12}C/^{13}C$ ratio using CN lines. Figure 2 compares the predictions for standard models (Dearborn 1990) with Gilroy's observed $^{12}C/^{13}C$ ratios. The new observations for cluster RGs show that:

• For M ≳ 2.5 $M_{\odot}$, the observed $^{12}C/^{13}C$ (≃ 26) is approximately constant. Dearborn's predictions are similarly constant, but at a lower $^{12}C/^{13}C$ ratio. This discrepancy is probably not severe: Dearborn, Eggleton and Schramm (1976) gave a higher ratio (≃ 25) at M = 2 $M_{\odot}$ for the same initial ratio ($^{12}C/^{13}C$ = 90) as assumed by Dearborn (1990).

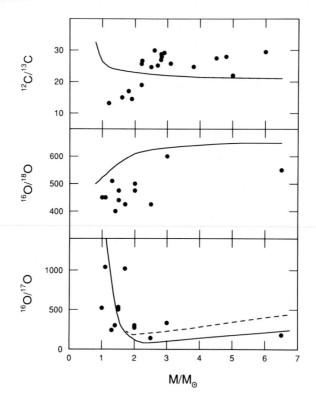

Figure 2. The $^{12}C/^{13}C$, $^{16}O/^{18}O$ and $^{16}O/^{17}O$ ratios for red giants. Gilroy's (1989) results for $^{12}C/^{13}C$ for giants in open clusters are shown in the top panel together with Dearborn's (1990) predictions for standard models. $^{16}O/^{18}O$ and $^{16}O/^{17}O$ measurements for field giants are taken from Harris, Lambert, and Smith (1988) with Dearborn's predictions given by the solid line. Predicted $^{16}O/^{17}O$ ratios for a higher rate for the $^{17}O(p,\alpha)$ reaction are shown by the broken line (see text).

• For M ≲ 2.5 $M_{\odot}$, the observed $^{12}C/^{13}C$ ratio decreases with decreasing mass: $^{12}C/^{13}C$ ≃ 10 (M/$M_{\odot}$). The standard models show the ratio to increase with increasing mass. Gilroy supplemented her results with representative points for old disk giants, field

Population II giants, and globular cluster giants. These metal-poor additions confirm the trend shown in Figure 2, but one should probably not mix giants of quite different metallicities

The discovery of the correlation between the $^{12}C/^{13}C$ ratio and mass of the low mass stars does not by itself appear to identify the failure of the standard models. Since the main sequence lifetime increases sharply with decreasing mass, it is tempting to suppose that internal mixing (rotationally induced?) operating at a rate that is approximately independent of mass will achieve a larger cumulative effect in the lower mass main sequence stars. Unfortunately, low mass stars also evolve more slowly up the RGB so that mixing and nuclear processing in RGB stars is likely to result in a larger cumulative effect in low mass stars. Evidence exists that, contrary to standard models of RGs, the RG's convective envelope is mixed continously at its base with CN-cycled material. Certainly, the Li abundance may decline as a star ascends the RGB - see Pilachowski's (1986) observations of RGs in the open cluster NGC 7789 and Pilachowski, Sneden and Hudek's (1990) report of a similar decline amongst field giants. Gilroy discusses some of the proposed mixing processes in main sequence and post-main sequence stars.

Two possible causes for the high ^{13}C abundances in low mass RGs may be dismissed. Substantial mass loss prior to the conclusion of the ascent of the RGB will lower the RG's $^{12}C/^{13}C$ ratio. But almost all of the cluster and some field giants having a low $^{12}C/^{13}C$ ratio have retained some Li in stark contradiction to the prediction for scenarios calling for severe mass loss; Li is retained by a main sequence star in a thin skin. Although the He-core flash that occurs in stars of $M \lesssim 2.2\ M_\odot$ may be a candidate for the origin of the additional mixing that leads to low $^{12}C/^{13}C$ ratios, it can probably be eliminated as the sole cause because some field giants of low $^{12}C/^{13}C$ ratio have luminosities below that of He-core burning and, in particular, giants on the RGBs of globular clusters show low $^{12}C/^{13}C$ ratios.

Proposed modifications to standard main sequence models may be tested against the O isotopic ratios. A main sequence star which is mixed at intermediate depths (*e.g.*, $M_r \sim$ 0.5-0.9 $M_\odot$ for a 1.5 $M_\odot$, where M_r is the mass within a radius r) evolves to a RG with a lower $^{12}C/^{13}C$ ratio but with O isotopic ratios little changed from those of an unmixed main sequence star. Mixing at deeper depths ($M_r < 0.5\ M_\odot$ for $M = 1.5\ M_\odot$) reduces the $^{12}C/^{13}C$ and $^{16}O/^{17}O$ ratios and increases the $^{16}O/^{18}O$ ratio of the RG.

The $^{16}O/^{18}O$ Ratio: Estimates of the $^{16}O/^{18}O$ ratio in field K giants are summarized in Figure 2 - note the estimated stellar masses are uncertain. It may be assumed that the giants' progenitors began with a $^{16}O/^{18}O$ ratio close to the solar (= 499) ratio. The first dredge-up is predicted to increase the surface $^{16}O/^{18}O$ ratio because ^{18}O is destroyed by protons in the deep interior of the main sequence star. The $^{18}O(p,\alpha)$ rate is given as uncertain by Harris *et al.* (1983) and only their 'low' (or a lower) rate fits the observed $^{16}O/^{18}O$ ratios (Harris and Lambert 1984). Laboratory measurements (Champagne and Pitts 1986) have confirmed that the rate is close to the 'low' rate. It appears that the observed $^{16}O/^{18}O$ ratios are systematically below the predictions based on the 'low' rate. Since the observational errors are $\pm$ 25% or so, the observations and predictions do overlap. A suspicion remains that the 'low' rate is higher than the true rate. Nonetheless, the $^{16}O/^{18}O$ ratios show that deep mixing in the main sequence star is unlikely to account for the low $^{12}C/^{13}C$ ratios.

The $^{16}O/^{17}O$ Ratios: The giant's convective envelope barely reaches the ^{17}O-rich core. For standard models, the enrichment of a low mass RG in ^{17}O is a steep function of mass - see Figure 2. For RGs more massive than 2 $M_\odot$, the predicted $^{16}O/^{17}O$ increases from the minimum (~ 150) than at 2 $M_\odot$. Dearborn's predicted $^{16}O/^{17}O$ ratios are based on the 'high' rate given by Fowler, Caughlan, and Zimmerman (1975) for $^{17}O(p,\alpha)$ which controls destruction of ^{17}O; the 'low' rate leads to low $^{16}O/^{17}O$ ratios ($\lesssim$ 50 for M $\gtrsim$ 2 $M_\odot$) in sharp disagreement with the observations. New measurements of the $^{17}O(p,\alpha)$ reaction report a rate above the 'high' rate (Bogaert *et al.* 1989, Landré *et al.* 1989) and lead to higher $^{16}O/^{17}O$ ratios for RGs of M $\gtrsim$ 2 $M_\odot$, in reasonable accord with the observations for the massive (M $\gtrsim$ 10 $M_\odot$) red supergiants (α Ori, α Sco, RS Per - Harris and Lambert 1984, Smith 1986). When the uncertainties in the assigned stellar masses of field giants are recognized, observed and predicted, $^{16}O/^{17}O$ ratios are in fair agreement (Figure 2) and, hence, deep mixing is an unlikely explanation for low $^{12}C/^{13}C$ ratios in RGs.

The final reconciliation of the observed and predicted compositions of giants which have undergone the first (and second) dredge-up will require even more observational attacks on RGs and their main sequence progenitors, as well as continued laboratory measurement of rates of certain nuclear reactions. Several key observational projects await the deployment of new telescopes and/or new spectrometers. I would note especially that the cryogenic infrared echelle spectrometers promised for large reflectors may enable the $^{12}C/^{13}C$ and $^{16}O/^{17}O/^{18}O$ ratios to be extended to cool main sequence stars (what are the initial ratios as a function of age and metallicity? Are these stars quite unmixed?) and may allow us to extend Gilroy's observations of open cluster giants to the $^{16}O/^{17}O/^{18}O$ ratios. Finally, the isotopic measurements must be integrated with determinations of the elemental (C,N,O) abundances.

5 THE THIRD DREDGE-UP

At the end of He-core burning, massive stars (M $\gtrsim$ 8 $M_\odot$) evolve through C-core burning to their end as a Type II supernova, but lower mass red giants (M $\lesssim$ 8 $M_\odot$) develop a C-O degenerate core and evolve along the AGB where they experience He-shell flashes (thermal pulses) separated by longer periods of quiescent H-shell burning. Models of intermediate mass (IM, 3-8 $M_\odot$) and some models of low mass (LM, < 3 $M_\odot$) AGB giants predict that, after the He-shell flash, products of He-burning (principally, ^{12}C and s-process nuclides) are mixed into the deep convective envelope and, hence, to the surface. This mixing is the third dredge-up. The cumulative effect of the dredge-up is to increase the C/O ratio and the s-process abundances of the atmosphere. When the third dredge-up was predicted, it and the associated He-shell flash were seen as the most likely explanation for the cool carbon stars and the s-process enrichments seen in oxygen-rich (spectral types MS and S) as well as carbon-rich red giants. In this concluding section, I shall sketch some of the observations that betray the nucleosynthesis achieved in thermally pulsing AGB stars.

The s-process: Merrill's (1952) discovery of technetium in the spectra of S stars came far in advance of even a crude understanding of nucleosynthesis within these red giants. Theoretical work by Iben (1975) on IM-AGBs showed that an s-process run by the neutron source ^{22}Ne$(\alpha,n)^{25}$Mg could lead to s-process elements (including Tc) with a solar-like abundance in the envelopes of these stars. Moreover, it was expected that IM-AGBs return sufficient mass to the interstellar medium to control the galactic abundances of the s-process nuclides (Truran and Iben 1977, Iben and Truran 1978).

The initial seeds of doubt about the IM-AGBs came when objective prism surveys of giants in the Magellanic Clouds (Blanco, McCarthy, and Blanco 1980) showed that the carbon stars were much less luminous than predicted by IM-AGB models: the carbon stars there and presumably also in our Galaxy seem to be LM-AGBs. What is the fate of the IM stars? There is evidence that these stars lose mass at rates so high that many evolve early away from the AGB (Hughes and Wood 1987). The relative contributions of IM-AGB and LM-AGB stars to galactic enrichment of s-process nuclides remains an open question. Here, I sketch the observational evidence that s-process enriched AGB stars are LM and not IM stars - a review is given elsewhere (Lambert 1991). I shall not discuss the theoretical work on the third dredge-up and the s-process in LM-AGBs: two experts on the internal upsets of these stars conclude a review by declaring "the observed enrichment of s-process elements in low mass stars remains a puzzle" (Sackmann and Boothroyd 1991)!

In IM-AGBs, the temperature in the He-burning shell exceeds that needed (T $\gtrsim$ 3 x 10^8 K) to ignite ^{22}Ne$(\alpha,n)^{25}$Mg. When this neutron source runs the s-process, the star's atmosphere is predicted to have a non-solar mix of the Mg isotopes; *e.g.*, ^{26}Mg/^{24}Mg > 1 is expected for a S star (Truran and Iben 1977; Scalo 1978; Malaney 1987). The Mg isotopic ratios are measureable through the MgH A-X lines. Analyses of s-process enriched O-rich giants show that the ratios are close to solar and quite unlike the predictions for IM-AGBs (see, for example, Tomkin and Lambert 1979; Smith and Lambert 1986). Although a survey of a large sample of s-process enriched giants should be made, the randomly selected sample of bright giants observed unsuccessfully for ^{25}Mg and ^{26}Mg excesses suggests that LM-AGBs greatly outnumber IM-AGBs in the solar neighbourhood. Unfortunately, the $^{13}C(\alpha,n)^{16}O$ reaction that is the leading candidate for the n-source in LM-AGBs does not result in a detectable spectroscopic signature.

Other observational evidence identifying LM-AGBs rather than IM-AGBs with Galactic S stars concerns the estimates of neutron density at the s-process site. In the limit that the neutron density, N(n), tends to zero, the s-process takes a unique path along the valley of stability because all unstable n-rich nuclei decay and do not capture a neutron. At the neutron densities expected in the He-burning shell, a competition occurs at certain long-lived unstable n-rich nuclides between decay and capture of a neutron and, hence, the s-process path has different paths at low and high N(n). By determining the abundances of nuclides on the alternate paths following the branch, one can estimate the N(n). This idea has been fully exploited in the analysis of solar system (meteoritic) abundances and several branches now provide concordant results: N(n) ~ 3 x 10^8 cm^{-3} for the 'main' component (Käppeler, Beer, and Wisshak 1989, Käppeler *et al.* 1990).

Operation of the $^{22}Ne(\alpha,n)$ source in IM-AGBs results in a neutron pulse with peak densities N(n) $\simeq$ 10^9-10^{11} cm^{-3}. Such densities are obviously incompatible with that derived from the solar system abundances. The decline of N(n) at the end of the pulse is rapid and the abundance pattern created at the peak N(n) in the pulse is not substantially altered in "the freeze out". Although the operation of the $^{13}C(\alpha,n)$ source in LM-AGBs is predicted to result in lower N(n), the predictions are dependent on the (unknown) details of the synthesis of a ^{13}C pocket at the top of the He shell: recipes for the formation of this pocket and its subsequent ingestion into the convective He shell are given by Hollowell and Iben (1988, 1989). The neutron density depends on the mass of ^{13}C, its rate of ingestion and the star's composition. Käppeler *et al.* (1990) describe a LM-AGB model whose He-shell leads to a near-solar abundance pattern except for mild deviations near several branches. These deviations show that the predicted N(n) exceeds slightly that gotten from the solar abundances. I presume that the discrepancy could be eliminated by adjustment of parameters describing the growth and utilization of the ^{13}C pocket.

Since the solar abundances are the fruits of several stellar generations of differing metallicity with a mass range within each generation, the classical analysis is not the ideal test of LM and IM-AGB models. Fortunately, two of the branches that provide estimates of N(n) are open to analyses in AGB stars:
  • the branch at ^{85}Kr that determines the Rb abundance relative to Sr and Y.
  • the branch at ^{95}Zr that controls production of the stable ^{96}Zr isotope.

For N(n) → 0, ^{96}Zr is not produced by the s-process. For N(n) → ∞, ^{96}Zr is produced with an abundance N(^{96}Zr)/N(^{94}Zr) $\simeq$ 1.8 for τ_0 $\simeq$ 0.3. The ratio N(^{96}Zr)/N(^{94}Zr) = 0.05 for N(n) $\simeq$ 2 x 10^8 cm^{-3} and 0.50 for N(n) $\simeq$ 3 x 10^9 cm^{-3} (Toukan and Käppeler 1990). Neutron densities within AGB stars are measureable via the ^{95}Zr branch provided that ZrO bands are present. We have obtained Reticon and CCD spectra of the B-X 0-1 and γ 1-2 R_1 heads near 6930 Å for a sample of S stars. Isotopic ratios are obtained from a comparison of synthetic and observed spectra (Smith 1988). Inspection of our spectra shows no evidence for a high abundance of ^{96}Zr. For two LPVs - R And and R Gem - examined in some detail, we estimate $N(^{96}Zr)/N(Zr) \lesssim 0.05$ or $N(n) \lesssim 5 \times 10^8$ cm^{-3} (Toukan and Käppeler 1990). These observations are incompatible with the high $N(n)$ predicted for IM-AGBs.

The branch at ^{85}Kr is not as simple as that at ^{95}Kr because a short-lived isomeric state is populated and reduces the effectiveness of the branch at the ^{85}Kr ground state. Thanks to the latter's 'long' lifetime, the products following the ^{85}Kr branch are sensitive to the duration (Δt) of the s-processing in the thermal pulse. Application of the ^{85}Kr branch to cool stars must rely on its effect on the Rb abundance. In the limit $N(n) \rightarrow 0$, the dominant path is through ^{85}Rb to ^{86}Sr. In the high density limit, the path through ^{86}Kr to ^{87}Rb and ^{88}Sr is opened and, thanks to the isomeric state of ^{85}Kr, runs in competition with the path through ^{85}Rb. The elemental abundance of Rb (relative to Sr and Y) is sensitive to the neutron density because the neutron capture cross-sections of ^{85}Rb and ^{87}Rb differ by a factor of about 10. (Note the Sr abundance, but not that of Y, is also sensitive to the ^{85}Kr branch.) The Rb/Sr ratio varies by more than an order of magnitude between the low and high density limits. The ^{85}Kr branch was exploited first by Tomkin and Lambert (1983) for the Barium star HR 774. Later analyses of three other Barium stars (Smith and Lambert 1984, Malaney and Lambert 1988) gave similar results: all values are close to the low density limit. Determinations of the Rb isotopic ratio have not been attempted. An analysis of the Rb I lines in Arcturus (Lambert and Luck 1976) suggests that such a determination is unlikely to be a more accurate indicator of neutron density than the Rb/Sr and Rb/Y ratios. At present we are examining Rb in a sample of bright M, MS, and S stars. Analyses of several stars and inspection of spectra of other stars show that the Rb/Sr ratios are solar-like in all cases. Of especial interest are HR 8714 with a high Li abundance and TV Aur, the most s-process enriched star of our sample.

These results for Rb show that the neutron density at the s-process site did not exceed about 10^9 cm^{-3}. The derived limit may depend on the form of the neutron pulse. Beer and Macklin (1989), who provide accurate measurements of the ^{85}Rb and ^{87}Rb neutron capture cross-sections, predicted Rb/Sr ratios for neutron pulses appropriate for IM-AGB models having C-O cores of mass 0.65 to 1.16 $M_\odot$. As expected for pulses having a peak density $N(n) \gtrsim 3 \times 10^9$ cm^{-3} and a rapid freezeout, the predictions exceed the ratios observed for Barium and MS/S stars by a large margin. Beer and Macklin show that pulses with a peak density $N(n) \sim 10^8$-10^9 cm^{-3} and pulse lengths $\Delta t \sim 30$-3 yr are required to fit the observations.

In summary, observations of ^{96}Zr and Rb in s-process enriched stars show that N(n) $\leq 10^9$ cm^{-3}. Such a limit is in conflict with models of IM-AGBs where the neutron source is ^{22}Ne(α,n).

Fluorine: The origin of ^{19}F, the only isotope of fluorine, has long been in question. Woosley *et al.* (1990) proposed that several rare nuclides are synthesized in Type II supernovae when neutrinos excite more abundant nuclei to unbound levels that decay by proton or neutron emission. This decay and reactions between the emitted protons and neutrons and abundant nuclei were suggested as a means ('the v-process') of synthesizing several minor nuclides. In particular, Woosley *et al.* predicted ^{19}F synthesis in the Ne-rich shell of the progenitor to yield F with a solar abundance relative to other species produced in this shell.

As this proposal of an effective site for ^{19}F nucleosynthesis appeared, we were completing an abundance analysis using the 2 μm vibration-rotation lines of the HF molecule in spectra of red giants (Jorissen, Smith, and Lambert 1991). Fluorine is overabundant in the s-process enriched giants such as the oxygen-rich S and carbon-rich N-type stars, as well as the Barium K giants. The overabundance is interpreted as evidence that ^{19}F is synthesized in (or near) the He-shell of an AGB star. Two obvious questions must be asked: How is ^{19}F synthesized? Which is the dominant site for ^{19}F synthesis - Type II SN or AGB stars? I discuss the first question, but leave the reader to find our preliminary answer to the second question in Jorissen *et al.* (1991): AGB stars appear to be the dominant site.

Nucleosynthesis of ^{19}F in an AGB star may be expected to occur in the He shell into which protons are mixed from the H-rich envelope. Such a mixing is required to raise the ^{13}C concentration in the He-shell to levels necessary to drive the s-process: ^{13}C is synthesized by the chain ^{12}C(p,γ)^{13}N(β^+,v)^{13}C which may continue to ^{14}N if the p/^{12}C ratio is not small. Neutrons are later released by ^{13}C(α,n)^{16}O. Several reaction chains are candidates for ^{19}F synthesis. All involve ^{15}N(α,γ)^{19}F as the final step with ^{15}N synthesis achieved in different ways. Unfortunately, no single path has yet been identified as readily synthesizing ^{19}F to the level required to account for the ^{19}F enrichment of AGB stars: the observations and an assumption that the dredged-up material has Z(^{12}C) = 0.2, as predicted by AGB models, lead to Z(^{19}F) ~ (1-4) x 10^{-4} for the material added to the envelope from the He-shell.

The most promising path for ^{19}F synthesis to the level inferred from observations is probably ^{14}N(α,γ)^{18}F(β^+,v)^{18}O(p,α)^{15}N(α,γ)^{19}F where the protons come from ^{14}N(n,p)^{14}C with the neutrons provided by the ^{13}C(α,n)^{16}O neutron source. This path may achieve the observed abundance of ^{19}F provided that adequate ^{13}C and ^{14}N are synthesized between the thermal pulses by protons mixed into the He shell. The ^{19}F that is synthesized by this and other schemes can be destroyed in the He shell by ^{19}F(α,p)^{21}Ne.

Destruction must be avoided, yet the temperature must be sufficiently high that ^{12}C is synthesized during the thermal pulse. This requirement can be satisfied as long as the temperature $T \lesssim 3 \times 10^8$ K. However, this limit, which is consistent with the identification of S and C stars as LM-AGBs, is dependent on the uncertain cross-section for ^{19}F(α,p).

Lithium: The origin of Li has stimulated much observational and theoretical discussion in recent years following the discovery by Spite and Spite (1982) that Li is present in halo stars with an abundance $\log \varepsilon(\text{Li}) \simeq 2.1$. I do not enter here into the debate on whether this abundance does or does not represent the stars' initial/primordial abundance and, hence, quite possibly the abundance from the Big Bang. If the observed Li abundance or a value close to it is the pregalactic value, there must be a source of ^{7}Li within the Galaxy that has raised the abundance to the present (local) value $\log \varepsilon(\text{Li}) \sim 3.3$. Very Li-rich AGB stars have been proposed as an effective source of ^{7}Li (Scalo 1976). Local examples of Li suppliers include the N (cool carbon) stars WZ Cas and WX Cyg and the S stars T Sgr and RY Sgr. Such stars are rare among N and S stars; for example, Catchpole and Feast (1976) surveyed 188 S stars and found 2 Li-rich stars. Recent abundance analyses using spectrum synthesis and model atmospheres show that except for the rare cases such as WZ Cas and T Sgr, the envelopes of AGB stars contain less Li than their main sequence progenitors (Kipper and Wallerstein 1990, Denn, Luck, and Lambert 1990) and, hence, ejection of these envelopes will reduce the interstellar medium's (ISM) Li abundance. Synthesis of Li is attributed to 'the ^{7}Be transport mechanism' (Cameron and Fowler 1971): ^{3}He$(\alpha,\gamma)^7$Be$(e^-,\nu)^7$Li where the ^{3}He was synthesized in the main sequence progenitor. Synthesis occurs at the hot base of the convective envelope (HBCE) of a luminous AGB star. Obvious questions arise: Do most or all of these AGB stars synthesize large amounts of Li before evolving off the AGB? Or are the super Li-rich stars a rare breed of special AGB stars? Are the Li-rich stars raising the ISM's Li abundance?

New insights into the first of these questions are being provided by high resolution spectroscopy of AGB stars in the Magellanic Clouds. As note above, an early finding from objective prism spectroscopy of Cloud AGB stars was that the carbon stars are much less luminous than predicted by IM-AGB models and, hence, may be LM-AGB stars. What is the fate of the IM-AGBs that were predicted to be luminous carbon stars? One possibility is that they develop a HBCE that, in the long interpulse interval, burns the dredged-up ^{12}C to ^{14}N and the IM-AGB is maintained as a S star. A proposal that a IM-AGB carbon star is shrouded by a thick circumstellar dust shell (*i.e.*, IRC+10216-like objects) may be rejected because too few luminous infrared sources have been found in the Clouds. One real possibility is that their evolution is terminated by severe mass loss before their envelopes are made C-rich.

The mass loss option has led to an occasional expression of the opinion that luminous IM-AGB stars do not exist in the Clouds. Certainly the mass loss rates appear to be so high that many stars evolved early off the AGB. In an important paper, Wood, Bessell, and Fox (1983, WBF) showed that both Clouds contain long-period variables (LPVs) with enhanced ZrO bands at luminosities close to maximum allowed luminosity for an AGB star. Our study (Smith and Lambert 1989, 1990) of WBF's stars shows that they are s-process enriched, as suspected by WBF, and, therefore, are S stars. An especially exciting result is that *all* stars within about 1 magnitude of the limiting luminosity for an AGB star ($M_{bol} \simeq -7.1$) are super Li-rich stars. More luminous stars that are core-burning supergiants show no detectable Li line. Lower luminosity AGB stars, mostly carbon stars, also do not show a Li enrichment; the observed sample is presently small, however. In short, all AGB stars that evolve close to the limit $M_{bol} \simeq 7.1$ synthesize Li in large quantities. We suppose that the rare super Li-rich S stars in the Galaxy are also luminous AGB stars and their apparent rarity arises simply because samples of Galactic S stars cannot be sorted by luminosity and so include many less luminous AGB stars.

WBF assigned current masses based on pulsation theory: the Li-rich stars have M ~ 4-8 $M_\odot$ if the LPVs are pulsating in the fundamental mode, but M ~ 2-4 $M_\odot$ if the pulsation is the first overtone mode. These estimates which do not allow for mass loss on and prior to the AGB may, perhaps, be considered lower limits to the main sequence masses. We are currently examining the Cloud and Galactic Li-rich S stars for evidence of high neutron-density s-processing. Our preliminary result is that we do not see the Rb and ^{96}Zr overabundances expected for the ^{22}Ne(α,n) source in IM-AGBs. Unless IM-AGBs can operate the s-process at low N(n), it appears that the Li-rich S stars are LM-AGBs, but they may have been IM main sequence stars. Our search for IM-AGBs in the Cloud and our Galaxy continues!

With the discovery that all luminous AGB S stars are Li-rich and with, of course, some additional assumptions not detailed here, it is possible to assess whether these stars contribute to galactic enrichment of Li: they probably did raise the Li from $\log \varepsilon(Li) \simeq 2.1$ to $\log \varepsilon(Li) \simeq 3$ at present (Smith and Lambert 1990, Lambert 1990).

In discussions of Li-rich cool giants, it is important to note that there are two types: S stars and carbon stars. The former are s-process enriched and plausibly identified as AGB stars that have undergone the third dredge-up and with a HBCE synthesizing Li. WZ Cas, the Li-rich carbon star, is not s-process enriched (Dominy 1985). If WZ Cas is representative of Li-rich carbon stars, their evolution must differ from that of the Li-rich S stars. Some LM-AGB models predict dredge-up of ^{12}C from a He-shell in which the s-process did not run (Lattanzio 1989). If these models are applicable, Li synthesis in cool carbon stars may be attributed to a HBCE. An alternative interpretation is to link the class of ^{13}C-rich (J-type) cool carbon stars to which WZ Cas belongs to the warm (R-type) carbon stars with luminosities less than those of AGB stars. Richer (1981) shows that Li-rich examples of low luminosity carbon stars exist in the Clouds. The contribution to galactic Li enrichment by these Li-rich carbon stars may be less than that from the S stars because the survey of the Clouds suggests that few carbon stars become Li-rich, but all luminous S stars may synthesize Li in large amounts.

6   CONCLUDING REMARKS

One goal of observational studies of nucleosynthesis in evolved stars is to answer the question - How, when, and where were the chemical elements synthesized? Stellar nucleosynthesis is a key part of the answer. Perhaps the first rational arguments in support of stellar nucleosynthesis were marshalled by Eddington who declared to the British Association in 1920 "I think that the suspicion has been generally entertained that the stars are the crucibles in which the lighter atoms which abound in the nebulae are compounded into more complex elements" (Eddington 1920). Convincing evidence - primarily theoretical - in support of stellar nucleosynthesis was assembled by Burbidge, Burbidge, Fowler, and Hoyle (1957) who legitimately claimed that "We have found it possible to explain, in a general way, the abundances of practically all the isotopes of the elements from hydrogen through uranium by synthesis in stars and supernovae". Today, as I hope this essay has shown, we can now address many difficult questions about nucleosynthesis, and so we expect to complete the identification of the sites within and outside stars at which the chemical elements are synthesized.

My research in nucleosynthesis is supported in part by the U. S. National Science Foundation and the Robert A. Welch Foundation of Houston, Texas.

REFERENCES

Becker, S. A. 1981, *Ap. J. Suppl.*, **45**, 475.

Beer, H., and Macklin, R. L. 1989, *Ap. J.*, **339**, 962.

Blanco, V. M., McCarthy, M. F., and Blanco, B. V. 1980, *Ap. J.*, **242**, 938.

Bogaert, G., André, V., Aguer, P., Barhoumi, S., Kious, M., Lefebvre, A., Thibaud, J. P., and Bertault, D. 1989, *Phys. Rev. C*, **39**, 265.

Brunish, W. M., and Truran, J. W. 1982, *Ap. J.*, **256**, 247.

Burbidge, E. M., Burbidge, G. R., Fowler, W. A., and Hoyle, F. 1957, *Rev. Mod. Phys.*, **29**, 547.

Cameron, A. G. W., and Fowler, W. A. 1971, *Ap. J.*, **164**, 111.

Catchpole, R. M., and Feast, M. W. 1976, *M.N.R.A.S.*, **176**, 501.

Champagne, A. E., and Pitts, M. L. 1986, *Nucl. Phys. A.*, **457**, 367.

Dearborn, D. S. P. 1990, *Phys. Repts.*, in press.

Dearborn, D. S. P., Eggleton, P. P., and Schramm, D. N. 1976, *Ap. J.*, **203**, 344.

Denn, G. R., Luck, R. E., and Lambert, D. L. 1990, *Ap. J.*, submitted for publication.

Dominy, J. F. 1985, *PASP*, **97**, 1104

Dufton, P. L., and Hibbert, A. 1981, *Astr. Ap.*, **95**, 24.

Dufton, P. L., Kane, L, and McKeith, C. D. 1981, *M.N.R.A.S.*, **194**, 85.

Eber, F., and Butler, K. 1988, *Astr. Ap.*, **202**, 153.

Eddington, A. S. 1920, *Observatory*, **43**, 341.

Gies, D. R., and Lambert, D. L. 1990, in preparation.

Gilroy, K. K. 1989, *Ap. J.*, **347**, 835.

Grevesse, N., Lambert, D. L., Sauval, A. J., van Dishoeck, E. F., Farmer, C. B., and Norton, R. H. 1991, *Astr. Ap.*, in press.

Harris, M. J., Fowler, W. A., Caughlan, G. R., and Zimmerman, B. A. 1983, *Ann. Rev. Astr. Ap.*, **21**, 165.

Harris, M. J., and Lambert, D. L. 1984, *Ap. J.*, **281**, 739.

_____. 1984, *Ap. J.*, **285**, 674.

Harris, M. J., Lambert, D. L., and Smith, V. V. 1988, *Ap. J.*, **325**, 768.

Hollowell, D., and Iben, I., Jr. 1988, *Ap. J. (Letters)*, **333**, L25.

_____. 1989, *Ap. J.*, **340**, 966.

Hughes, S. M. G., and Wood, P. R. 1987, *Proc. Astr. Soc. Australia*, **7**, 147.

Iben, I., Jr. 1975, *Ap. J.*, **196**, 525 and 549.

Iben, I., Jr., and Truran, J. W. 1978, *Ap. J.*, **220**, 980.

Jorissen, A., Smith, V. V., and Lambert, D. L. 1991, *Ap. J.*, to be submitted.

Kane, L, McKeith, C. D., and Dufton, P. L. 1980, *Astr. Ap.*, **84**, 115.

Käppeler, F., Beer, H., and Wisshak, K. 1989, *Repts. Prog. Phys.*, **52**, 945.

Käppeler, F., Gallino, R., Busso, M., Picchio, G., and Raiteri, C. M. 1990, *Ap. J.*, **354**, 630.

Kipper, T., and Wallerstein, G. 1990, *PASP*, **102**, 574.

Lambert, D. L. 1990, *Mat.-fys. Medd.*, **42**:4, 75.

_____. 1991, in *The Evolution of Stars - The Photospheric Abundance Connection*, ed. G. Michaud, (Dordrecht: Kluwer), in press.

Lambert, D. L., and Luck, R. E. 1976, *Observatory*, **96**, 100.

Landré, V., Prantzos, N., Aguer, P., Bogaert, G., Lefebvre, A., and Thibaud, J. P. 1990, *Astr. Ap.*, in press.

Lattanzio, J. C. 1989, in *Evolution of Peculiar Red Giants*, ed. H. R. Johnson and B. Zuckerman (Cambridge: CUP), p. 161.

Luck, R. E. 1978, *Ap. J.*, **219**, 148.

Luck, R. E., and Lambert, D. L. 1981, *Ap. J.*, **245**, 1018.

Lyubimkov, L. S. 1984, *Astrofizika*, **20**, 475.

_____. 1988, *Astrofizika*, **29**, 479.

_____. 1989, *Astrofizika*, **30**, 99.

Malaney, R. A. 1987, *Ap. J.*, **321**, 832.

Malaney, R. A., and Lambert, D. L. 1988, *M.N.R.A.S.*, **235**, 695.

Merrill, P. M. 1952, *Ap. J.*, **116**, 21.

Mihalas, D. 1972, *Non-LTE Model Atmospheres for B and O Stars*, NCAR-TN/STR-76.

Nissen, P. E. 1976, *Astr. Ap.*, **50**, 343.

Pilachowski, C. A. 1986, *Ap. J.*, **300**, 289.

Pilachowski, C. A., Sneden, C., and Hudek, D. 1990, *A.J.*, **99**, 1225.

Richer, H. B., 1981, *Ap. J.*, **243**, 744.

Sackmann, I.-J., and Boothroyd, A. 1991, in *The Evolution of Stars - The Photospheric Abundance Connection*, ed. G. Michaud, (Dordrecht: Kluwer), in press.

Scalo, J. M. 1976, *Ap. J.*, **206**, 795.

_____. 1978, *Ap. J.*, **221**, 627.

Smith, V. V. 1986, *B.A.A.S.*, **18**, 679.

_____. 1988, in *The Origin and Distribution of the Elements*, ed. G. J. Mathews, (Singapore: World Scientific), p. 535.

Smith, V. V., and Lambert, D. L. 1984, *PASP*, **96**, 226.

_____. 1986, *Ap. J.*, **311**, 843.

_____. 1988, *M.N.R.A.S.*, **235**, 695.

_____. 1989, *Ap. J. (Letters)*, **345**, L75.

_____. 1990, *Ap. J. (Letters)*, **361**, L69.

Spite, F., and Spite, M. 1982, *Astr. Ap.*, **115**, 357.

Tomkin, J., and Lambert, D. L. 1979, *Ap. J.*, **227**, 209.

_____. 1983, *Ap. J.*, **273**, 722.

Toukan, K. A., and Käppeler, F. 1990, *Ap. J.*, **348**, 357.

Truran, J. W., and Iben, I., Jr. 1977, *Ap. J.*, **216**, 797.

Wolff, S. C., and Heasley, J. N. 1985, *Ap. J.*, **292**, 589.

Wood, P. R., Bessell, M. S., and Fox, M. W. 1983, *Ap. J.* **272**, 99.

Woosley, S. E., Hartmann, D. H., Hoffman, R. D., and Haxton, W. C. 1990, *Ap. J.*, **356**, 272.

Abundance Distribution in Galactic Stellar Populations

P. E. NISSEN

Institute of Astronomy
University of Aarhus
DK-8000 Aarhus C, Denmark.

SUMMARY. Some recent results on the kinematics, chemical compositions, and ages of halo and disk stars in the Solar neighborhood are reviewed. Evidence for the existence of three discrete stellar components: the halo, the thick disk and the thin disk, is presented. The thick disk appears to be as old as the halo. The change from a high α-element/Fe ratio in halo stars to the Solar ratio occurs in connection with the thick-disk/thin-disk transition, i.e. at $[Fe/H] \simeq -0.6$, and not in connection with the halo/thick-disk transition at $[Fe/H] \simeq -1.0$ as usually assumed.

1. INTRODUCTION

Abundance distributions in stellar populations provide important information about Galactic evolution and nucleosynthesis of the elements, especially if abundance ratios of CNO-elements, α-elements, odd-even elements, iron-peak elements, and s- and r-process elements can be determined. Ideally, one would like to study these abundance ratios for *in situ* samples of stars in the halo, the bulge, the thick disk and the thin disk. Given, however, that we have not yet been able to determine detailed abundance ratios for such distant samples of stars, we are bound to rely on information provided by the stars that happen to be relatively close to the Sun. Fortunately, as we shall see, it seems possible to separate these stars in kinematically discrete groups representing the halo, the thick disk and the thin disk.

With the present contribution I do not intend to give a broad review of our knowledge of abundance ratios in Galactic stellar populations. Instead, I will concentrate on some recent surveys on kinematics, chemical compositions and ages of F and G stars in the Solar neighborhood, which have provided new information about the formation and early evolution of the Galaxy, and in which I have been engaged. The results to be discussed are of particular interest in connection with the recent debate on the classical collapse model of Eggen, Lynden-Bell & Sandage (1962) versus the accretion model of Searle & Zinn (1978).

2 KINEMATICS OF HALO AND DISK STARS; V_{rot} VERSUS [Fe/H]

In order to understand better the formation and evolution of the Galaxy it is important to know whether the halo and disk stars in the Solar neighborhood are discrete groups or whether there is a continuous kinematic transition from halo to disk stars. In this connection the relation between Galactic rotational velocity, V_{rot}, and metallicity, [Fe/H], is of particular importance, and an extensive debate about this relation has arisen in the last few years. Sandage & Fouts (1987) argue that there is a smooth increase of the average V_{rot} from about zero at [Fe/H]≈-2.0 to about 200 kms^{-1} at [Fe/H]≈0.0. Norris (1986) and later Norris & Ryan (1989) find that $\langle V_{rot} \rangle$ is constant at about 25 kms^{-1} for -2.5 < [Fe/H] < -1.5 and then raises steeply to about 200 kms^{-1} at [Fe/H]≈-0.5.

Recently, the V_{rot}-[Fe/H] diagram for high-velocity stars in the Solar neighborhood has been the subject of two accurate studies. The first study by Carney, Latham & Laird (1990) has emerged from an impressive survey of almost 1000 stars taken from the Lowell Proper Motion Catalogue. Radial velocities, accurate to ± 1 kms^{-1}, and metallicities with a mean error of 0.15 dex have been obtained from *low* S/N, high-resolution spectra. Distances have been determined from UBV(RI) photometry by assuming that all stars are dwarfs. It is estimated that only 10% of the sample are subgiants.

The second study of the V_{rot}-[Fe/H] diagram is based on *uvby-β* photometry of 711 high-velocity stars by Schuster & Nissen (1988). The interstellar reddening of a star (if any) and T_{eff} are determined from $β$ and $b-y$, and the metallicity, accurate to 0.20 dex, is determined from m_1 (Schuster & Nissen 1989a). Evidence for a cosmic age dispersion of ± 2.5 Gyr among halo stars is presented by Schuster & Nissen (1989b), and in the latest paper (Nissen & Schuster 1991) distances, kinematics and ages are derived. The distances are based on an absolute magnitude calibration that includes an evolutionary correction derived from c_1 and $b-y$.

The V_{rot}-[Fe/H] diagram of Nissen & Schuster (1991) is shown in Fig.1. Just by looking at this diagram it seems evident that there is *not* a continuous transition from the metal-poor halo stars to the metal-rich disk stars. Rather, two discrete populations are seen: the halo around V_{rot}≈0 kms^{-1} and [Fe/H]≈ -1.6 and the disk around V_{rot}≈175 kms^{-1} and [Fe/H]≈-0.4. It is striking that the corresponding V-[Fe/H] diagram of Carney et al. (1990) looks very much the same as Fig.1. A similar distribution of stars is also found by Morrison et al. (1990) from a study of K-giants. In the metallicity range -1.4 < [Fe/H] < -0.6 there are slow rotating halo stars and fast rotating disk stars, but a scarcity of stars with [Fe/H]≈-1.0 and V_{rot}≈100 kms^{-1}. Nissen & Schuster corroborate this overlapping by studying histograms of V_{rot} for different intervals in [Fe/H], and suggest that a straight line through the points ([Fe/H], V_{rot}) = (-0.3, 0 kms^{-1}) and (-1.5, 175 kms^{-1}) separates the two populations rather well. In the following I shall use this line of separation to classify stars as halo or disk.

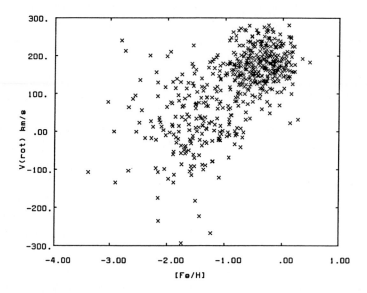

Fig.1. The V_{rot}-[Fe/H] diagram for the stars of Nissen & Schuster (1991).

The sample of Schuster & Nissen (1988) was selected as stars having a total space velocity with respect to the LSR greater than about 80 km s^{-1}. Hence, their group of disk stars includes only the high velocity component of the Galactic disk. The asymmetric drift of the group is 50 km s^{-1} and the velocity dispersion in W is 46 km s^{-1}. These values are close to those currently adopted for the thick disk (Gilmore et al. 1989). On the other hand the average metallicity, [Fe/H]$\approx$-0.4, is somewhat higher than the value of [Fe/H]$\approx$-0.6 normally quoted for the thick disk. Therefore, the group of disk stars seen in Fig.1 probably contains a mixture of metal-poor thick-disk stars and a high-velocity tail of more metal-rich thin-disk stars.

3 CHEMICAL ABUNDANCES OF HALO AND DISK STARS

3.1 Description of Two Extensive Programs

The advent of RETICON and CCD detectors in combination with efficient high resolution spectrographs has resulted in several accurate studies of abundance ratios in halo and disk stars. The most extensive of these studies has been carried out by Edvardsson et al. (1991) and concerns abundance ratios in 189 Solar-type disk stars ranging in metallicity from [Fe/H]$\approx$-1.0 to +0.4. In a complementary program Magain et al. (1991) are studying about 40 dwarfs and subgiants with [Fe/H]$\lesssim$-1.0.

The stars included in the two programs were selected from magnitude limited catalogues of *uvby-β* photometry (see Olsen 1983), and the Strömgren indices *β*, *b-y*, c_1 and m_1 were used to determine effective temperatures, surface gravities and metallicities. High resolution, high S/N spectra were obtained at the ESO 1.4 m CAT telescope and the McDonald Observatory 2.7 m telescope with their respective Coudé Echelle Spectrometers.

Abundances were determined from each measured atomic line by the aid of a synthetic line profile program, which calculates the radiative transfer through a model atmosphere under the assumption of LTE. The oscillator strengths were determined by means of equivalent widths measured in the Solar spectrum. The model atmospheres were calculated using basically the methods and program of Gustafsson et al. (1975), but including new extensive semi-empirical line data calculations of Kurucz (1989, private communication). These new models have flux distributions which are very much closer to the observed fluxes for the Sun and other stars than that of the previous generation of models. The main improvement is due to the greatly increased line opacity in the UV spectral range, which actually removes the long discussed problem of the "missing Solar UV opacity".

In the case of iron about 20 neutral absorption lines and 2 ionic lines are available for the determination of [Fe/H]. The agreement between the abundances derived from neutral and ionic lines is better than 0.1 dex except for the most metal-rich stars, suggesting that possible non-LTE effects are rather negligible for [Fe/H]<0.0. Although the accurate empirical study of non-LTE effects in metal-poor dwarfs by Magain and Zhao (1990) indicates that the effects can be somewhat larger than 0.1 dex for low excitation FeI states, their study confirms that the non-LTE corrections are negligible, when high excitation FeI lines are used to derive [Fe/H] as is the case in the work of Edvardsson et al. We conclude that differential values of [Fe/H] for stars with similar metallicity have accuracies better than ± 0.05 dex. The absolute values of [Fe/H] may have somewhat larger errors, say ± 0.10 dex, mainly due to uncertainties in the determination of T_{eff} as a function of [Fe/H] (Nissen 1990).

3.2 [O/Fe] versus [Fe/H]

The oxygen abundances are derived from the equivalent widths of the OI triplet at 7773 Å and the rather faint OI line at 6158.2 Å. These lines were observed for the southern stars only. The results are shown in Fig.2.

The data of Abia & Rebolo (1989) in Fig.2 have been obtained from an LTE analysis of the OI triplet in unevolved field stars and agree well with the results of Edvardsson et al. although the scatter is somewhat larger. According to Abia & Rebolo [O/Fe] reaches very high values for the most metal deficient stars in conflict with the results of Barbuy (1988) and Barbuy &

Erdelyi-Mendez (1989) obtained from the forbidden oxygen line at 6300 Å in the spectra of K giants and G subgiants.

What is the reason for the large discrepancy in [O/Fe] derived from neutral OI lines in dwarf stars and from the [OI] line in giants? One suggestion is departures from LTE for the high excitation states corresponding to the OI lines. However, Abia & Rebolo (1989) estimate the maximum non-LTE corrections for metal-poor stars to be 0.2 dex only. This is confirmed by recent computations of Kiselman (1990, private communication), who also estimates the non-LTE corrections to the data by Edvardsson et al. to be less than 0.1 dex. On the other hand, Magain (1988) finds the oxygen abundance of the metal-poor Solar-type star HD76932, as derived from the triplet, to be 0.6 dex higher than the oxygen abundance derived from the [OI] line. Barbuy & Erdelyi-Mendez (1989) determined a corresponding difference of 0.15 dex for the somewhat cooler, metal-poor subgiant HD10700. Although the forbidden line is very weak, these results seem quite significant and suggest that the non-LTE computations for the oxygen atom are not adequate.

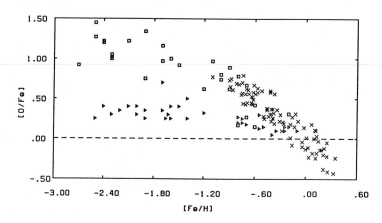

Fig.2. [O/Fe] versus [Fe/H]; ×: Edvardsson et al. (1991); □: Abia & Rebolo (1989); ∇: Barbuy (1988) and Barbuy & Erdelyi-Mendez (1989).

Another reason for the large discrepancy between the results shown in Fig.2 may be problems with the effective temperature scale for metal-poor stars. When the oxygen abundance is derived from high-excitation OI lines and the iron abundance from FeI lines, [O/Fe] becomes very sensitive to errors in T_{eff}. An increase of T_{eff} with 100 K decreases the derived [O/Fe] by almost 0.2 dex. Furthermore, [O/Fe] of giants derived from the forbidden oxygen line is sensitive to errors in the atmospheric parameters, T_{eff} and g.

As discussed by Gilmore et al. (1989) the trend of [O/Fe] with [Fe/H], in particular the possible existence of a change of slope at [Fe/H]≈-1.0, provides very interesting information on the time scale for Galactic formation

and evolution. The reason is that oxygen comes from massive Type II supernovae, while part of the iron is coming from Type I supernovae with a time
delay of about 1 Gyr. However, due to the discrepancies displayed in Fig.2,
it is impossible to draw any conclusions from the present data.

3.3 [Ca/Fe] versus [Fe/H]

According to current theories of Galactic chemical evolution α-elements like
Mg, Si and Ca are expected to depend on [Fe/H] in much the same way as
oxygen (Pagel 1989). Hence, it should also be possible to use [α/Fe] versus
[Fe/H] as a probe of Galactic evolution. This has the advantage that abundance ratios between α-elements and iron can be determined with higher
accuracy than O/Fe. Among the α-elements Ca offers the best case. Neutral
calcium exhibit many unblended absorption lines in the spectra of F and G
stars suitable for very accurate abundance studies. Precise atomic data are
available for these lines from laboratory work (Smith 1988) and from the
Solar spectrum.

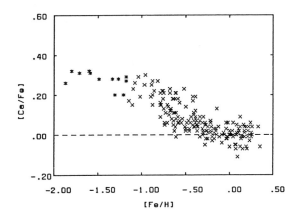

Fig.3.   [Ca/Fe]   versus
[Fe/H] for stars in the
programs of Edvardsson
et al. (1991) and Magain
et al. (1991). X,   disk
stars. *, halo stars.

A detailed study of CaI has recently been carried out by Drake (1990, preprint), who finds that non-LTE effects in differential abundance studies of
Solar-type stars are less than 0.05 dex as was the case for iron. Furthermore, [Ca/Fe] is nearly independent of possible errors in T_{eff} and g (Nissen
1990). The dominant error source is the uncertainty of the equivalent widths
and from the scatter of abundances derived from individual lines we estimate
the error of the mean value of [Ca/Fe] to be as small as 0.04 dex.

Fig.3 shows [Ca/Fe] versus [Fe/H] for the stars in the survey of Edvardsson
et al. (1991) supplemented with some preliminary data from the program of
Magain et al. (1991). The stars have been divided into halo and disk stars
according to their position in the V_{rot}-[Fe/H] diagram as explained in Sect.
2. As seen the transition from a high ratio of [Ca/Fe]$\approx$0.3 among the metal-
poor stars to the Solar ratio, [Ca/Fe]=0.0, starts at [Fe/H]$\approx$-0.8, whereas

the halo/disk transition occurs at [Fe/H]≃-1.1.

3.4 Chemical Composition and Kinematics of Disk Stars

Calcium is not the only element for which the abundance changes steeply as a function of [Fe/H]. Fig.4 shows [Mg/Fe] versus [Fe/H] for those stars in the survey of Edvardsson et al. (1991), for which at least two Mg lines were used for the abundance determination. As seen [Mg/Fe] has a value of about 0.4 dex for [Fe/H]<-0.8 in fairly good agreement with the plateau of [Mg/Fe]≃0.5 for halo stars found by Magain (1989). In the range -0.8<[Fe/H]<-0.4 [Mg/Fe] decreases and reaches a constant level of about 0.1 dex for [Fe/H]>-0.4. No obvious explanation has been found for this systematic offset from zero at Solar metallicities, but it may be due to unknown blends in the MgI lines used (8712.68 and 8717.82 Å), because the stars have in average somewhat higher effective temperatures than the Sun.

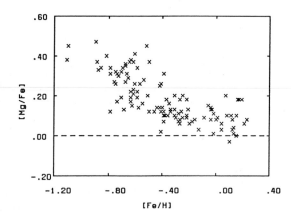

Fig.4. [Mg/Fe] versus [Fe/H] for the stars of Edvardsson et al. (1991).

It is interesting to compare Fig.4 with the plots of the Galactic velocity components V and W versus [Fe/H] (Fig.5). In both diagrams the velocity dispersion in the range -0.8<[Fe/H]<-0.4 is about a factor of two higher than the dispersion in the range -0.4<[Fe/H]<0.0. In this connection we note that there is no kinematic bias in the selection of stars in the program of Edvardsson et al. (1991). The F5-G2 stars in the magnitude limited catalogues of Strömgren photometry were divided into 9 metallicity groups according to the δm_1 index and the 20 brightest slightly evolved main sequence stars in each group were observed.

The discontinuity in the kinematic properties of disk stars at [Fe/H]≃-0.4 suggests that we are dealing with two discrete populations of disk stars: the thick-disk stars with [Fe/H]≲-0.4 and the thin-disk stars with mainly [Fe/H]≳-0.4 but probably overlapping in [Fe/H] with the thick disk. According to this interpretation the transition from the high Ca/Fe and Mg/Fe ratios to the Solar ratios occur in connection with the thick-disk/thin-disk transition.

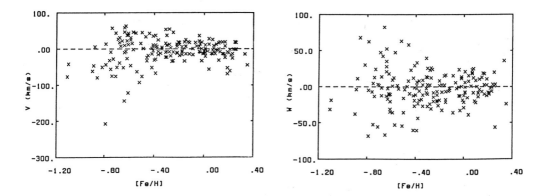

Fig.5. The Galactic velocity components V and W (corrected for Solar motion) versus [Fe/H] for the stars of Edvardsson et al. (1991).

The dispersion of [Mg/Fe] in the range -0.4<[Fe/H]<0.0 is 0.05 dex. In the range -0.8<[Fe/H]<-0.4 the dispersion is 0.10 dex suggesting that a cosmic scatter in Mg/Fe is present. As seen from Fig.6 there is even a marginal correlation between [Mg/Fe] and V for the thick-disk stars. A possible interpretation of this correlation is discussed in Sect. 5.

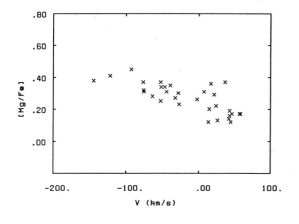

Fig.6. [Mg/Fe] versus the Galactic velocity component V for stars from Edvardsson et al. (1991) having -0.8 < [Fe/H] < -0.4 and ages A > 10 Gyr.

4 AGES OF HALO AND DISK STARS

Ages of turn-off stars have been determined by comparing their position in the $\delta M_V - \log T_{eff}$ diagram with isochrones from VandenBerg (1985). $\delta M_V = M_V(ZAMS) - M_V(star)$ and T_{eff} are derived from the position of the star in the $c_1 - (b-y)$ diagram. The technique and the calibrations used are described in more detail by Nissen & Schuster (1991) and Edvardsson et al. (1991). Earlier, the same method was used by Twarog (1980) to derive an age-metallicity relation for F-type stars and by Strömgren (1987) in a discussion of the relations between age, chemical composition and kinematics for a large sample of F stars within 100 pc.

For turn-off stars with a given metallicity differential ages can be determined to an accuracy of about 25%, i.e. $\sigma(\log A)=0.10$, provided that they are older than about 3 Gyr. As shown by Andersen et al. (1990) from a study of binary stars the error may be larger for younger stars due to the neglect of convective overshooting in the models of VandenBerg. Furthermore, the absolute ages derived may be affected by systematic errors in the calibrations and in the stellar models. These errors could well be a function of [Fe/H].

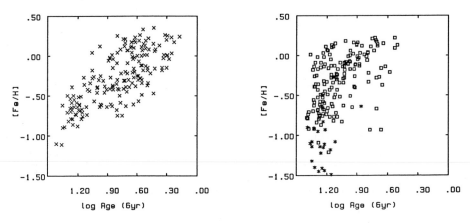

Fig.7. [Fe/H] versus $\log A$, where A is the stellar age in Gyr. Left fig.: disk stars from Edvardsson et al. (1991). Right fig.: high-velocity stars from Nissen and Schuster (1991) belonging to the disk (□), and to the halo, *.

Fig.7 shows the [Fe/H]-$\log A$ diagram for the Edvardsson et al. disk stars (selected as the 20 brightest F5-G2 stars in 9 metallicity groups) and the corresponding diagram for the Nissen & Schuster high-velocity stars (selected as stars having a total space velocity with respect to the LSR greater than 80 km s^{-1}). As seen from the left figure there is an overall age-metallicity relation among disk stars. However, the cosmic scatter in this relation is considerably. At a given [Fe/H] the rms deviation of $\log A$ is 0.20 dex, twice the estimated error of the age determinations. This scatter becomes even more pronounced when we consider the high-velocity component of the disk. The high-velocity stars are generally older than the normal disk stars, and some high-velocity stars with [Fe/H]≈0.0 are nearly as old as the halo stars.

The most interesting result of Fig.7 is that the bulk of high-velocity stars with [Fe/H]<-0.4, i.e. the thick-disk stars, have about the same high age as the halo stars. The classification into halo or disk was done on the basis of the position of the stars in the V_{rot}-[Fe/H] diagram as described in Sect.2.

Due to the overlap in [Fe/H] between the halo and the disk the conclusion of similar ages for the two components is not sensitive to possible systematic errors in the age determination. The absolute value derived for the age of the halo stars (18 Gyr in average) may however be affected by these errors. We also note that the data in Fig.7 do not exclude an age spread of about 2 Gyr among halo stars as advocated by Schuster & Nissen (1989b) and as recently found by VandenBerg et al. (1990) from a study of color-magnitude diagrams of globular clusters.

Due to the large spread in [Fe/H] at a given age one has to be very careful when deriving an *average* age-metallicity relation for a given region in the Solar neighborhood. The age-metallicity relation in the so-called Solar cylinder of the disk will be different from the age-metallicity relation in a sphere, because the scale height of stars depends on [Fe/H]. Consequently, no attempt has been made to derive an average age-metallicity relation from the data shown in Fig.7. Instead models should be tested by comparing directly with Fig.7 using the appropriate selection criteria for the stars.

5 DISCUSSION AND CONCLUSIONS

From the classical collapse model of Galactic formation by Eggen, Lynden-Bell & Sandage (1962) continuous relations between kinematic properties of stars and the metallicity parameter [Fe/H] have been predicted at variance with the discontinuity between halo and disk stars discussed in Sect.2. Some collapse models may however be compatible with the data presented in this paper. As discussed by Gilmore et al. (1989) a rapid increase in the dissipation and star formation rates due to enhanced cooling once [Fe/H] has passed -1.0 would introduce an apparent kinematic discontinuity. In this scenario the age difference between the halo and disk may well be small enough to agree with the age data discussed in Sect.4.

Although all of the data presented in the present paper probably can be explained in terms of Galactic collapse models, accretion models seem more appealing. They are usually based on the idea of Searle and Zinn (1978) that globular clusters and halo stars form in small satellite galaxies, which are then accreted by a thin, fast rotating Galactic disk. As suggested by Freeman (1990) the heating of the early, partly stellar disk, resulting from dynamical friction of the satellite galaxies could produce the thick disk. The kinematics of the halo, on the other hand, depends on the dynamics of the merging galaxies. After the accretion period new formed disk stars are accelerated by encounters with molecular clouds and/or spiral arms. Hence, we have three different mechanisms being responsible for the kinematics of the halo, the thick disk and the thin disk and explaining the discontinuities seen in Fig.1 and Fig.5.

In Sect.3 we found that the change from a high α-element/Fe ratio to the Solar ratio takes place at the [Fe/H] value for the thick-disk/thin-disk transition. Assuming that the change in the element ratio is due to the occurrence of Type I supernovae having a time delay of about one Gyr with respect to Type II's, it follows that halo and thick-disk stars have similar high ages, confirming the direct age determinations discussed in Sect.4. Finally, we note that the increased dispersion of [Mg/Fe] and the possible correlation between [Mg/Fe] and V in the range -0.8<[Fe/H]<-0.4 may be due to different rates of chemical evolution in the disk as a function of galactocentric distance.

References

Abia, C., Rebolo, R. 1989, ApJ, 347, 186

Andersen, J., Nordström, B., Clausen, J.V. 1990, ApJ, 363, L33

Barbuy, B. 1988, A&A, 191, 121

Barbuy, B., Erdelyi-Mendez, M. 1989, A&A, 214, 239

Carney, B.W., Latham, D.W., Laird, J.B. 1990, AJ, 99, 572

Edvardsson, B., Andersen, J., Gustafsson, B., Lambert, D.L., Nissen, P.E., Tomkin, J. 1991, (in preparation)

Eggen, O.J., Lynden-Bell, D., Sandage, A.R. 1962, ApJ, 136, 748

Freeman, K.C. 1990, in *Astrophysics - Recent Progress and Future Possibilities, Invited Reviews at a Symposium in Honour of Bengt Strömgren*, Eds. B. Gustafsson and P.E. Nissen, The Royal Danish Academy of Sciences and Letters, Mat.Fys.Medd. 42:4, p.187

Gilmore, G., Wyse, R.F.G., Kuijken, K. 1989, ARA&A, 27, 555

Gustafsson, B., Bell, R.A., Eriksson, K., Nordlund, Å. 1975, A&A, 42, 407

Magain, P. 1989, A&A, 209, 211

Magain, P., Zhao, G. 1990, ApJ (submitted)

Magain, P. 1988, Proc. IAU Symp. 132, eds. G. Cayrel de Strobel and M. Spite, Klüwer, Dordrecht, p. 485

Magain, P., Edvardsson, B., Gustafsson, B., Nissen, P.E. 1991, (in preparation)

Morrison, H.L., Flynn, C., Freeman, K.C. 1990, AJ, 100, 1191

Nissen, P.E. 1990, Proc. XI IAU European Astronomy Meeting, ed. M. Vazques, Cambridge University Press, (in press)

Nissen, P.E., Schuster, W.J. 1991, A&A (submitted)

Norris, J. 1986, ApJS, 61, 667

Norris, J.E., Ryan, S.G. 1989, ApJ, 340, 739

Olsen, E.H. 1983, A&AS, 54, 55

Pagel, B.E.J. 1989, Rev. Mex. Astr. Astrof., 18, 161

Sandage, A., Fouts, G. 1987, AJ, 93, 74

Schuster, W.J., Nissen, P.E. 1988, A&AS, 73, 225

Schuster, W.J., Nissen, P.E. 1989a, A&A, 221, 65

Schuster. W.J., Nissen, P.E. 1989b, A&A, 222, 69

Searle, L., Zinn, R. 1978, ApJ, 225, 357

Smith, G. 1988, J.Phys.B., 21, 2827

Strömgren, B. 1987, The Galaxy, Proc. NATO Adv. Study Inst., eds. G. Gilmore and R. Carswell, Reidel, Dordrecht, p.229

Twarog, B.A. 1980, ApJ, 242, 242

VandenBerg, D.A. 1985, ApJS, 58, 711

VandenBerg, D.A., Bolte, M., Stetson, P.B. 1990, AJ, 100, 445

Questions

D. Lynden-Bell: In the diagram in which you showed quite a clear separation between disk and halo stars I would like to know how the stars are selected. If there is a bias towards high velocity stars so as to show some then the way that bias acts clearly affects what is seen in the diagram. High proper motion clearly biases us towards high velocity stars <u>and</u> towards nearby disk stars. Is this bias less effective for stars of intermediate velocities and distances?

P.E. Nissen: The stars were indeed selected as a sample of high-velocity stars. However, I do not think this can explain the deficiency of stars with intermediate rotational velocities $V_{rot} \simeq 100 \, \mathrm{kms^{-1}}$ and [Fe/H]$\simeq$-1.0. Such stars would have about the same proper motions as stars with $V_{rot} \simeq 100 \, \mathrm{kms^{-1}}$ and [Fe/H]$\simeq$-1.6 of which we see many.

B.E.J. Pagel: Do you think there is a discrete division between thin and thick disk?

P.E. Nissen: Yes, as seen from our kinematical diagrams, V vs. [Fe/H] and W vs. [Fe/H], the velocity dispersion changes rather abruptly at a metallicity of [Fe/H]$\simeq$-0.4. In earlier diagrams of this type the sharp change in velocity dispersion may have been smeared out because of less accurate [Fe/H] determinations.

M.G. Edmunds: You showed the correlation [Mg/Fe] against V velocity. If you plot [Fe/H] against V for <u>exactly</u> the same stars do you get the same correlation?

P.E. Nissen: No, there is no obvious correlation.

B. Barbuy: Given the number of discussions on the discrepancies on the oxygen abundances it is important to point out that a result [O/Fe]$\simeq$0.4 for metal-poor stars, derived from the forbidden line, was also obtained by Sneden et al. (1990) (Proc. of the VIth Cambridge Workshop on Cool Stars, Stellar Systems and the Sun, Ed. G. Wallerstein).

Discovery of the Most Metal-Poor Stars

T.C. BEERS

Michigan State University

1 A SURVEY FOR EXTREMELY METAL-POOR STARS

In the four decades following the report of field stars with sub-solar abundance by Chamberlain and Aller (1951), numerous dedicated searches for metal-deficient stars in the halo of the galaxy have resulted in the detection of several hundred halo stars in the abundance range $-2.0 \leq [Fe/H] \leq -1.0$. Although recent proper-motion selected surveys (Laird, Carney, and Latham 1988; Norris and Ryan 1989) have discovered on order 100 stars in the abundance range $-3.0 \leq [Fe/H] \leq -2.0$, with only a few notable exceptions stars with $[Fe/H] \leq -3.0$ have rarely been identified. This paucity of the most metal-poor stars has led some to argue that either (a) stars of sufficiently low mass to survive to the present could not form from the metal-deficient gas of the early galaxy (Jones 1985), (b) self-enrichment of early globular clusters followed by their disruption into the halo led to a lack of presently observable stars below $[Fe/H] = -3.0$ (Cayrel 1986), or (c) an as-yet poorly understood early nucleosynthesis event, such as that due to supermassive $(M > 100~M_\odot)$ stars led to an early boost in the ambient metallicity of the galaxy (Carr *et.al.* 1984).

As we discuss below, a new survey, optimized to detect the most metal-poor stars within 1-5 kpc of the sun, has detected extremely metal-deficient stars in numbers which are consistent with predictions of simple models for galactic chemical evolution in which star formation proceeds from a zero (or nearly so) initial abundance of metals in the early galaxy. The discovery of large numbers of extremely metal-deficient stars will allow strong constraints to be placed on the low end of the metallicity distribution function for halo stars, opens the possibility of "reading back" the mass distribution of early Type-II supernovae, and will eventually enable direct tests of recently proposed non-standard big bang cosmological scenarios.

1.1 The Objective-Prism/Interference-Filter Technique

As discussed in Beers, Preston, and Shectman (1985), our survey technique is a modification of traditional objective-prism methods used in the past to discover a limited number of relatively bright low-metallicity stars. Candidate stars in the present sample are identified from wide-field objective-prism plates obtained with the Curtis Schmidt telescope at CTIO. An interference-filter placed near the focal plane restricts the bandpass of our survey to approximately 150 Å centered near the CaII K feature at 3933 Å. Because of the interference-filter, long (90 minute) exposures can be taken without undue sky fog or confusion of the spectra. The faintest stars in our survey reach $B = 16$, several magnitudes fainter than previous surveys. The plates are visually inspected with a low-power binocular microscope. The vast majority of images exhibit strong CaII H and K absorption typical of solar abundance F, G, and K stars. A small fraction of stars (typically 1 in 1000) exhibit weak CaII H *and* K lines, which identifies them as metal-poor (MP) candidates. Roughly five times this number of spectra exhibit weak CaII K and strong Hϵ, and are classified as AB- or A-type stars. This category is dominated by members of the field horizontal-branch (FHB) populations, but also includes main sequence A stars and other miscellaneous hot stars.

Medium-resolution (1Å) digital spectra are obtained for the MP candidates, over the wavelength range $\lambda\lambda 3700 - 4500$ Å. Roughly sixty percent of the MP candidates are revealed to be metal-deficient; the remainder are a mix of hot stars, FHB stars, or late-type M stars with emission filling in the H and K features. Broadband UBV photometry is obtained for the stars verified as metal-poor.

1.2 Present Status of the Survey
To date, 193 plates have been obtained in the southern survey. A recent extension of the survey in the north with the Burrell Schmidt at KPNO has resulted in 92 acceptable survey plates. Fifteen repeat plates have been obtained to allow for a statistical quantification of expected temperature-related selection effects. A total of 6750 square degrees of sky has been surveyed to date. We plan to double this sky coverage, eventually obtaining plates of the entire sky above $b = \pm 45°$. Fifty percent of the extant survey plates have been visually scanned so far. On order 4500 MP candidates have been identified. One third of these candidates (1500 stars) have had the required spectroscopic follow up. Thus the sample discussed herein comprises roughly 20 percent of the extant survey, or 10 percent of the eventual survey.

2 ESTIMATION OF METAL ABUNDANCE FROM MEDIUM-RESOLUTION SPECTRA

To obtain estimates of metal abundance from medium-resolution spectra, we compare the strength of the observed CaII K feature to models of the predicted line strength as a function of the broadband $B-V$ color and [Fe/H] . Clearly, this method relies on the existence of a one-to-one correlation (not necessarily a linear one, but one free of cosmic scatter) between CaII and the overall stellar metal abundance, as parameterized by [Fe/H] , over its range of applicability. Fortunately, recent work by Magain (1987), Gratton and Sneden (1988), Magain (1989), and others, indicates that [Ca/Fe] is remarkably free of scatter at a given [Fe/H] .

2.1 Model Atmosphere Calculations

We employ spectral synthesis calculations for the predicted CaII K line strength kindly carried out for us by Norris. A fixed ratio [Ca/Fe] = 0.38 is adopted for stars in the range $-5.0 \leq [Fe/H] \leq -1.0$. The Revised Yale Isochrones are then used to obtain estimates of the expected $B - V$ colors over the range of abundance. Deficiencies in the model calculations, including the as-yet poorly constrained variation of [Ca/Fe] as a function of [Fe/H], and the mapping of $B - V$ colors from the theoretical to the observation plane are calibrated and removed as described below.

2.2 Calibration of Model Predictions

To facilitate a comparison between our model estimates of metal abundance and other independent measures of [Fe/H] , we have obtained observations on our system of a sample of 80 stars with metallicity estimates derived from high-resolution spectroscopy. We construct a residual map of the difference in metal abundance as derived from the Norris models and the adopted standard metallicities ($RES_1 = [Fe/H]_K - [Fe/H]_o$) in the KP' (an optimized CaII K equivalent width corrected for interstellar contamination) $(B-V)_o$ plane. A multiple linear regression fit to RES_1 is used to remove correlated effects with KP' and $(B-V)_o$, which account for on order 70 percent of the variance in RES_1. A comparison of the corrected estimates of metal abundance with the available standard stars indicates that our method has a scatter of roughly 0.15–0.2 dex over the abundance range $-4.5 \leq [Fe/H] \leq -1.0$. For further discussion of this method, see Beers *et.al.* (1990).

Broadband photometry is available for only about one third of our program objects. For those stars without $B - V$ colors, we adopt a calibration of $B - V$ with Balmer line strength, parameterized as HP (a weighted mean equivalent width of the Hγ and Hδ lines). In Figure 1 we plot KP' *vs.* HP for 1066 stars. Contours of constant metal abundance are indicated by the dashed lines. The placement of the $[Fe/H] = -4.0$ contour is rather poorly constrained due to a lack of available calibration stars. This difficulty should be removed shortly, as high-resolution spectroscopy of our extremely metal-poor stars is being actively pursued. Confirmation of several ultra metal-poor stars discovered in our survey is already in hand. Bonifacio, Castelli, and Molaro (1990) obtain $[Fe/H] = -4.23 \pm 0.31$ for the giant CS 22885-96; Molaro and Castelli (1990) report $[Fe/H] = -4.29 \pm 0.19$ for the turnoff star CS 22876-32.

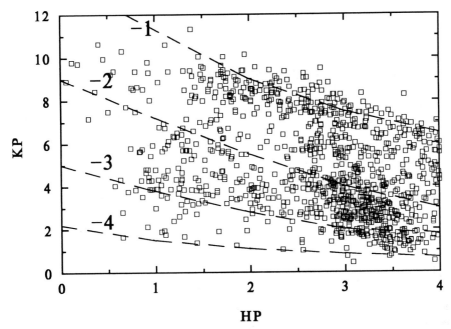

Fig. 1 1066 Stars with [Fe/H] < –0.5

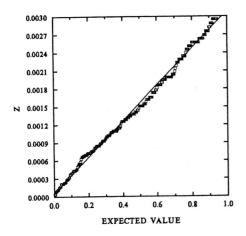

Fig. 2(a) Stars with [Fe/H] < –2.5

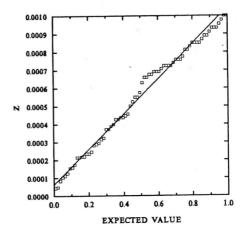

Fig. 2(b) Stars with [Fe/H] < –3.0

3 THE METALLICITY DISTRIBUTION FUNCTION OF EXTREME POPULATION II STARS

With the large numbers of metal-poor stars obtained in the HK survey, the crucial question of the form of the metallicity distribution function (MDF) of the most metal-deficient stars can now be investigated.

3.1 Predictions of the Simple Model of Galactic Chemical Evolution

Simple models (Searle and Sargent 1972; Pagel and Patchett 1975; Hartwick 1976) for galactic chemical evolution predict that, in the limit of complete conversion of gas to stars, the MDF takes the form:

$$f(z) = \frac{1}{y}e^{-z/y} ,\qquad(1)$$

where $z = 10^{[Fe/H]}$, and y is the effective yield parameter, which has value between 0.03 and 0.06. In the low-metallicity limit ($z \ll y$), this function approaches a constant. Thus, the MDF of the most metal-poor stars is predicted to be flat.

3.2 Quantile Plots for Stars with $[Fe/H] \leq -2.5$

A simple check on the observed shape of the MDF at low abundance can be obtained in the following manner. The metallicities of stars selected from a flat MDF should be consistent with random draws from a distribution which is uniform on the interval $0 \leq z \leq z_{upp}$, where z_{upp} is the upper limit of the metallicity sample. In Figure 2(a) we show a quantile plot (where the expected value is obtained from the assumption of a uniform distribution) based on 232 stars from our survey with $[Fe/H] \leq -2.5$, corresponding to a $z_{upp} = 0.003$. The linear behavior of this plot indicates that a uniform distribution is a remarkably good description of the data in hand. A similar plot for the 72 stars with $[Fe/H] \leq -3.0$ ($z_{upp} = 0.001$) is shown in Figure 2(b). Clearly, there is no evidence in the available data for a deficiency of the most metal-poor stars as compared to the form predicted by the simple models.

4 FUTURE PROSPECTS

Much work remains, but with it will come great opportunity for refinement of our picture of galactic chemical evolution. Here we summarize several interesting directions for future research based on stars discovered by the HK survey.

4.1 The Eventual Survey

Assuming a future discovery rate similar to that so far, we estimate that a dedicated spectroscopic follow up of the 10,000 MP candidates to be identified will result in the detection of on order 5000 stars with $[Fe/H] \leq -2.0$, 500 stars below $[Fe/H] = -3.0$, 50 stars below $[Fe/H] = -4.0$, and perhaps 5 stars with abundances below $[Fe/H] = -5.0$. If stars as metal-deficient as $[Fe/H] = -6.0$ exist, and the MDF stays flat to this abundance level, we have a 50 percent chance of finding at least one star at this extreme level of metal poverty. In addition to the great improvement in constraining the MDF, the non-kinematic selection of these stars assures they will be of great utility for quantification of the kinematics of the inner halo as a function of abundance.

4.2 The Mass Distribution of Early Type-II Supernovae

The relative abundances of heavy elements in metal-deficient stars summarize many aspects of the cumulative history of star formation, stellar evolution, and nucleosynthesis processes which have taken place in the galaxy. Thielemann, Hashimoto, and Nomoto (1990) concisely describe the nucleosynthesis sites expected to dominate the early chemical evolution of the Galaxy, primarily intermediate- to high-mass Type-II supernovae. By matching the relative abundances of various nuclear species (in particular the alpha elements such as Ca, Mg, and O) in successively lower-metallicity stars to the predicted abundances of Type-II supernova precursors, we may be able to obtain an estimate of the masses for these objects. At the lowest metallicities we are probing such an early chemical time that we should begin to see variations due to local enhancements from individual supernovae, and thus place limits on the time that the galaxy could be considered well-mixed.

4.3 Tests of Non-Standard Big Bang Models

The most metal-deficient stars provide one of the only available tests of the predictions of non-standard big bang cosmologies. If the neutron exposures suggested by recent calculations are correct, isotopes of the light elements Be and B are expected to be 10–1000 times greater than their standard big bang abundances (Kajino and Boyd 1990). The rather large range of allowed parameters in these models permits a definitive test to be carried out by measuring the lines of Be and B in stars with abundances below $[Fe/H] = -3.5$. Although faint, the large numbers of stars in this abundance range discovered by the HK survey should prove useful.

This work received partial support from grants AST 86-17265 and AST 90-01376 from the National Science Foundation, and an All University Research Initiation Grant from Michigan State University.

REFERENCES

Beers, T.C., Preston, G.W., and Shectman, S.A. 1985, *A.J.*, **90**, 2089.

Beers, T.C., Preston, G.W., Shectman, S.A., and Kage, J.A. 1990, *A.J.*, **100**, 849.

Bonifacio, P., Castelli, F., and Molaro, P. 1990, preprint.

Carr, B.J., Bond, J.R., and Arnett, W.D. 1984, *Ap.J.*, **277**, 445.

Cayrel, R. 1986, *A.A.*, **168**, 81.

Chamberlain, J.W., and Aller, L.H. 1951, *Ap.J.*, **114,**, 52.

Gratton, R.G., and Sneden, C. 1988, *A.A.*, **204**, 193.

Hartwick, F.D.A. 1976, *Ap.J.*, **209**, 418.

Jones, J.E. 1985, *P.A.S.P.*, **97**, 593.

Kajino, T., and Boyd, R.N. 1990, *Ap.J.*, **359**, 267.

Laird, J.B., Carney, B.W., and Latham, D.W. 1988, *A.J.*, **95**, 1843.

Magain, P. 1987, *A.A.*, **179**, 176.

Magain, P. 1989, *A.A.*, **209**, 211.

Molaro, P., and Castelli, F. 1990, A.A., (in press).

Norris, J.E., and Ryan, S.G. 1989, *Ap.J.*, **340**, 739.

Pagel, B.E.J., and Patchett, B.E. 1975, *M.N.R.A.S.*, **172**, 13.

Searle, L., and Sargent, W.L.W. 1972, *Ap.J.*, **173**, 25.

Thielemann, F.K., Hashimoto, M., and Nomoto, K. 1990, *Ap.J.*, **349**, 222.

[O/Fe] vs [Fe/H] AT LOW METALLICITY

C. ABIA

Institut d'Astronomie, d'Astrophysique et de Géophysique
Bruxelles, Belgium

R. CANAL

Departament d'Astronomia i Meteorologia, Universitat de Barcelona, Spain
Laboratori d'Astrofísica, Institut d'Estudis Catalans

J. ISERN

Centre d'Estudis Avançats de Blanes, C.S.I.C., Girona, Spain
Laboratori d'Astrofísica, Institut d'Estudis Catalans

1 INTRODUCTION

Recent observations (Abia and Rebolo 1989) have extended our knowledge of the $[O/Fe]$ vs $[Fe/H]$ relationship down to very low metallicities: almost to $[Fe/H] \simeq -3.5$. Based upon the (permitted) lines of the OI infrared triplet at 7773 Å in old disk and halo population dwarfs, the new results show a monotonous increase in $[O/Fe]$ with decreasing metallicity: from $\simeq 0.0$ at $[Fe/H] \simeq 0.0$, up to $\simeq +1.1$ at $[Fe/H] \simeq -2.0$. In addition, there is some hint as to that for $[Fe/H] \leq -2.0$ it might remain constant at $\simeq +1.1$. Such a trend is in contrast with previous results (see Wheeler, Sneden and Truran 1989) based, at metallicities $[Fe/H] \leq -1.0$, on the forbidden line of OI at 6300 Å in old giants and subgiants.

In the present paper we have tried to answer two questions:

a) Can the new results (controversial as they might be) still be explained from "standard" hypotheses concerning galactic chemical evolution?

b) Since chemical abundance ratios at low metallicities reflect the early evolution of the Galaxy, starting from the nucleosynthesis by the shortest-lived stars and progressively including the contributions from longer-lived ones as metallicity increases: could these results give us some insight on nucleosynthesis by supernovae of different types and on the nature of their progenitor stars?

To this end we have constructed a fairly "standard" galactic evolution model.

2 MODEL HYPOTHESES

Our model is based on the following hypotheses (see Abia, Canal, and Isern 1990 for more details):

1) The star life times as a function of their mass, $\tau_M(M)$, are taken into account as in Talbot and Arnett (1971).

2) A star formation rate (SFR) of the form:

$$\Psi(t) = \alpha \, \sigma_g^n(t) \qquad (\alpha = 0.6 \text{ Gyr}^{-1}; \ n=1) \qquad (1)$$

is adopted.

3) Infall of unenriched gas, exponentially decreasing in time, with an e-folding constant of 4.5 Gyr, is assumed. This gives a present infall rate, $t_{present} \simeq 0.2 \text{ M}\odot \text{ pc}^{-2} \text{ Gyr}^{-1}$, and a present total mass density $\sigma_T \simeq 70 \text{ M}\odot \text{ pc}^{-2}$.

4) The initial mass function (IMF) is first taken constant in space and time and it is of the form:

$$\Phi(M) \propto M^{-(1+x)} \qquad (x = 1.35; \ 0.8 \text{ M}\odot \le M \le 100 \text{ M}\odot) \qquad (2)$$

(we will later study the effects of modifying this hypothesis).

3 SUPERNOVA RATES

3.1 Type Ia (SNIa) rates

They are computed through the formulation of Greggio and Renzini (1983), and we consider two different hypotheses as to their progenitors:

1) A binary system made of a carbon-oxygen white dwarf plus a red giant companion. There, the presupernova life time is that of the secondary star: $\tau^* = \tau_M^{second}$.

2) A double degenerate system made of two carbon-oxygen white dwarfs (Iben and Tutukov 1984). In this case, the presupernova life time is that of the secondary plus the time necessary for it to fill its Roche lobe due to energy loss by gravitational wave radiation: $\tau^* = \tau_M^{second} + \tau_{GR}$. For calculating the last, a distribution function of the semimajor axes, a_0, of the binary is adopted:

$$h(a_0) \simeq 1/a_0 \qquad (10 \text{ R}\odot \le a_0 \le 10^4 \text{ R}\odot) \qquad (3)$$

(Tutukov and Yungelson 1980).

In both cases SNIa are assumed to originate from intermediate-mass stars:

$M_{bin} \simeq 4 - 14$ M$\odot$. In order to fit the present galactic SNIa rate, it has to be assumed that only $\sim 2\%$ of all appropriate binaries do become SNIa. In the case where SNIb are equally assumed to be binaries (CO white dwarf + He star), no distinction between them and SNIa is made.

3.2 Type Ib/Ic (SNIb/c) rates

Here we also consider two different hypotheses as to the progenitors:

1) Very massive stars (M $\geq$ 30 M$\odot$) in the Wolf–Rayet stage.

2) Intermediate–mass stars (7 M$\odot$ $\leq$ M $\leq$ 10 M$\odot$) having lost their hydrogen–rich envelopes (the SNI1/2 of Iben and Renzini 1983).

3.3 Type II (SNII) rates

SNII are assumed to result from core collapse of massive (M $\geq$ 10 M$\odot$) stars. When hypothesis 1 (in **3.2**) for SNIb/c is adopted, the SNII presupernova mass range is reduced to 30 M$\odot$ $\geq$ M $\geq$ 10 M$\odot$.

The SNIb/c and the SNII rates are computed as in Matteucci and Tornambè (1988) and the present–time galactic rates for the different types are in the proportions:

$$(SNII/SNIa/SNIb/c)_{present} = 1.4/0.4/0.7 \text{ SN per 100 yr} \qquad (4)$$

It must be stressed that the maximum SNIa rate is attained at later times for hypothesis 2 than for hypothesis 1 (in **3.1**).

4 NUCLEOSYNTHESIS PRESCRIPTIONS

4.1 Single stars

In the mass range 0.8 M$\odot$ $\leq$ M $\leq$ 8 M$\odot$, they are taken from Renzini and Voli (1981), with the parameters: Y = 0.11, Z = 0.02, α = 1.5, and η = 0.33.

In the M $\geq$ 8 – 10 M$\odot$ range, the ^{16}O yields are from Woosley and Weaver (1986) for the ^{12}C$(\alpha,\gamma)^{16}$O rate from Kettner et al. (1982).

For the stars with masses M $\geq$ 50 M$\odot$, the ^{16}O yield due to mass loss is included according to the prescription of Prantzos et al. (1986). As for ^{56}Fe, it is assumed that each SNII ejects $\simeq$ 0.07 M$\odot$, as in SN 1987A (Woosley 1988; Hashimoto, Nomoto, and Shigeyama 1989; Arnett, Schramm, and Truran 1989). SNIb/c eject $\simeq$ 0.15 M$\odot$ of ^{56}Fe, both in cases 1 and 2 (in **3.2**).

4.2 Binaries (SNIa)

They are assumed to eject $\simeq$ 0.14 M$\odot$ of ^{16}O and $\simeq$ 0.6 M$\odot$ of ^{56}Fe (Nomoto, Thielemann, and Yokoi 1984).

5 RESULTS AND DISCUSSION

Results from different combinations of hypotheses are displayed in Fig. 1 (a, b, and c), where the observational data are also plotted. From Fig. 1a we see that the combination of hypotheses 2 (from 3.1) and 1 (from 3.2) (solid line), that fits both ^{16}O and ^{56}Fe in the Solar System plus the current gas fraction, somewhat underpredicts [O/Fe] for [Fe/H] ≤ -1.0. Such underproduction is more pronounced for combination 1 (3.1) plus 1 (3.2) (dotted line), and still more so in the calculation of Matteucci and Greggio (1986) (dashed line). From Fig. 2b we see that the combination 2 (3.1) plus 2 (3.2) (solid line) is clearly incompatible with the observations, whereas this same combination plus the assumption of null ^{56}Fe production in SNII (dashed line) gives a strong rise in [O/Fe] for [Fe/H] ≤ -2.0 that is not observed. This indicates that some ^{56}Fe has to be produced in SNII, although less than in the models of Woosley and Weaver (1986). Finally, in Fig. 2c we have tried to increase [O/Fe] in halo stars by modifying our initial hypothesis 4 (in 2). The continuous line (that gives the best fit) corresponds to adopting a slope x = 1.00 in the IMF for t ≤ 5x10^8 yr and x = 1.35 for t > 5x10^8 yr, and hypotheses 2 (3.1) plus 1 (3.2). The dashed line shows the result of allowing "supermassive" (100 M⊙ ≤ M ≤ 300 M⊙) star formation, also only for t ≤ 5x10^8 yr and hypotheses 2 (3.1) plus 1 (3.2). The supermassive stars would yield ≈ 60% of their mass as ^{16}O, acording to Ober, El Eid, and Fricke (1983). The dash-dotted line gives the result of adopting x = 1.0 in the IMF, but also for t ≥ 5x10^8 yr: it clearly gives an ^{16}O excess over the observations for [Fe/H] ≥ -1.5. Another interesting point is that adopting a galactic SNIa rate as low as 1 SNIa per 100 years (Evans, van den Bergh, and McClure 1989) gives an ^{56}Fe abundance in the Solar System that is too low by a factor of 3.

Since the ^{12}C$(\alpha,\gamma)^{16}$O rate determines the C/O ratio in the presupernova stars and SNIa are not important sources of ^{12}C, one expects the [C/O] ratio in very metal-poor stars to be representative of the relative production of these elements in SNII + SNIb/c (in the case that the last ones have massive progenitors). We have thus equally computed the evolution of the C abundance

Fig.1.– Predicted evolution of [O/Fe] *vs* [Fe/H] in different cases:

(**a**) *Solid line* corresponds to hypotheses 2 (**3.1**) and 1 (**3.2**). *Dotted line* corresponds to hypotheses 1 (**3.1**) and 1 (**3.2**). *Dashed line* shows the result of Matteucci and Greggio (1986).

(**b**) *Solid line* shows the results from hypotheses 2 (**3.1**) and 2 (**3.2**). *Dashed line* corresponds to the same hypotheses plus no production at all of ^{56}Fe in SNII.

(**c**) *Solid line* is the result from hypotheses 2 (**3.1**) and 1 (**3.2**) plus an IMF with flatter slope durig the first 5×10^8 yr. *Dashed line* shows the prediction for the same combination of hypotheses but allowing "supermassive" star formation during the first 5×10^8 yr. *Dash-dotted line* gives instead the result of adopting a "flat" IMF during the whole life of the Galaxy.

Numbered hypotheses correspond to different assumptions as to the progenitors of SNIa and SNIb/c (see text).

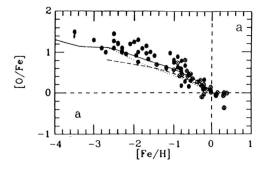

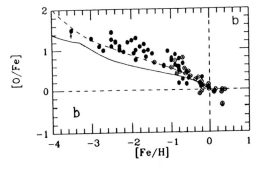

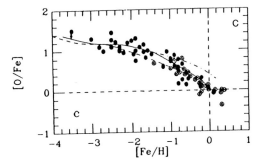

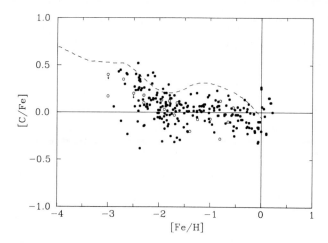

Fig. 2.- Predicted evolution of the $[C/Fe]$ ratio as a function of $[Fe/H]$.

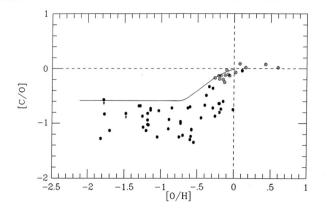

Fig. 3.- Predicted evolution of the $[C/O]$ ratio as a function of $[O/H]$.

and we compare the predicted $[C/Fe]$ *vs* $[Fe/H]$ and $[C/O]$ *vs* $[O/H]$ with the observed ones in Figs. 2 and 3, respectively. ^{12}C yields in low- and intermediate-mass stars are from Renzini and Voli (1981). In massive stars, from Woosley and Weaver (1986) (Kettner et al. 1982 $^{12}C(\alpha,\gamma)^{16}O$ rate). It is $\simeq 0.05$ M$\odot$ per SNIa explosion (Nomoto, Thielemann, and Yokoi 1984). We see that the predicted $[C/Fe]$ is too high at any metallicity (as in Matteucci 1988 and in Andreani et al 1988). These authors suggest that the problem comes from incorrect yields for low- and intermediate-mass stars. The model reproduces well the increase in $[C/Fe]$ in stars with $[Fe/H] \leq -1.8$. It also predicts nearly constant $[C/Fe]$ for $[Fe/H] \leq -2.5$, but there are too few observations to either confirm or disprove this point. Fig. 3 suggests that the $^{12}C(\alpha,\gamma)^{16}O$ rate cannot be significantly lower than the one given by Kettner et al. (1982): the fit would improve if it were even higher.

6 CONCLUSIONS

a) To explain the observed $[O/Fe]$ ratio at low metallicities, only small amounts of ^{56}Fe should be produced by SNII. Best agreement with observations is obtained with a "flat" IMF for $t \leq 5\times10^8$ yr and a steeper one for $t > 5\times10^8$ yr. SNIa do produce $\geq 80\%$ of all ^{56}Fe in the Galaxy.

b) The SNIa scenario that is most compatible with the observations is the "double degenerate" one (Iben and Tutukov 1984). For SNIb/c, either very massive stars (M $\geq$ 30 M$\odot$) or simply a fraction of massive stars (M $\geq$ 10 M$\odot$) are compatible.

Acknowledgements This work has been supported in part by CICYT grants N^{os} PB87-0147 and PB87-0304.

REFERENCES

Abia, C., and Rebolo, R. 1989, *Astrophys. J.*, **347**, 186

Abia, C., Canal, R., and Isern, J. 1990, *Astrophys. J.*, **366**, in press

Andreani, P., Vangioni-Flam, E., and Audouze, J. 1988, *Astrophys. J.*, **334**, 698

Arnett, W.D., Schramm, D.N., and Truran, J.W. 1989, *Astrophys. J. (Letters),*

339, L25

Evans, R., van den Bergh, S., and McClure, R.D. 1989, *Astrophys. J.*, **345**, 752

Greggio, L., and Renzini, A. 1983, *Astron. Astrophys.*, **118**, 217

Hashimoto, M., Nomoto, K., and Shigeyama, Y. 1989, *Astron. Astrophys.*, in press

Iben, I. Jr., and Renzini, A. 1983, *Ann. Rev. Astr. Astrophys.*, **21**, 271

Iben, I., and Tutukov, A.V. 1984, *Astrophys. J. Suppl.*, **54**, 335

Kettner, K.V., Becker, H.W., Buchmann, L., Görres, J., Krawinkel, H., Rolfs, C., Schamalbrock, P., Trautvetter, H.P., and Vlieks, A. 1982, *Zs. Phys.*, **A308**, 73

Matteucci, F. 1988, in *ESO Workshop on Stellar Evolution and Dynamics in the Outer Halo of the Galaxy*, ed. F. Matteucci and M. Azzopardi (Garching: ESO), p. 609

Matteucci, F., and Greggio, L. 1986, *Astron. Astrophys.*, **154**, 279

Matteucci, F., and Tornambè, A. 1985, *Astron. Astrophys.*, **142**, 13

Nomoto, K. Thielemann, F.K., and Yokoi, K. 1984, *Astrophys. J.*, **286**, 644

Ober, W.W., El Eid, M., and Fricke, K. 1983, *Astron. Astrophys.*, **119**, 61

Prantzos, N., Doom, C., Arnould, M., and de Loore, C. 1986, *Astrophys. J.*, **304**, 695

Renzini, A., and Voli, M. 1981, *Astron. Astrophys.*, **94**, 175

Talbot, R.J., and Arnett, W.D. 1971, *Astrophys. J.*, **170**, 409

Tutukov, A.V., and Yungelson, L.R. 1980, in *Close Binary Stars*, ed. M.J. Plavec, D.M. Popper, and R.K. Ulrich (Dordrecht: Reidel), p. 15

Wheeler, J.C., Sneden, C., and Truran, J.W. 1989, *Ann. Rev. Astr. Astrophys.*, **27**, 279

Woosley, S.E. 1988, *Astrophys. J.*, **330**, 218

Woosley, S.E., and Weaver, T.A. 1986, in *Radiation Hydrodynamics in Stars and Compact Objects*, ed. D. Mihalas and K.H.A. Winkler (Berlin: Springer), p. 91

Barium abundance in metal poor stars *

P. François

Observatoire de Paris, DASGAL, 61 Av. de l'Observatoire, 75014 PARIS, FRANCE

1 Introduction

The subject of this talk deals with the abundance of s-process elements in metal-poor stars. However, this presentation will be restricted to the Barium . After a short historical introduction, I shall present new results pointing out the difficulty of deriving accurate barium abundances. These data coupled with the one of Zhao and Magain (1990, private communication) will then be critically compared with the results of Gilroy et al. (1988).

2 What do we know about Barium in metal-poor stars?

Apart from the nearby halo star HD122563, little was known concerning the behaviour of [Ba/Fe] as a function of [Fe/H] before the paper of Spite and Spite (1978). They studied a sample of 11 halo stars both dwarfs and giants and they showed that the [Ba/Fe] ratio departed from the solar ratio in stars with a metallicity smaller than about -1.5 . I should stress to avoid any confusion that metallicity means all over this talk the iron abundance faithfully to the spectroscopists' tradition and not the general content in heavy elements starting from the carbon as used in models of chemical evolution of galaxies (Matteucci and François 1989).

In 1985, Luck and Bond determined the abundance of s-process elements in a sample of 3 metal-poor field red giants. They confirmed the results of Spite and Spite (1978) : the heavy s-process element Ba behaves like a secondary element ($[s/Fe] \propto [Fe/H]$) .

In 1988, Gilroy et al. confirmed this trend showing however a large spread with respect to the secondary-like behaviour for stars with $[Fe/H] < -1.5$

A new determination (Zhao and Magain 1990) of the ratio [Ba/Fe] in a sample of 14 dwarfs with a metallicity ranging from -1.4 to -2.4 led to a contradictory result. They found the [Ba/Fe] ratio to be solar with a significative dispersion. The last results from Zhao and Magain (private communication) based on a sample of stars with [Fe/H] lower than -2 show a secondary-like behaviour of Barium.

---

* based on observations collected at ESO, La Silla

We decided to observe Barium in a new sample of stars with a metallicity lying between -1.5 and -2.2 in order to detect at which metallicity Barium starts to have a primary-like behaviour and to understand why there is such a spread in the data.

3 Observations and Reductions:

High resolution, high signal to noise ratio spectra were acquired at the 1.4m CAT ESO telescope feeding the Coudé Echelle Spectrgraph (CES) equipped with the short camera and the CCD detector. We used the new facility offered by ESO (Remote Control Observations) where the instrument and the telescope are driven directly from the ESO headquarters in Garching (Germany). For each star, we took two spectra centered on two different transitions of Barium.

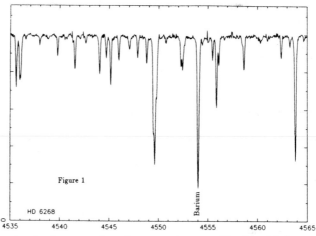

Figure 1

HD 6268

Figure 1 shows a typical example of a spectrum. It should be noted that the barium line is clearly visible and that the spectral region of the spectrum does not contain too many lines : the continuum is then well defined. The error made on the abundance determination coming from the continuum placement uncertainty is then small. The spectra were reduced using the reduction package ASRETI running on the VAX computer of the PARIS OBSERVATORY. The observed stars have been previously studied by different authors ; the main characteristics of their atmosphere are known, from the "Catalogue of [Fe/H] determinations " (Cayrel et al. 1985). With these parameters, we have interpolated the models in the grid of Gustafsson ,1981) computed with the same hypotheses as in Gustafsson et al. (1975). The oscillator strengths of the lines have been determined by fitting the profile of the lines of a solar Atlas (Delbouille et al. 1973) with a computed profile (using the Holweger-Müller model). For these computations, the solar abundances of Anders and Grevesse (1989) have been adopted. The main assumptions ans procedures used for computing the abundances are are described in François and Raiteri (1991).

4 Errors :

The barium abundance can be measured in metal poor stars by the mean of 3 transitions ($\lambda4454.03\mathring{A}$, $\lambda6141.72\mathring{A}$, $\lambda6496.91\mathring{A}$) from the first ionized state (BaII).In extreme metal-poor dwarfs, only the resonnance line of BaII is used. This means that we have to rely on one single to derive the abundances. This transition coming from an ionized state is moreover sensitive to the choice of the gravity. However, the most important source of error comes from the uncertainty on the microturbulent velocity. The lines of Barium lie on the shoulder of the curve of growth and are highly sensitive on this parameter. A variation of 0.5 Km/s in the microturbulent velocity changes the [Ba/Fe] ratio by a amount of about 0.2 dex.

5 Results and Discussion :

On figure 2 are plotted the [Ba/Fe] ratios as a function of [Fe/H] for our sample of stars as well as previous determination by other observers. In our sample of star, we find a solar [Ba/Fe] ratio which confirms what was found by Zhao and Magain (1990).For the star HD 6268, we found a high [Ba/Fe] ratio of about +0.4 dex. This star is suspected to be on the AGB (Luck and Bond 1985) and we will not take it into account for our discussion.

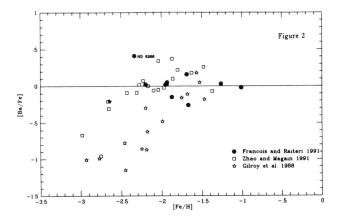

Most of the data concerning the stars fainter than [Fe/H]= - 2 are data coming from Gilroy et al. (1988) and it should be questioned whether the spread in the observed [Ba/Fe] for a given metallicity is real or not. We showed in the previous section how sensitive were strong lines to the microturbulent velocity. But, it should be added that the error made on the measurement of the equivalent width or on the placement of the continuum which can the consequence of low signal to noise spectra leads also to significant errors in the abundance determination. This last type of error can be detected by looking at the abundance determination for other species for which we have some confidence. We know that the [Ca/Fe] ratio exhibits an overabundance of about 0.3 dex in halo stars (See for example Zhao and Magain 1990). The Calcium abundance has been also determined by Gilroy et al. (1988) and we propose to use it as a check . We plotted in Figure 3a the [Ca/Fe] ratio coming from Gilroy's paper as a function of metallicity and in Figure 3b the data coming from Zhao and Magain (1990). A comparison of these two figures shows that part of the dispersion in the data of Gilroy et al. (1988) may come from errors in atmospherical parameters. The consequence is that part of the spread found for [Ba/Fe] comes from uncertainties in the derivation of the abundance .

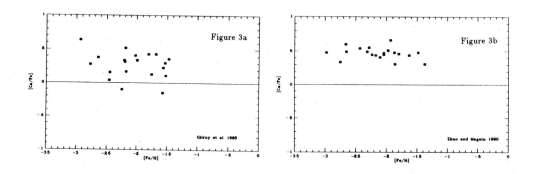

We could have an idea of the average trend of [Ba/Fe] as a function of [Fe/H] in binning the data in order to minimize the intrinsic errors. In Figure 4, we plotted the binned data [Ba/Fe] in order to have five objects per bin centered on the average [Fe/H] found. This figure shows that [Ba/Fe] remains solar until a [Fe/H] of about -1.7 and then becomes sub-solar showing a clear secondary-like behaviour.

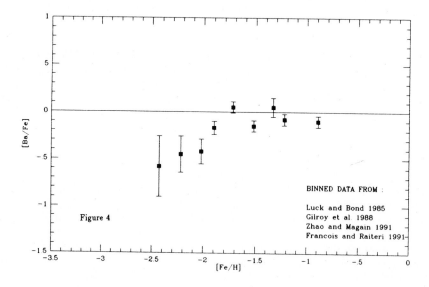

References :

Anders E., Grevesse, N. : 1989, *Geochim. Cosmochim. Acta*, **53**, 197

Cayrel de strobel, G., Bentolila, C., Hauck, B. , Duquennoy, A. : 1985, *Astron. Astrophys. Suppl.*,**59**, 149

Delbouille, L., Neven, L., Roland G. : *Photometric Atlas of the Solar spectrum from 3000 to 10000 Å*, Institut d'Astrophysique, Université de Liège 1973)

François, P., Raiteri, C.M. : 1991, in preparation Gilroy, K., Sneden, C., Pilachowski, C., Cowan, J. : 1988, *Astrophys. J.* , **327**, 298

Gustafsson, B., Bell, R.A., Eriksson, K., Nordlund, A. : 1975, *Astron. Astrophys.*, **42**, 407

Gustafsson, B. : 1981, private communication

Luck, R.E., Bond, H.E. : 1985, *Astrophys. J.* , **292**, 559

Matteucci, F., François, P. : 1989, *Month. Not. R. ast. Soc.*, **239**, 885

Spite, F., Spite, M. : 1978, *Astron. Astrophys.* ,**67**, 23

Zhao, G., Magain, P. : 1990, *Astron. Astrophys.* ,**238** , 242

DISCUSSION:

Grevesse :
You could perhaps look at the BaI line. Although it is blended in the solar spectrum, it might be blended by a metallic line. And you could eventually see it in your low metallicity star.

Pagel :
Gilroy et al. claimed a more smooth Eu, Ba relation than Fe,Ba . Do you agree ?

François :
Yes, I agree . I think the reason comes from the fact that the errors made on the Barium and on the Europium abundance determination are about the same and the effect is that the true position is shifted along the axis of secondary-like behaviour. Thus, their relation Eu,Ba seems real.

Precision Spectroscopy and Stellar Parameters for Bright Stars

G. SMITH

Department of Physics, University of Oxford

1 INTRODUCTION

Several years ago, thanks to the generosity of Professor Donald Blackwell, I was able to measure some atomic data for neutral calcium lines using the carbon tube furnace at Oxford (Blackwell et al. 1979). My initial purpose in making these measurements was to tackle a problem in atomic physics but, once this project had been completed (Smith 1988; Smith and Raggett 1981), I began to consider what applications this data might have to the interpretation of stellar spectra. I had precise oscillator strengths and some knowledge of collisional broadening parameters for lines of moderately high excitation, 2–3 eV above the ground level, covering a range in strength from weak to strong in the solar spectrum. My idea was to see how much I could learn about the surface properties of bright stars using just a limited amount of high-quality data. One obvious requirement was for stellar spectra of high resolution and high photometric accuracy covering the calcium-line regions of interest. Here again I was fortunate in being invited by Professor David Lambert to use the excellent spectroscopic facilities at McDonald Observatory, Texas. With the help of Professor Lambert and his colleagues, I've gradually been able to accumulate sets of spectra for stars covering quite a wide range of surface temperature, gravity and metallicity.

2 INSTRUMENTATION

At McDonald Observatory we have been using the 2.7 m telescope and Coudé spectrograph which for most of the time was fitted with an 1872-element (15μ) Reticon detector (Vogt et al. 1978). The echelle grating gives about 25Å of spectrum per integration at a dispersion of 1Å per mm with a limit of resolution of 0.05Å. Under normal seeing conditions, we could achieve a signal/noise ratio of about 100 in about 2 min for a star of 0.0 mag or about 2 hr for a star of 4.0 mag. The Reticon has now been replaced by a more efficient CCD system based on an 800 ×

800 (15μ) chip. The shorter chip means that we only get about 10Å per integration but the greater efficiency allows us to obtain spectra of similar quality in about 10 min for a star of 4.0 mag or about 60 min for a star of 6.0 mag. Wider spectral ranges can be recorded at lower dispersion using various conventional gratings.

3 RESULTS

3.1 G-type Dwarfs
I began by looking at stars not too different from the Sun. The calcium-line regions contained Fe I and Fe II lines with relatively unblended profiles and these gave useful information on the metallicity and temperature of the atmosphere. My model atmospheres were based on the temperature distribution of the Holweger-Müller solar atmosphere (Holweger and Müller 1974) scaled for the desired effective temperature. The variation of pressure with optical depth was obtained by integrating the hydrostatic equation for the desired surface gravity and metallicity. Since only the iron abundance relative to the Sun was needed, atomic data for iron was derived from the solar flux spectrum of Kurucz *et al.*, 1984. My procedure was as follows:

(i)   Derive [Fe/H], iron abundance relative to the Sun, as a function of effective temperature (initially assuming solar surface gravity). The region of intersection of curves (loci of constant equivalent width) arising from Fe I and Fe II lines provides initial estimates of the metallicity and effective temperature. The result is fairly insensitive to the adopted microturbulence parameter.

(ii)   Using these initial estimates, derive Ca/H, the calcium abundance, as a function of microturbulence. Loci of constant equivalent width for weak to medium-strong calcium lines normally intersect in a narrow region or "neck" which provides initial estimates of the calcium abundance and microturbulence.

(iii)   Compare the observed profile of the strong Ca I 6162Å line with profiles calculated for different values of the surface gravity (using the initial estimates of the other parameters). The collisionally-broadened line wings are sensitive to surface gravity and provide an improved estimate of this quantity.

(iv)   Iterate the previous three stages until a self-consistent set of parameters is obtained. In fact the procedure converges very rapidly.

Observations of the extremely strong Ca II 8542Å line profile were used to provide additional constraints on the stellar parameters. Smith and Drake (1987) have shown that the depth of absorption in the wings of this line in solar-type stars is sensitive mainly to metallicity. The same paper reports the following results for the stars η Cas A (G0) and τ Ceti (G8):

	η Cas A			τ Ceti		
Effective temperature (K)	5890	$\pm$	50	5380	$\pm$	50
log g	4.4	$\pm$	0.1	4.5	$\pm$	0.1
[Fe/H]	-0.24	$\pm$	0.05	-0.53	$\pm$	0.05
[Ca/H]	-0.24	$\pm$	0.05	-0.34	$\pm$	0.05
Microturbulence (km s^{-1})	1.45	$\pm$	0.05	1.25	$\pm$	0.05

The values of effective temperature are close to those given by the Infrared Flux Method (Blackwell *et al.* 1990) and those adopted in the recent extensive investigation of G and K star temperatures by Bell and Gustafsson (1989). The binary nature of η Cas A allows an independent check on the surface gravity. The integrated flux measured by Blackwell *et al.* (1990) can be combined with the known parallax to yield the luminosity. This can then be combined with the known mass and the effective temperature to give the gravity. The result is log g = 4.4 $\pm$ 0.1 in excellent agreement with the spectroscopic value. It is also interesting to note that in η Cas A [Ca/H] = [Fe/H] whereas in τ Ceti [Ca/H] exceeds [Fe/H] by about 0.19 dex. This is an example of a much more general feature of metal-deficient stars which has been confirmed by the work of several groups in recent years (see, for example, the review by Lambert, 1987). The abundances relative to the Sun of the alpha-particle elements, such as calcium, exceed that of iron by an amount increasing from zero to about 0.3 dex as the iron abundance falls from about one-half solar to about one-tenth solar. In fact the first indications of this phenomenon were obtained by Wallerstein (1962) in a survey of field G-dwarf abundances. Evidence for the enhancement in relative abundance of alpha-particle elements over iron in τ Ceti was found shortly afterwards by Pagel (1964). I am currently analysing spectra of the G8 subdwarf Gmb1830 where [Fe/H] $\simeq$ -1.3 and preliminary results indicate that [α/Fe] $\simeq$ $+0.3$.

3.2 Late-type Giants

The spectroscopic analysis of giant stars is complicated by the fact that their surface layers are generally cooler than those of the Sun and, as a consequence, their spectra are more complex making the blending problem more severe. Furthermore the lower atmospheric pressure means that the collisionally-broadened strong-line wings are less well developed. There is now strong evidence that non-LTE effects are present at least in some spectral features. It is not obvious what are the best model atmospheres to use for giants. J.J. Drake and I (Drake and Smith, 1991) have recently completed an analysis of calcium and iron lines in the K0 giant, Pollux. The strong wings of the Ca II 8542Å line, which are likely to be formed in LTE, provided a useful constraint on the temperature structure of the atmosphere. We found that this temperature structure closely resembled that of a MARCS model atmosphere (Bell *et al.*, 1976). A program to generate MARCS models has kindly been made available to UK astronomers through the STARLINK network by Professor B. Gustafsson. Since the most recent and comprehensive analysis of the spectrum of Pollux (Ruland *et al.*, 1980) had found empirical evidence for departures from LTE in low-excitation lines of Fe I, and subsequent calculations by Steenbock (1985) had indicated that high-excitation lines were most likely to be formed in conditions close to LTE, we determined the effective temperature using only Fe II lines and weak, high-excitation Fe I lines. We found that the wings of the Ca I 6162Å line were still sufficiently sensitive to gravity so that our iterative procedure for determining a self-consistent set of stellar parameters could proceed as described for solar-type stars. In addition Drake (1991) has investigated the effects of departures from LTE in calcium lines by means of a 20-level model calcium atom. These effects were found to be mostly small and restricted to the spectrum of the neutral atom. Our final results corrected for departures from LTE are:

Effective temperature (K)	4865	$\pm$	50
log g	2.75	$\pm$	0.15
[Fe/H]	-0.04	$\pm$	0.05
[Ca/H]	-0.11	$\pm$	0.10
Microturbulence (km s^{-1})	1.4	$\pm$	0.1

The effective temperature is very close to the value 4842 ± 50 K found by Blackwell *et al.* (1990) using their Infrared Flux Method. The surface gravity, together with the radius derived from recent measurements of the integrated flux and parallax, implies a mass for Pollux of 1.7 ± 0.6 $M_\odot$. The position of Pollux on the H-R diagram is compatible with calculated evolutionary tracks for masses within this range on the assumption that Pollux is either undergoing its first ascent of the red-giant branch or has reached the clump-giant stage. We have also investigated spectra of the somewhat metal-poor K1 giant Arcturus where our results essentially confirm those found by Bell *et al.* (1985). In particular we find that [Ca/H] exceeds [Fe/H] by about 0.2 dex.

4 CONCLUSION

Provided spectroscopic data of good quality are available, it seems that the procedure outlined above is capable of providing reliable stellar atmospheric parameters.

The method is limited to spectral types between about F5 and K2. In stars hotter than F5 the Ca I 6162Å line no longer has well developed wings so is unsuitable as a gravity indicator. In stars cooler than K2 the presence of molecular features prevents the accurate determination of equivalent widths and wing profiles. The method also fails for low gravity stars, log g < 1.5, where the absorption coefficient in the strong line wings is no longer sensitive to collisional broadening. Within these limitations the procedure leads to parameters that are well constrained and self-consistent though, in the case of giant stars, many more iterations are required to achieve self-consistency than in the case of dwarf stars. There is a definite advantage to be gained from using strong-line profiles of neutral and singly-ionised lines of the same element to determine both the surface gravity and the temperature structure of the atmosphere since only one, initially unknown, abundance is involved. The temperature structure is best determined from singly-ionised lines since these arise from the dominant ionisation stage and their wings are most likely to be formed in LTE. If good profiles of the Ca I 6162Å line and of the Ca II 8542Å line are available, together with an effective temperature determined by the Infrared Flux Method, a "quick" approach to stellar parameters may be possible. Taking the effective temperature as known, one can determine the metallicity, [M/H], and the calcium abundance from the 8542Å line wings, assuming [M/H] = [Ca/H]. The surface gravity is then determined from the 6162Å line profile. I have also investigated the use of the wings of the strong Mg I 8806Å line for determining the magnesium abundance. In dwarf stars [Mg/H] = [Ca/H] is indicated, as might be expected since both are alpha-particle elements, but in giant stars anomalous results are obtained possibly due to departures from LTE. Similar conclusions were reached by Bell *et al.* (1985) in their study of Arcturus.

5 REFERENCES

Bell, R.A., Eriksson, K., Gustafsson, B. and Nordlund, A.: 1976, Astron. Astrophys. Suppl. Ser., **23**, 37.

Bell, R.A., Edvardsson, B. and Gustafsson, B.: 1985, Mon. Not. R. astr. Soc., **212**, 497.

Bell, R.A. and Gustafsson, B.: 1989, Mon. Not. R. astr. Soc., **236**, 653.

Blackwell, D.E., Ibbetson, P.A., Petford, A.D. and Shallis, M.J.: 1979, Mon. Not. R. astr. Soc., **186**, 633.

Blackwell, D.E., Petford, A.D., Arribas, S., Haddock, D.J. and Selby, M.J.: 1990, Astron. Astrophys., **232**, 396.

Drake, J.J.: 1991, submitted to Mon. Not. R. astr. Soc.

Drake, J.J. and Smith, G.: 1991, submitted to Mon. Not. R. astr. Soc.

Holweger, H. and Müller, E.A.: 1974, Solar Phys., **39**, 19.

Kurucz, R.L., Furenlid, I., Brault, J.W. and Testerman, L.: 1984, Solar Flux Atlas from 596 to 1300 nm, National Solar Observatory, Kitt Peak, Arizona.

Lambert, D.L.: 1987, J. Astrophys. Astron., **8**, 103.

Pagel, B.E.J.: 1964, Roy. Obs. Bull., **87**, E227.

Ruland, F., Holweger, H., Griffin, R., Griffin, R. and Biehl, D.: 1980, Astron. Astrophys., **92**, 70.

Smith, G. and Raggett, D.St.J.: 1981, J. Phys. B: At. Mol. Phys., **14**, 4015.

Smith, G. and Drake, J.J.: 1987, Astron. Astrophys., **181**, 103.

Smith, G.: 1988, J. Phys. B: At. Mol. Phys., **21**, 2827.

Steenbock, W.: 1985 in Cool Stars with Excesses of Heavy Elements, eds. M. Jaschek, P.C. Keenan, Reidel, Dordrecht, p231.

Vogt, S.S., Tull, R.G. and Kelton, P.: 1978, Appl. Optics, **17**, 574.

Wallerstein, G.: 1962, Astrophys. J. Suppl. Ser., **6**, 407.

QUESTION

Blackwell: Have you determined the microturbulence for your solar-type stars using both Ca I and Fe I lines? I am concerned that such measurements may be affected by non–LTE to a considerable extent.

Smith: I have found reasonable agreement between microturbulence values determined from Fe I and Ca I lines in solar-type stars. However, only the value determined from Ca I lines is based on oscillator strengths measured in the laboratory. For Fe I lines I have been using oscillator strengths derived from the solar spectrum, and this requires the assumption of a particular solar abundance and microturbulence. Any non–LTE effects present in the solar spectrum will be absorbed into the "solar" oscillator strengths. The iron lines therefore determine the microturbulence in solar-type stars relative to the Sun. In the case of late-type giant stars I only get consistent microturbulence values when I restrict my sample of Fe I lines to lines of excitation greater than about 3 eV.

Theoretical Models and Spectra for Arcturus – a Step toward Extragalactic Globular Cluster Abundances

R.C. PETERSON AND C.M. DALLE ORE

University of Arizona and Harvard-Smithsonian Center for Astrophysics

1 SPECTRAL OBSERVATIONS OF ANDROMEDA CLUSTERS

In a comprehensive comparison of observational spectra, Burstein *et al.* (1984) showed that spectra of Andromeda globular clusters were generally similar to those of Galactic globulars but that certain line strength indices behaved anomalously. In particular, CN was too strong and Hβ too weak for a given strength of the Mg b + MgH index in spectra of several metal-rich Andromeda clusters. They argued that this was most probably due to a substantial younger population in the anomalous Andromeda clusters, but that no one explanation was fully satisfactory.

High-resolution spectra of the Andromeda clusters might assist in identifying the nature of the anomaly. The velocity dispersion of an Andromeda cluster is usually < 40 km s^{-1} (Peterson 1989), small enough that individual features may be discerned. From the differences between red spectra where cool stars dominate versus blue-green where the hotter stars become more influential, and from the relative strengths in each spectral region of various features arising from different elements, one may hope to disentangle the effects of the possible presence of hotter main-sequence turnoff stars or metal-rich blue horizontal-branch stars versus the abundance ratios of species such as magnesium that are overabundant in Galactic metal-poor clusters.

To this end we have begun to obtain high S/N spectra of Andromeda globular clusters in the region of the Mg b lines at 5200 Å and also in the near infrared, 7500 – 9000 Å, which includes the Ca II triplet and many weaker lines such as Mg I at 8806 Å. The similarity of the near IR spectra to that of an individual giant in the Galactic globular M71 is shown in Figure 1. The noisy line is the observed spectrum of the integrated light of cluster Bo 193 in Andromeda; the smooth line is the observed spectrum of star 21 in M71, after broadening by a Gaussian of $\sigma = 14$ km s^{-1} to account for the stellar velocity dispersion of the cluster. Agreement is very good in lines as diverse as the Ca II line at 8662 Å, the moderately strong Fe I line at 8688 Å, and the weak Ti I line at 8683 Å. Once abundances and abundance ratios are found from this region, the bluer region should yield the population mix.

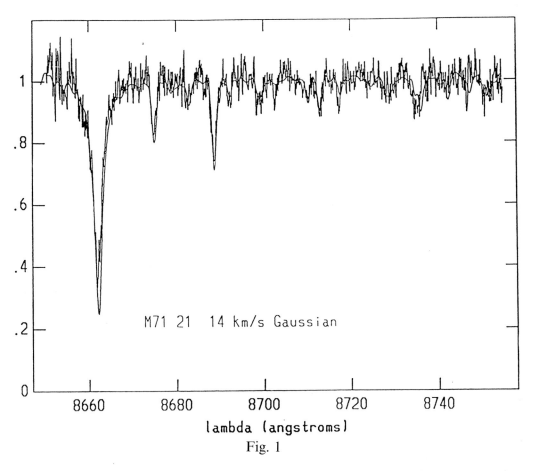

Fig. 1

2 CALCULATION OF THEORETICAL MODELS AND SPECTRA

The quantitative determination of abundances depends on comparisons of spectra observed for a particular star with those calculated for a model atmosphere appropriate for that star. For stars in M71, with an abundance approaching that of the Sun, has been a difficult task (Leep, Oke, and Wallerstein 1987). Until now two problems have arisen: the accuracy of the transition probablilities of individual atomic lines has been inadequate to derive small differences in elemental abundances, and the stellar models have failed to reproduce the flux distributions spectrum of either late-type stars or the Sun (Gustafsson and Bell 1979).

These problems have now largely been solved with the completion by Kurucz (1990) of a greatly improved and expanded list of atomic and molecular line parameters, one which includes more than forty million atomic lines and updated data for molecules such as CO, CN, C_2, H_2, TiO, SiO, and five hydrides. The new Kurucz solar model based on this list agrees with the flux measurements of Labs *et al.* (1987), with no indication of missing opacity.

3 RESULTS FOR ARCTURUS

As a first step toward the application of this list to cool giants, we have reanalyzed Arcturus. The work below is described by Peterson, Dalle Ore, and Kurucz (1991).

We first adjusted *gf* values of individual lines to match those in the Sun. The spectrum calculated using the solar model of Maltby *et al.* (1986) was plotted on top of the solar intensity spectrum at disk center observed by Brault and digitized by Kurucz, which has a resolution of 522,000 and a signal-to-noise (S/N) ratio of 2000 – 9000. Based on discrepancies in the comparison we refined the *gf* values and repeated the procedure until the agreement was satisfactory. Except for extremely weak lines, lines blended with telluric features, lines affected by hyperfine splitting, and lines which are sensitive to the surface temperature structure of the Sun, we expect the revised values to be good to 0.05 dex = 12%. External comparisons with the laboratory *gf* values of Ti I lines measured by Blackwell *et al.* (1986 and references cited therein), and the solar *gf* values of Thévenin (1989, 1990), support this estimate. To date ~3000 *gf* values in the 5000 – 6500 Å and 7500 – 9000 Å regions have been changed.

Our modified line list was applied to calculate the spectrum of Arcturus for a given input model. We first used the Kurucz line opacity distribution functions (LODF's) to calculate a model with the parameters found for Arcturus by Johnson *et al.* (1977), $T_{eff} = 4250$ K, $\log g = 1.7$, and [Fe/H] $= -0.5$ with scaled-solar abundances. We then employed the revised list of line parameters to compute a theoretical spectrum. This was compared to the Griffin atlas, digitized by Kurucz. From the spectral line fit in the 5000 – 5500 Å, 6000 – 6500 Å, and 7500 – 8000 Å regions, we estimated better values of T_{eff}, $\log g$, and the elemental abundances. A new model with parameters $T_{eff} = 4300 \pm 30$ K, $\log g = 1.5 \pm 0.2$, $v_t = 1.7 \pm 0.3$ km s^{-1}, and [Fe/H] $= -0.5 \pm 0.1$, and enhanced light-element abundances provided the best fit. With respect to the solar abundances of Grevesse and Anders (1989), [O/Fe] = [Mg/Fe] = [Si/Fe] = 0.4; [N/Fe] = [Al/Fe] = [Ca/Fe] = [Ti/Fe] = 0.3; [Sc/Fe] = 0.2; and [C/Fe] = [Cr/Fe] = [Ni/Fe] = 0.0. The uncertainty in the relative abundances is $\pm$ 0.1 dex for N, Ca, and Al, and less than 0.1 dex for the other elements listed, which have sufficiently many weak atomic lines that they are not sensitive to v_t.

Using our $\log g = 1.5$ the mass of Arcturus is found to be 0.72 M$_\odot$ (see Trimble and Bell 1981). The uncertainty of $\pm$ 0.1 in $\log g$ leads to a 1σ mass range of 0.57 – 0.91 M$_\odot$, the same mass range inferred for giants in globular clusters (Rood 1973). The enhancement of light elements in Arcturus is also the same as that seen in halo stars of much lower metallicity. As discussed by Wheeler, Sneden, and Truran (1989), virtually all stars more metal-poor than [Fe/H] $= -1$ show an overabundance of about 0.3 – 0.5 dex in the light elements O, Mg, Si, Ca, and probably Ti.

Our $T_{eff} = 4300$ K agrees with that of Di Benedetto and Foy (1986), who found 4294 $\pm$ 30 K from the integrated flux and angular diameter. The relative flux distribution of our model is also in accord with the observations of Honeycutt *et al.* (1977), as shown in Figure 2. In that figure we also plot the flux distributions for our model with parameters from Johnson *et al.* and the model tabulated by Frisk *et al.* with $T_{eff} =$ 4375 K, log $g = 1.5$, and [Fe/H] $= -0.8$ with an enhanced magnesium abundance. These fluxes are all calculated with the new Kurucz LODF's rather than by the authors. The heavy line plots the fluxes observed by Honeycutt *et al.* (1977). Both our model and that of Frisk *et al.* reproduce the fluxes of Honeycutt *et al.* over the range indicated to within the errors of the observations. Ours was calculated *ab initio*, while Frisk *et al.* had to invoke extra *ad hoc* opacity blueward of 5000 Å.

The goodness of the spectral fit is shown in Figs. 3 and 4. In each figure the upper panel is the solar spectrum; the lower, Arcturus. The wavelength scale in nanometers is indicated along the bottom. The heavy line with noise in these panels indicates the observed spectrum. The light line represents calculations using our modified line list. Above each panel, the species identification appears following the last three digits of the wavelength, then lower energy level or molecular transition, intrinsic central depth, and adopted *gf* value.

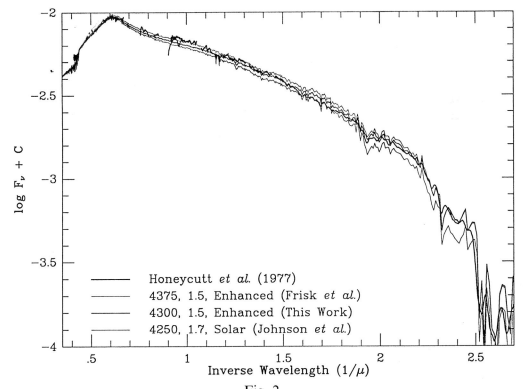

Fig. 2

Fig 3

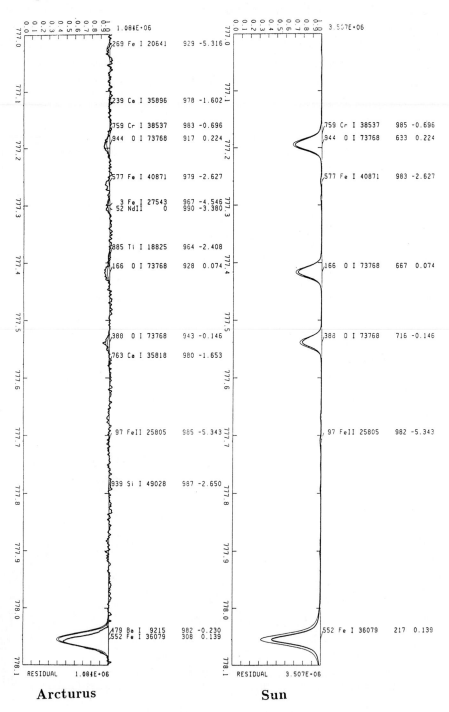

Arcturus   **Sun**

Fig 4

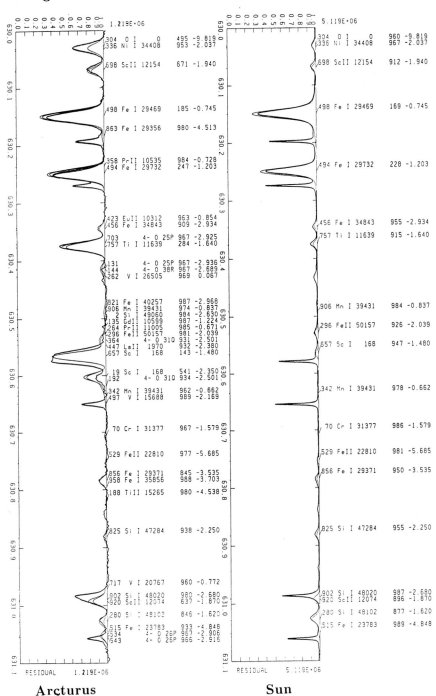

Arcturus   **Sun**

The two light lines in the lower panels show calculations for our two models: the Johnson *et al.* one and our final one. The line which shows better agreement is that based on our own model. Everywhere that a line is well matched in the Sun, it is well matched in Arcturus. Lines of Fe I from low and high excitation levels agree well, as do those of Fe II, Mg I/Mg II, Ti I/Ti II, and Sc I/Sc II. The strengths of molecular features are all well reproduced. The strengths and profiles of both the [O I] 6300.3 Å line and the oxygen triplet at 7771 – 7775 Å are reproduced exactly. This indicates that the temperature structure is correct in the line-forming region. It also suggests that nonLTE effects are no larger in Arcturus than in the Sun.

4 REFERENCES

Blackwell, D. E., Booth, A. J., Menon, S. L. R., and Petford, A. D. 1986, *M.N.R.A.S.,* **220,** 289.

Burstein, D., Faber, S. M., Gaskell, C. M., and Krumm, N. 1984, *Ap. J.,* **287,** 586.

Di Benedetto, G. P., and Foy, R. 1986, *Astr. Ap.,* **166,** 204.

Frisk, U., Bell, R. A., Gustafsson, B., Nordh, H. L., and Olofsson, S. G. 1982, *M.N.R.A.S.,* **199,** 471.

Grevesse, N., and Anders, E. 1989, in *AIP Conf. Proc. 183, Cosmic Abundances of Matter,* ed. C. J. Waddington (New York: AIP), p. 1.

Griffin, R. 1968, *A Photometric Atlas of the Spectrum of Arcturus* (Cambridge: Cambridge University Press).

Gustafsson, B., and Bell, R. A. 1979, *Astr. Ap.,* **74,** 313.

Honeycutt, R. K., Ramsey, L. W., Warren, W. H., and Ridgway, S. T. 1977, *Ap. J.,* **215,** 584.

Johnson, H. R., Collins, J. G., Krupp, B., and Bell, R. A. 1977, *Ap. J.,* **212,** 760.

Kurucz, R. L. 1990, *Transactions of the IAU,* Vol. XXB, ed. M. McNally (Dordrecht: Kluwer), p. 168.

Labs, D., Neckel, H., Simon, P. C., and Thuillier, G. 1987, *Solar Phys.,* **107,** 203.

Leep, E. M., Oke, J. B., and Wallerstein, G. 1987, *A. J.,* **93,** 338.

Maltby, P., Avrett, E. H., Carlsson, M., Kjeldseth-Moe, O., Kurucz, R. L., and Loeser, R. 1986, *Ap. J.,* **306,** 284.

Peterson, R. C. 1989, in *Dynamics of Dense Stellar Systems,* ed. D. Merritt (Cambridge: Cambridge Unversity Press), p. 161.

Peterson, R. C., Dalle Ore, C. M., and Kurucz, R. L. 1991, *Ap. J.,* submitted.

Rood, R. T. 1973, *Ap. J.,* **184,** 815.

Thévenin, F. 1989, *Astr. Ap. Suppl.,* **77,** 137.

Thévenin, F. 1990, *Astr. Ap. Suppl.,* **82,** 179.

Trimble, V., and Bell, R. A. 1981, *Q.J.R.A.S.,* **22,** 361.

Wheeler, J. C., Sneden, C., and Truran, J. W. 1989, *Ann. Rev. Astr. Ap.,* **27,** 279.

RR Lyrae Stars as Tracers of the Galactic Halo

T.D.KINMAN

National Optical Astronomy Observatories [1]
Kitt Peak National Observatory, Tucson, Arizona 85726-6732

After the globular clusters, the RR Lyrae variables are probably the best known tracers of the galactic halo. Their absolute brightness and characteristic light variations allow them to be detected unambiguously not only in the halo of our own galaxy but also in the Magellanic Clouds and M31 system. A particular advantage of these variables in tracing the halo is that there are a great many more *metal-poor halo* RR Lyrae stars than *metal-rich disk* RR Lyrae stars in any sight-line that is out of the galactic plane. Consequently, the adulteration of a halo sample with disk stars, a very important matter when determining the shape of the halo, is much less of a problem with the RR Lyrae stars than with other possible tracers such as red giants or non-variable horizontal branch stars. Also, both the absolute magnitudes and unreddened colors of these variables are comparatively well known so that they are amongst the best distance indicators (Pritchet, 1988). Figure 1 shows the spatial density distribution as a function of galactocentric distance (R_g) of both the type ab RR Lyrae stars in the Lick Astrograph fields and the halo ([Fe/H] $\leq$ −0.8) globular clusters (Zinn, 1985). Outside the solar circle, both types of object follow either a de Vaucouleur's law or an inverse power law (exponent $\sim$ −3.5). The spatial distribution of the RR Lyrae stars closely follows that of the clusters and since the variables are some thousand times more numerous per cubic kpc than the clusters, their value for delineating the halo is obvious. The way in which the instabilty strip is populated depends *inter alia* on metallicity; it follows that the RR Lyrae stars cannot be perfect tracers of the whole range of metallicities in the halo population. Recently, Suntzeff, Kinman and Kraft (1991) (hereafter SKK) completed a metallicity survey of 177 field RR Lyrae stars (over 90% of which come from the Lick surveys) and discussed the distribution of these variables in globular clusters as a function of both metallicity and galactic location. This SKK data is used here to compare the metallicity distributions of these field RR Lyrae stars with those of other metal-poor halo objects in order to see how good a tracer the RR Lyrae stars are in this respect.

---

[1]The National Optical Astronomy Observatories are operated by the Association of Universities for Research in Astronomy, Inc., under cooperative agreement with the National Science Foundation

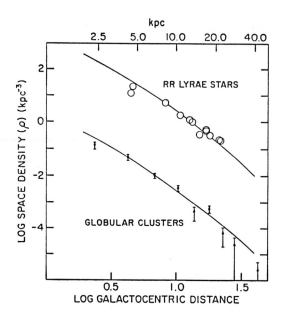

Figure 1. The ordinate is the logarithm of the space density (objects per cubic kpc) and the abscissa is the logarithm of the galactocentric distance (kpc). The open circles refer to type ab RR Lyrae stars discovered in the surveys with the Lick Astrograph. The small filled circles refer to the metal-poor ([Fe/H]$\leq$ −0.8) halo globular clusters.

Figure 2 (left) shows the metallicity distribution as a function of R_g of the 177 field RR Lyrae stars that are discussed by SKK. There is clearly a trend of the [Fe/H] distribution with R_g. We need to isolate as large and as homogeneous a sample of the field RR Lyrae stars as possible. I have therefore chosen only the 112 stars that are outside the solar circle but within 30 kpc (i.e. $8 \leq R_g \leq 30$ kpc). The surveys in this volume are reasonably complete for the larger amplitude variables (Δ B $\geq$ 0.75 mag.) and *in this region* there is no gradient of the metallicity with R_g. It is possible of course, that the stars in this sample have a spread in age. Figure 2 (right) compares the metallicity distribution of these 112 RR Lyrae stars with that of the 36 globular clusters that are in the same range of R_g. The data for the globular clusters comes from the compilation given in SKK. The two distributions are rather similar with a maximum at [Fe/H] $\sim$ −1.6. There are, however, proportionately fewer very metal-poor RR Lyrae stars than very metal-poor clusters. The number of clusters is small, however, and the most metal-poor part of the cluster metallicity function is not well determined.

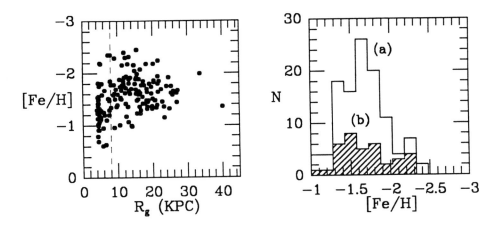

Figure 2.   The diagram to the left gives the abundance [Fe/H] *vs.* the galactocentric distance (R_g) in kpc for 177 field RR Lyrae stars according to SKK. Outside the solar circle (indicated by the vertical dashed line), this population shows no abundance gradient. The diagram to the right gives the number of objects (N) per 0.15 dex interval of abundance [Fe/H] for (a) 112 field RR Lyrae stars with $8 \leq R_g \leq 30$ kpc from SKK and (b) (hatched) the 36 globular clusters in the same volume of space.

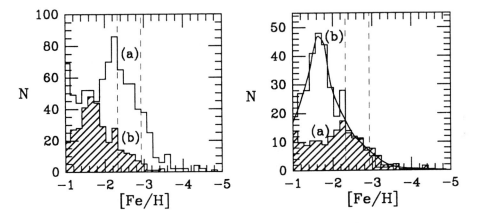

Figure 3.   The diagram to the left shows the number of stars (N) per 0.15 dex interval of abundance [Fe/H] for (a) the metal-poor stars in the catalogue of Beers *et al.* (BPS) and (b) (hatched) the metal-poor high space-motion dwarfs in the catalogue of Ryan and Norris (RN). In the diagram to the right, the distribution of the stars in the BPS catalogue (a) has been normalized to the RN distribution using the abundance interval between the two vertical dashed lines. The composite distribution is shown by the full curve.

There are two other sources of relatively local halo objects which can be used to derive a complete metallicity distribution for the halo. The first is a survey for very metal-poor stars by Beers, Preston and Shectman (1985) (BPS); a new unpublished catalogue of nearly a thousand of these stars was kindly made available to me by Beers. This source can be used to establish an upper limit to the relative numbers of the most metal-poor stars. There is, however, observational selection in this BPS survey in the sense that it progresssively excludes stars that are more metal-rich than [Fe/H] $\sim$ −2.3. The second source is the catalogue of 373 kinematically-selected dwarf halo stars of Ryan and Norris (1990) (RN). This contains stars whose space velocities are greater than 250 km s^{-1} with respect to the local standard of rest; 238 of these come from their southern subdwarf survey and the remainder from the northern survey of Laird, Carney and Latham (1988). The existence of "thick disk" stars (of as yet ill-defined kinematics) with metal abundances as low as [Fe/H]$\sim$ −1.6, (Morrison, Flynn and Freeman 1990) suggests that disk contamination might be a problem among the more metal-rich RN stars. *Because of this uncertainty, no attempt is made to compare the RR Lyrae metallicities with those of these local stars at the metal-rich end of the distribution.* The metallicity distributions of the two local surveys are given with a 0.15 dex binning in Figure 3 (left). The two distributions have similar slopes in the abundance range −2.3 $\geq$ [Fe/H] $\geq$ −2.9; the BPS distribution was therefore normalized to the RN distribution in this abundance range. A composite halo (CH) metallicity distribution (shown by the full curve in Figure 3 (right)) was then derived; this follows the RN distribution for [Fe/H]$\geq$ −2.5 and the normalized BPS distribution for more metal-poor objects. The RR Lyrae and globular cluster metallicity distributions of Figure 2 (right) are compared to normalized versions of this composite distribution in Figure 4 (left) and Figure 4 (right) respectively. The normalization in this case was made in the metallicity range −1.425 $\geq$ [Fe/H] $\geq$ −1.825, i.e. around the peak of the three distributions.

The RR Lyrae stars, the globular clusters and the composite curve all show a peak in their metallicity distributions at [Fe/H]$\sim$ −1.6. The full-width at half maximum of the three distributions are also similar (the CH and globular cluster distributions have a width of $\sim$ 0.9 dex while that of the RR Lyrae stars is $\sim$ 0.8 dex). Both clusters and RR Lyrae stars show a lack of stars with respect to the CH distribution at low metallicities ([Fe/H]$\leq$ −2.0). These deficiencies amount to some 5 clusters and 28 RR Lyrae stars respectively. Laird, Rupen, Carney and Latham (1988) also compared a similar sample of halo globular clusters ($R_g \geq$ 7 kpc) with respect to their local subdwarf metallicity distribution; they found a deficiency of about 4 metal-poor clusters and concluded that the difference was of marginal statistical significance. The deficiency in the number of metal-poor RR Lyrae stars is clearly more significant: *the RR Lyrae stars trace about 80 % of the outer halo population and fail to trace the most metal-poor 20 %.* This is a rough estimate that could be changed by future improvements in the CH and RR Lyrae metallicity scales.

The SKK RR Lyrae sample is incomplete for amplitudes less than about 0.75 mag.; this produces a bias against the discovery of metal-poor stars (Butler, Kinman and Kraft, 1979). A new technique for finding RR Lyrae stars amongst stars of spectral types A and F is being investigated. This should allow the discovery of metal-poor variables of much lower amplitude and hence improve the value of the RR Lyrae stars as tracers of the galactic halo.

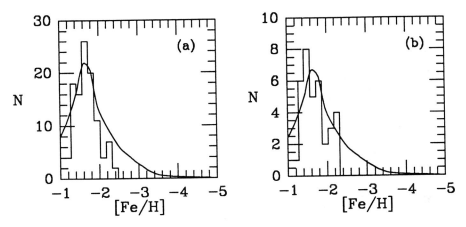

Figure 4.   The ordinate (N) is the number of objects per 0.15 dex interval of the abundance [Fe/H] (abscissa). Plot (a) is for 112 RR Lyrae stars and plot (b) for 36 globular clusters that have $8 \leq R_g \leq 30$ kpc. The full curve is the composite metallicity distribution described in the text.

Sandage (1990) and Renzini (1983) have suggested that the Oosterhoof effect arises in part from the absence of RR Lyrae stars in globular clusters of intermediate metallicity ($-1.3 \geq [Fe/H] \geq -1.6$). In the SKK data for objects with $8 \leq R_g \leq 30$ kpc, there are 38 RR Lyrae stars in the abundance range $-1.3 \geq [Fe/H] \geq -1.6$ and 74 RR Lyrae stars in the range $-1.6 \geq [Fe/H] \geq -2.0$; the corresponding numbers of stars in the RN catalogue are 74 and 108 respectively for these two abundance ranges. The ratio of the numbers in the two abundance ranges is almost the same for the two classes of objects; there is no suggestion of a relative deficiency in the numbers of intermediate metallicity *field* RR Lyrae stars. SKK define n_{RR} as the number of RR Lyrae stars that a cluster would have if scaled to an absolute visual magnitude of -7.5. The average value of n_{RR} for the clusters in the range $-1.3 \geq [Fe/H] \geq -1.6$ is 34, while the average value of n_{RR} for the clusters outside this range is 26. A deficiency of RR Lyrae stars therefore does not occur in clusters with $8 \leq R_g \leq 30$ kpc and $-1.3 \geq [Fe/H] \geq -1.6$ in using the SKK data. It is found, however, that the clusters in the abundance range $-1.25 \geq [Fe/H] \geq -1.45$ do have the low average n_{RR} of 7. We therefore consider the case where $-1.25 \geq [Fe/H] \geq -1.45$ defines the intermediate metallicity range. Again, considering SKK RR Lyrae stars with $8 \leq R_g \leq 30$ kpc, there are 21 RR Lyrae stars with $-1.25 \geq [\text{Fe/H}] \geq -1.45$ and 72 with $-1.45 \geq [\text{Fe/H}] \geq -2.0$. The corresponding numbers of stars in the RN catalogue are 44 and 148 respectively for the two abundance ranges. Again, there is good agreement between the ratios found for the RR Lyrae stars and for the stars in the RN catalogue. *The SKK data therefore show no evidence for a lack of field RR Lyrae stars of intermediate abundance.* It should be emphasized that this result refers specifically to the *field* RR Lyrae stars that lie beyond the solar circle ($8 \leq R_g \leq 30$ kpc). The possibility that there is a narrow range (~ 0.1 dex) of intermediate abundances that lacks field RR Lyrae stars cannot be excluded because such a gap in the distribution would be smoothed out by the observational errors.

It is a pleasure to thank Dr. T.C. Beers and also Dr. S.G. Ryan for allowing me to discuss their data before publication. I would also like to thank both Dr. H.L. Morrison and Dr A. Saha for helpful discussions.

REFERENCES

Beers, T., Preston, G. and Shectman, S. (1985), Astron. J., **90**, 2089.
Butler, D., Kinman, T. and Kraft, R. (1979), Astron. J., **84**, 993.
Laird, J., Carney, B. and Latham, D. (1988), Astron. J., **95**, 1843.
Laird, J., Rupen, M., Carney, B. and Latham, D. (1988), Astron. J., **96**, 1908.
Morrison, H., Flynn, C. and Freeman, K. (1990), Astron. J., **100**, 1191.
Pritchet, C. (1988), in The Extragalactic Distance Scale, eds. S. van den Bergh and C. Pritchet, A.S.P. Conference Series.
Renzini, A. (1983), Mem. Soc. Astr. Italiana, **54**, 335.
Ryan, S. and Norris, J. (1990), Astron. J., (in press).
Sandage, A. (1990) Ap. J., **350**, 631.
Suntzeff, N., Kinman, T. and Kraft, R. (1991), Ap. J., (in press).
Zinn, R. (1985), Ap. J., **293**, 424.

Pagel: Morrison, Flynn and Freeman have identified a group of red giants called by them the 'metal-weak thick disk', which is of great interest to me because I predicted such a thing, but they claim that this population has no RR Lyraes which is a problem. Could you comment on this?

Kinman: The Morrison, Flynn and Freeman paper is very interesting. It is, however, difficult at the moment to isolate the 'thick disk' amongst the nearby RR Lyrae stars because the necessary data (radial velocities and spectroscopic abundances) are incomplete. This incompleteness is particularly severe at low galactic latitudes where the 'thick disk' RR Lyrae stars should be most prominent. I have just completed the observations for a photometric study in B,V and H for 150 of the nearest RR Lyrae stars. We now have up-to-date ephemerides for these stars and in the near future Morrison and I hope to obtain spectroscopic abundances and radial velocities for the variables that currently lack them. The isolation of the 'thick disk' RR Lyrae stars should then be less difficult.

The S-Process in Massive Stars with Different Metallicities

N. PRANTZOS

Institut d' Astrophysique de Paris, and
Service d' Astrophysique, CEN Saclay, France

ABSTRACT

Slow neutron capture nucleosynthesis during core Helium burning in massive stars is rein-vestigated, with a complete set of stellar models (i.e. covering the whole mass range for massive stars) and the most recent nuclear data. For the first time, the production of secondary nuclei as a function of stellar metallicity is investigated in detail, and important effects are revealed at low metallicities. It turns out that 1) massive stars are, most proba-bly, the site of the so called "weak" s-process component (i.e. s- nuclei with mass number $60 < A < 90$) and 2) the yield of s- nuclei in that site is not a linear function of metallicity as usually assumed in galactic chemical evolution considerations.

1. INTRODUCTION

In recent years, secondary nucleosynthesis accompanying central He burning in massive stars has been investigated by several authors (e.g. Busso and Gallino 1985, Prantzos et al. 1987, Langer et al. 1989). This site is thought to be the principal candidate for the production of the weak s-process component (e.g. Prantzos 1989; Kappeler et al. 1989, for reviews). The ^{22}Ne(α,n) reaction is the main neutron source in that site, being activated towards the end of He burning, at temperatures $T \sim 2$-3 10^8 K (as originally suggested by Peters, 1968).

Despite recent investigations of the subject, we revisited that nucleosynthetic site for the following reasons (the last one being the most important):

i) The range of stellar masses covered in previous studies is limited; since different stellar evolution and nucleosynthesis codes are used in those works, the results (although com-plementary) do not allow for a complete, self-consistent picture of the s-process over the whole mass range of massive stars (i.e. from 10 to 70-100 $M_\odot$).

ii) Many important nuclear data have been revised very recently (e.g. the neutron capture cross section of ^{22}Ne, Beer et al. 1989; see Sec. 2), making necessary a revision of all the recent s-process calculations.

iii) Despite the limited information available on s-elements in low metallicity stars, it is expected that observational activity in the years to come will considerably improve our knowledge on the production and evolution of those elements in the Galaxy. It is, therefore, important to have accurate predictions of the yields of s-nuclei as a function of metallicity, in order to be able to compare chemical evolution models with observational data. Notice that most s-nuclei (A>90), belonging to the "main" s-process component, are thought to be produced in a different site, namely during thermal pulses in low and/or intermediate mass stars. However, the study of that site is plagued with various uncertainties up to now, even for solar metallicity stars; on the contrary, core He burning in massive stars is a well understood stage of stellar evolution, allowing for an accurate study of the metallicity effects.

In this work we shall try to answer the following questions: 1) can massive stars produce the weak s-process component?, 2) in that site, is the yield of s-nuclei proportional to metallicity (expressed either by iron or oxygen), as usually assumed?, and 3) did massive stars effectively produce the weak s-component at low metallicity (e.g. $z/z_\odot \leq 0.01$), or is it mainly a contribution from the r-process? We shall show that the answer to those questions is, 1)*yes* (but..), 2)*no*, and 3)*perhaps*, respectively. In Sec. 2 we present the stellar and nuclear physics inputs of this study, along with some assumptions concerning the abundance evolution of various elements in the Galaxy. In Sec. 3 the results for solar metallicity stars are presented, whereas in Sec. 4 results for lower than solar metallicities (down to $z/z_\odot = 10^{-4}$) are presented.

2. INPUT PHYSICS

The phase of core He-burning in massive stars was studied with the stellar evolution code developed by Sugimoto and Nomoto (1980). The evolution of He cores with masses $M_\alpha =$ 2.5, 4, 6, 8, 10, 16, and 32 $M_\odot$, respectively, was studied until He exaustion; the obtained temperature and density profiles as a function of time and mass coordinate were used then for "post-processing", i.e. studying nucleosynthesis with a large network, appropriate for the s-process (see below). Indeed, since He core nucleosynthesis is not affected by the evolution of the stellar enveloppe, and since the resulting s-process yields are not expected to be modified by the subsequent advanced evolutionary phases of the star (except in the innermost regions, which will be "locked" in the neutron star or black hole, anyway), this "post-processing" approximation is quite satisfactory.

The above He cores (2.5 to 32 $M_\odot$) correspond to main sequence stellar masses in the range $M_{ms} = 8(11)$ to 55(78) $M_\odot$, the exact value depending on the poorly known effects of overshooting and mass loss (their importance increases with increasing stellar mass); numbers in parenthesis above correspond to no overshooting at all, i.e. the standard Schwarschild criterion is used for convection. The M_α-M_{ms} relationship can be approximated by: $M_\alpha = \alpha \log(M_{ms}/M_\odot) M_{ms}$, with α ranging from 0.24 to 0.40, depending on the amount of overshooting adopted.

The s-process network used in this work contains 440 nuclei from ^{12}C to ^{210}Bi ($\sim$30% of them unstable), linked with 720 reactions. The rates for the various alpha- or neutron-induced reactions are taken from the compilation of Caughlan and Fowler (1988; CF88 in the follwing) or previous versions of it, while radiative neutron capture cross sections are taken from the compilation of Bao and Kappeler (1987). Finally, for beta decays and electronic captures the (temperature and electronic density dependent) rates of Takahashi and Yokoi (1987) are adopted.

Notice the important recent modification of the $^{22}Ne(n,\gamma)$ cross-section: $\sigma_{22}=0.06\pm0.005$ mb (Beer et al. 1989), w.r.t. $\sigma_{22}=0.9$ mb in Bao and Kappeler; this revision (which is not included in all previous recent calculations of the s-process in massive stars) considerably reduces the importance of neutron "poisoning" by ^{22}Ne. Notice also that the rate of the main neutron source, the $^{22}Ne(\alpha,n)$ reaction, is still subject to considerable uncertainties, and could be either higher or lower w.r.t. its CF88 value by a factor of $\sim$10 (Wolke et al. 1989); obviously, a large modification of that rate will change drastically the results reported here.

In order to mimic as realistically as possible the chemical composition at the beginning of core He burning for stars of various metallicities, account should be taken of the evolution of the abundances of various elements in our Galaxy. For our study, the most important abundances are those associated with the *seed* nuclei (i.e. ^{56}Fe) and the *neutron source* (i.e ^{22}Ne). For the first one, observations exist down to very low metallicities, [Fe/H]=-4 (where [M1/M2] = log(M1/M2)$_{star}$ - log(M1/M2)$_{\odot}$, in the standard notation). The second one is not directly observable, but its abundance should be associated to the one of ^{16}O, since it is assumed that the previous H burning phase transforms all CNO nuclei (mostly ^{16}O) in ^{14}N, which in the beginning of He burning turns into ^{22}Ne. The observational situation concerning the evolution of O vs. Fe in the Galaxy is not clear as yet (e.g. Wheeler et al. 1989), and for our purposes we adopted three different "histories" for those elements covering the whole range of possibilities (cases A, B, and C, respectively, in Fig. 1: case A is clearly an unrealistic one, but allows for a very useful test of a *constant source/seed ratio*; see Prantzos et al. 1990, for a thorough discussion of the importance of a realistic source/seed ratio).

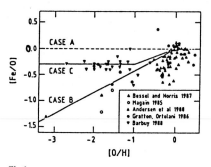

Fig 1. Observed abundance ratios of [Fe/O] (=log(Fe/O) −log(Fe/O)$_{\odot}$) vs. metallicity, expressed here as [O/H] (adapted from Wheeler et al., 1989). The dotted line (case A) corresponds to solar Fe/O ratio, independent of metallicity, and is studied here for purely illustrative purposes. Case B is based on the results of Bessel and Norris for 2 stars only, and seems to be supported by recent observations of Abia and Rebolo (1989), not shown here; it is a rather extreme case. Case C is our standard senario

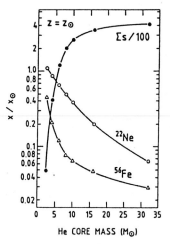

Fig 2. Ratio of final over initial abundances for ^{22}Ne, ^{56}Fe, and the sum of pure s-nuclei at He exhaustion, as a function of the mass of the He core. In the case of ^{22}Ne, the ^{14}N abundance resulting from the CNO cycle (see text) is considered as initial one (when ^{14}N is completely transformed in ^{22}Ne, the maximum mass fraction of the latter nucleus is $X_{22} = X_{14} 22/14$)

3. NUCLEOSYNTHESIS IN SOLAR METALLICITY STARS

With the CF88 rate, ^{22}Ne(α,n) becomes efficient at rather high temperatures, that is only towards the end of He burning (at He mass fraction $X_{\alpha} <0.03$). Because of their lower central temperatures, stars with $M_{\alpha} <6\ M_{\odot}$ burn only a small fraction of their ^{22}Ne (as can be seen in Fig. 2), contrary to more massive ones; for obvious reasons, the depletion of ^{56}Fe (the seed nucleus for the s-process) as a function of stellar mass follows the behaviour of ^{22}Ne. The effect of the higher temperatures on the *efficiency* of the neutron source is clearly seen in Fig. 3, where the neutron densities in the center of the studied He cores are shown as a function of X_{α}: a difference by more than a factor of 5 exists between the peak neutron densities of the 4 $M_{\odot}$ and 16 $M_{\odot}$ He cores.

Besides the efficiency of the neutron *source* (depending on both temperature and source abundance), the neutron economy depends also on the efficiency of the various neutron *poisons*. This effect (a very important one, especially in view of the low metallicity case) can be quantified by introducing the quantity: $P_i(t) = \Sigma Y_i(t)\lambda_{n,i}(t)$, where $Y_i(t)$ is the number fraction of the "poison", and $\lambda_{n,i}(t)$ is the rate of the neutron induced reaction on it. We distinguished 6 main classes of neutron poisons: 1) ^{12}C, 2) ^{20}Ne (those two are "primary" nuclei, i.e. they are produced always to similar amounts by He burning, irrespectively of metallicity; ^{16}O belongs to the same class, but its neutron capture cross-section is much smaller than the ones of those two), 3) the region between ^{21}Ne and ^{26}Mg (referred to as Ne-Mg), 4) the region Al-Fe, 5) ^{56}Fe, the s-process seed, and 6) all nuclei heavier than ^{56}Fe ("heavies" in the following). It turns out that in solar metallicity stars, $P_t(t) = \Sigma P_i(t)$ is always dominated by Ne-Mg towards the end of He burning (when s-process is occuring); however, the "primary" neutron poisons (^{12}C, in $M_\alpha < 6$ $M_\odot$ He cores, and ^{20}Ne in more massive ones), play a very important role also. One may easily anticipate that in lower metallicity stars (with less Ne-Mg or Fe during He burning, but the same amounts of ^{12}C and ^{20}Ne, since their production does not depend on metallicity), the role of ^{12}C and ^{20}Ne as neutron poisons will be even more important (see Sec. 4).

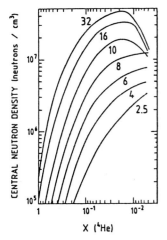

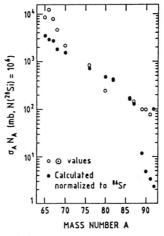

Fig 3. Central neutron densities as a function of ^{4}He mass fraction (X_4) in the He cores of our study. Notice that the efficiency of the s-process depends rather on the average neutron density in the convective core, a quantity with no physical significance since neutrons are in local equilibrium in each mass shell

Fig. 4. Empirical (circles) and theoretical (dots) $\sigma_A N_A$ curves vs. mass number A for the most overproduced nuclei in the $65 < A < 90$ region. Empirical data on neutron capture cross sections σ_A are taken from Bao and Kappeler (1987) and solar system abundances N_A from Anders and Grevesse (1989). Theoretical data correspond to the last column of Table 2, normalised to ^{86}Sr. For $A < 90$ the fit is excellent for pure s-nuclei (^{76}Se, ^{82}Kr, ^{86}Sr, ^{87}Sr) but clearly fails at heavier ones (notice the overproduction of ^{80}Kr)

The obtained yields of s- nuclei present the same qualitative features in all the He cores studied in this work (significant quantitative differences exist, however, due to different neutron densities). The results, averaged over an Initial Mass Function (IMF) of the Salpeter type $[\Phi(M) \propto M^{-2.35}]$, present the following characteristics: (1) all the pure s-nuclei of the weak component have similar overproduction factors, except ^{89}Y and ^{87}Rb; (2) some Zn isotopes, as well as the s-r isotopes of Ga, Ge, Sr also present significant (but lower than the pure s- nuclei) enhancements; (3) ^{80}Kr is enhanced twice as much as any other isotope. All those results are attributed essentially to the relatively low neutron densities encountered in the He burning cores of massive stars. Point (3), first discussed by Prantzos et al. (1987), presents for the moment a serious difficulty to the senario of the weak s-process component originating in massive stars. One may ignore it, hoping that some, up to now overlooked, nuclear physics subtlety will eventually help solving it, without perturbing too much the pattern for the other nuclei; in that case, the solar system weak s- process component is well reproduced, as can be seen from Fig. 4, where theoretical data is compared to observations.

4. THE S-PROCESS IN LOW METALLICITY STARS

Calculations have been performed for metallicities $z/z_\odot = 10^{-1}$, 10^{-2}, 10^{-3}, and 10^{-4}, respectively, and for cases A (solar source/seed ratio at all metallicities), B, and C (source/seed ratio is given by galactic chemical evolution in the last two cases).

In case A one might naively expect that, since the source/seed ratio remains constant, the efficiency of the s-process (expressed, for instance, by the number n_c of neutrons captured per initial ^{56}Fe seed nucleus) would also remain constant, leading to a production pattern similar to the one obtained in solar metallicity stars. This is not the case, however, as can be seen in Fig. 5a, the reason being the following: as the metallicity gets smaller, the abundances (not the ratio) of the source and seed nuclei and the secondary neutron poisons become also smaller, but the abundances of the *primary neutron poisons*, like ^{12}C and ^{20}Ne, remain always the same, i.e. their relative contribution to neutron poisoning increases as metallicity decreases; obviously, at very low metallicities, those primary neutron poisons capture almost all available neutrons, leaving insignificant amounts to be captured by ^{56}Fe seed nuclei for the s-process.

This situation is seriously modified if one takes into account the *differential* evolution of the source/seed ratio shown in Fig. 1. As can be seen in Figs. 5b (for the rather "extreme" case B) and 5c (case C, our "standard" one), n_c *increases* w.r.t. case A at intermediate metallicities ($z/z_\odot \sim 10^{-1} - 10^{-2}$, depending on stellar mass), and decreases below the corresponding results of case A at low metallicities. This quite interesting pattern clearly reflects the competition at low metallicities between two opposite factors: the primary neutron poisons (which tend to *decrease* n_c with decreasing z, as shown in Fig. 5a), and the enhanced source/seed ratio (which tends to *increase* n_c with decreasing z); at very low metallicities the former factor always dominates, but at intermediate ones interesting enhancements in the efficiency of the s-process are obtained.

It is clear from the above that the "s-yield vs. metallicity" relationship *cannot be a linear one*, as is often invoqued (e.g. Tinsley 1980, Lambert 1987). The exact form of that relationship can be seen in Figs. 6, where metallicity is expressed by iron (Fig. 6a) and oxygen (Fig. 6b): in both cases the production of ^{86}Sr, a pure s-nucleus of the weak component, is found to be enhanced (w.r.t. the expectations based on a linear relationship) at intermediate metallicities, and considerably reduced at low metallicities. The same pattern is obtained for the other pure s-nuclei and the other stellar masses considered.

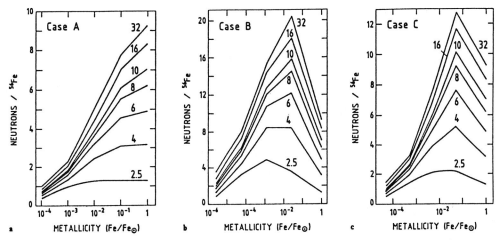

Fig 5 a-c. Number of neutrons n_c captured per initial ^{56}Fe nucleus as a function of Fe metallicity, in the 7 He cores of this study and for cases A (a), B (b) and C (c), respectively (see also Fig. 1). The lines are drawn only to guide the eye and each corresponds to one of the 7 He cores. The effect of primary neutron poisons at low metallicities is clearly seen in each figure (reduction of n_c), while the effect of a large (w.r.t. solar) source/seed ratio is also seen between a, b, and c (increase of n_c)

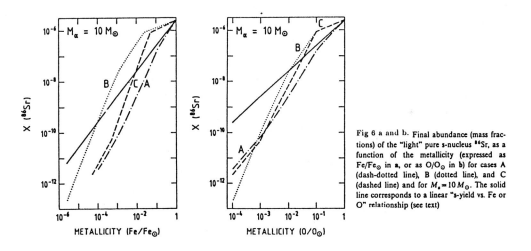

Fig 6 a and b. Final abundance (mass fractions) of the "light" pure s-nucleus ^{86}Sr, as a function of the metallicity (expressed as Fe/Fe$_\odot$ in a, or as O/O$_\odot$ in b) for cases A (dash-dotted line), B (dotted line), and C (dashed line) and for $M_\alpha = 10\,M_\odot$. The solid line corresponds to a linear "s-yield vs. Fe or O" relationship (see text)

5. CONCLUSIONS

The main results of this work are as follows:

(1) The operation of the ^{22}Ne(α,n) source during central He burning in massive stars may have produced the solar system weak s-process component, *if* the problem of the associated overproduction of ^{80}Kr is solved.

(2) In lower than solar metallicities, two (previously unexplored) factors are found to affect

significantly the results of slow neutron capture nucleosynthesis: the "primary" neutron poisons, like ^{12}C and ^{20}Ne (which tend to decrease the s-process efficiency with decreasing metallicity), and the enhanced in the past (because of galactic chemical evolution considerations) source/seed ratio (which tends to increase the s-process efficiency with decreasing metallicity).

(3) The relationship "s-yield vs. metallicity" is not a linear one, because of the two factors mentionned in (2) above. Although this result is obtained only for the operation of $^{22}Ne(\alpha,n)$ in massive stars, similar effects should be expected for its operation in low and/or intermediate mass stars.

AKNOWLEDGEMENTS: I wish to thank my collaborators in this work, M. Hashimoto and K. Nomoto, as well as M. Arnould, for their help. I gratefully aknowledge the generous allocation of computer time in the CRAY XMP of CEN Saclay.

REFERENCES

Bao Z., Kappeler F. 1987, *Atom. Data Nucl. Data Tables*, **36**, 411

Beer H., Rupp G., Voss F., Kappeler F. 1989, *Proceedings of 5th Workshop on Nuclear Astrophysics*, eds. W. Hillebrandt, E. Muller, Max Planck Institute, p. 6

Busso M., Gallino R. 1985, *A. A.*, **151**, 205

Caughlan G., Fowler W. 1988, *Atom. Data Nucl. Data Tables*, **40**, 283

Jorissen A., Arnould M. 1990, *A. A.*, in press

Kappeler F., Beer H., Wisshak K. 1989, *Rep. Prog. Phys.*, **52**, 945

Langer N., Arnould M., Arcoragi J.P. 1989 *A. A.*, **210**, 187

Lambert D. 1987, *J. Ap. A.*, **8**, 103

Peters J. 1968, *Ap. J.*, **154**, 224

Prantzos N. 1989, in *Nuclear Astrophysics*, Eds. M. Lozano, M. Gaillardo, J. Arias, Springer-Verlag, p. 1

Prantzos N., Arnould M., Arcoragi J.P. 1987, *Ap. J*, **315**, 209

Prantzos N., Arnould M., Casse M. 1988, *Ap. J. Let*, **331**, L15

Prantzos N., Hashimoto M., Nomoto K. 1990, *A. A.*, **234**, 211

Sugimoto D., Nomoto K. 1980, *Sp. Sci. Rev.*, **25**, 155

Takahashi K., Yokoi K. 1987, *Atom. Data Nucl. Data Tables*, **36**, 375

Tinsley B. 1980, *Fund. Cosm. Phys.*, **5**, 287

Wheeler J.C., Sneeden G., Truran J. 1989, *A. R. A. A.*, **27**, 516

QUESTIONS

Ph. Podsiodlowski: Did you compare your s-element distributions with the claimed s-process anomaly in SN1987A?

N. P.: We made specific calculations for SN1987A (Prantzos et al. 1988; actually the only quantitative calculation for that case), but with the *old* rate for ^{22}Ne(n,γ). It seemed difficult (although not impossible) to reproduce the Ba overabundance claimed by Williams, by core He burning in the progenitor of SN1987A (the *new*, smaller, rate for that reaction should make it easier). However, Williams' claim has not been confirmed, and one cannot be sure if this was a real effect or not.

B. Pagel: Do you think similar (metallicity) effects can apply to the the main s-process component also?

N. P.: We do not know yet what is the neutron source for the main s-process component. Although only one other relevant calculation, parametrized and not conclusive, has been done up to now (Malaney and Fowler 1988, M.N.R.A.S.), my guess is that similar effects should be expected whenever the ^{22}Ne(α,n) source operates; I am much less certain about the metallicity effects on the ^{13}C(α,n) source.

B. Pagel (to N. P. and D. Lambert): I see no reason why the Solar System s-process distribution would differ from that in a single AGB star, after several dredge-up cycles.

N. P.: I see no reason why the Solar System s-process distribution would *not* differ from that in a single AGB star, after several dredge-up cycles! To my knowledge, theory up to now managed to produce solar system s-process distributions only in small and very low metallicity stars, under really special conditions (e.g. calculations by Hollowell and Iben, or Gallino, Kappeler and collaborators). More massive and/or normal metallicity AGB stars failed (up to now) to give solar system s-process distributions.

THE COOL END OF THE WHITE DWARF LUMINOSITY FUNCTION

J. Isern[1], E. García-Berro[2], M. Hernanz[1], R. Mochkovitch[3] and A. Diaz[4]

[1]Centre d'Estudis Avançats de Blanes (CSIC) and Grup d'Astrofísica (IEC).
[2]Departament de Física Aplicada (UPC) and Grup d'Astrofísica (IEC).
[3]Institut d'Astrophysique de Paris
[4]Departament de Física i Enginyeria Nuclear (UPC)

1. INTRODUCTION

Since white dwarfs are the final remnants of low and intermediate mass stars and since they can be very old, it is possible, at least in principle, to extract information both on the white dwarfs themselves and on the structure and evolution of the Galaxy (García-Berro et al., 1988b, Mochkovitch et al., 1990). The necessary tool to extract such information is the luminosity function of white dwarfs. The first reliable luminosity function observationally obtained is due to Liebert, Dahn and Monet (1989), and it is characterized by a monotonic increase in the number of white dwarfs with magnitude, followed by an abrupt shortfall at a visual magnitude $M_V \simeq 16$. Due to the lack of reliable model atmospheres, the bolometric correction is uncertain, and the luminosity at which the shortfall appears can only be placed somewhere in the range $-4.2 \leq \log(L/L_\odot) \leq -4.6$. The cool and faint end of this function was obtained from the LHS catalogue (see Liebert, Dahn and Monet 1989 for a detailed discussion), and the region $M_V > 18$ has not yet been explored.

For the last decades, plasma physics developments have led to a better understanding of the physical conditions in white dwarf interiors, where the elements are completely ionized and form what is called a Coulomb plasma. The importance of Coulomb interactions comes not only from the changes introduced in the specific heat, but also because they lead to a first-order transition from liquid to solid that releases latent heat. Detailed descriptions of such a plasma can be found in Hansen, Torrie and Vieillefosse (1977) and Ichimaru, Iyetomi and Ogata (1988). For a typical white dwarf ($0.6M_\odot$ made of carbon and oxygen) central solidification appears at $\log(L/L_\odot) \approx -3.5$, thus implying that the interior of white dwarfs belonging to the the faint end of the luminosity function are completely dominated by the physics of solid Coulomb plasmas.

White dwarf interiors, however, are not only made of one chemical element but of a mixture of elements and, as the solidification starts, there is always some degree of chemical separation, even in the case where the components are miscible in all proportions. This effect is not negligible. For instance, if carbon and oxygen were not miscible in the solid phase, pure carbon flakes would form an rise towards the surface, melting there. Meanwhile, pure oxygen flakes would sink and settle at the center. If this process would continue all the star would become completely differentiated and $\approx 10^{47}$ erg of gravitational energy would be released. As this would happen at $L \approx 10^{-4} L_\odot$, the cooling process would be delayed ≈ 8 Gyr (García-Berro *et al.*, 1988a).

2. INFLUENCE OF THE PHASE DIAGRAM IN THE COOLING PROCESS

2.1 Influence of the carbon oxygen mixture

Obviously, the cooling process not only depends on the thermal contents of white dwarfs, but also on the structure of the envelope (mass of the helium layer, metal contents and so on). As here we are only interested in the behaviour of the interior, we can mimic the properties of the envelope by adopting a relationship between the temperature of the core (which is assumed to be isothermal) and the luminosity like that of Lamb and Van Horn (1975) or that of Wood and Winget (1989). The nature of the adopted envelope is an open problem and any change in the relationship connecting the core with the envelope can introduce dramatic changes in the cooling time (D'Antonna and Mazzitelli 1990).

In order to make comparisons, we have adopted as standard model a white dwarf of $0.6 M_\odot$ composed of an homogeneous mixture of carbon and oxygen with mass fractions $x_C = x_O = 0.5$. The relationship connecting the luminosity with the central temperature was obtained from Lamb and Van Horn (1975). The adopted phase diagram for the liquid-solid transition was that of Barrat, Hansen and Mochkovitch (1988). This diagram, which is of the spindle form, predicts a slight oxygen enrichment of the central regions of the star. The time needed to reach a luminosity of $10^{-4.5} L_\odot$ is 10.9 Gyr. Here it is necessary to be careful because any small change in the luminosity translates into a big change in the age of the white dwarfs. The uncertainties introduced by the bolometric correction amount to 4 Gyr, for instance.

The quoted values depend on the detailed chemical composition of the core. As a result of the different temperatures at which it is built, the core is already stratified at the birth of white dwarf, being the oxygen abundance much higher in the central regions than in the outer ones (D'Antonna and Mazzitelli 1990). As a consequence of the higher average molecular weight per particle, the higher freezing temperature and the smaller release of gravitational energy, the white dwarf cooling is more rapid in this last case. The time necessary to arrive at $L = 10^{-4.5} L_\odot$ is 9.8 Gyr.

It is useful to compare these results with those obtained in the case in which the sedimentation induced by freezing is neglected. In this case, the time necessary to reach a luminosity $L = 10^{-4.5} L_\odot$ is 10.1 Gyr, 0.8 Gyr less than in the first case. This is not an enormous difference, if compared with the uncertainties introduced by the poorly known properties of the envelope and the values of the bolometric correction, but it is a positive correction that must be always taken into account in order to use white dwarfs as indicators of the age of the galactic disk. This is especially important in view of the present discrepancy with the age of the globular clusters.

Recently, Ichimaru *et al.* (1988) have discovered that the internal energy of a mixture due to Coulomb effects is a linear combination of the Coulomb energies of each species weighted by the abundance by number of each species. Furthermore, they have found that the entropy is smaller in the liquid phase that in the solid phase. As the liquid phase is favored, the phase diagram adopts an azeotropic form (the solidification temperature versus the chemical composition goes through a minimum). The azeotropic point for the carbon oxygen mixture is $T/T_C = 0.94$ and $X_A = 0.16$, where T_C is the freezing temperature for pure carbon and X_A is the oxygen abundance by number. As the abundance of oxygen is larger than the azeotropic one, its distribution after freezing is similar to that obtained with the Barrat, Hansen and Mochkovitch (1988) diagram. Due to the smaller freezing temperatures involved, the delay introduced in the cooling process is larger. The time necessary to reach $L = 10^{-4.5} L_\odot$ is 11.5 Gyr, a quantity that would not be so far from the age of globular clusters provided that the present knowledge of the properties of the envelope were reliable.

2.2 The Role of the Minor Species
The existence of an azeotrope in the phase diagram opens, however, new and very interesting possibilities for obtaining additional sources of energy (García-Berro *et al.*, 1988b). Since any binary mixture with an atomic number ratio larger than that of the carbon and oxygen mixture will also display an azeotropic behavior at freezing, the phase diagram will have a negative slope in the region of the smaller abundances of the heavier element. In this case, the liquid remnant will be denser than the solid and the last will migrate outwards, thus a chemical stratification of the star, with the heaviest elements being concentrated in the innermost regions, will result.

White dwarfs are made of carbon, oxygen, and a mixture of minor components reflecting the initial metallicity of the parent star. Among them, ^{22}Ne is the most important one. It comes from helium burning of the ^{14}N left by the CNO cycle. Its total amount as well as its distribution throughout the white dwarf interior depend on the initial metallicity and mass of the progenitor. Despite of its low abundance, the importance of ^{22}Ne as a source of gravitational energy comes from the fact that white dwarf structures are very sensitive to the shape of the electron mole number distribution Y_e (Mochkovitch 1983). A white dwarf with all the neon concentrated in the innermost regions is far more compact than a white dwarf with the same amount of this element uniformly distributed throughout the entire star.

The phase diagrams for C-Ne and O-Ne mixtures as computed by Isern et al (1990) show the existence of an azeotropic point at $X_{Ne} = 0.053, T/T_C = 0.982$ and $X_{Ne} = 0.04, T/T_O = 0.995$ respectively (or $x_{Ne}^a = 0.09$ and $x_{Ne}^a = 0.05$, if mass fractions are preferred), where T_C and T_O are the solidification temperatures for pure carbon and pure oxygen. As it was mentioned in Isern *et al.*, (1990), these results are strongly dependent upon the values taken by the entropy of the solid and the fact that in white dwarfs the mixtures are not binary but at least ternary, and they must be considered as preliminary.

As the neon abundance is smaller than the azeotropic value, a neon poor solid forms which, being lighter, rises towards the surface and melts somewhere. As a consequence, the liquid mixture becomes progressively more neon rich. When the liquid reaches the azeotropic abundance, a solid with this last composition forms. This solid, being denser than the mixture, sinks and a neon rich core progressively forms. The final structure is a core containing all the ^{22}Ne of the star surrounded by a mantle free of neon. The energy released by this process ranges from 3.2 to 5.8×10^{46} erg (depending on the assumed chemical composition) and the corresponding delay introduced in the cooling ranges from 0.6 to 3.0 Gyr.

This delay is not negligible and must be taken into account in the calculation of the luminosity functions. Notice that any increase of the azeotropic composition and/or a decrease of the solidification temperature of the azeotropic mixture would result in a dramatic enhancement of the cooling time. It is difficult, however, to predict how this effect translates into the white dwarf luminosity function, because the abundance of ^{22}Ne depends on the metallicity of the interstellar medium from which stars form. Therefore, white dwarfs coming from metal poor stars will cool down more quickly than white dwarfs coming from metal rich populations.

3. THE LUMINOSITY FUNCTION

In order to compare the evolution of white dwarfs with the observations it is necessary to construct theoretical luminosity functions. The method we have followed is that of Iben and Laughlin (1989), with the same input (IMF, lifetimes in the main sequence...) but cooling times obtained from the standard model quoted in Section 2.1 for a grid of stars covering a mass range of 0.6 to 1.2 $M_\odot$ and assuming either a uniform distribution of C and O or a stratified distribution like those of D'Antonna and Mazzitelli (1990).

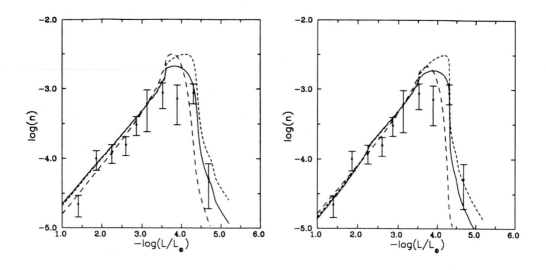

Figure 1. Luminosity functions obtained for three different SFRs: (a) The distribution of C and O has been assumed constant through all the white dwarf, (b) The distribution of C and O was obtained from D'Antonna and Mazzitelli (1990). Case (1), solid line, case (2), dotted line, case (3), dashed line.

One of the ingredients in the construction of the luminosity function is the SFR. In order to study its influence on the luminosity function we have considered three cases: (1) Constant SFR (figures 1a and b). Adopted as a standard, the corresponding luminosity functions were obliged to fit the observed cutoff. The age of the galactic disk obtained in this way was 10.5 and 9.5 respectively. The slowing down in the cooling induced by solidification translates into a prominent bump. (2) Exponentially decreasing SFR. Taken from Canal *et al.*, this volume. The age of the disk was assumed to be the same as in case (1). In this case, the position of the cutoff remains unchanged but the bump is more pronounced. (3) Arbitrary SFR. This SFR has the form $\psi \propto t^2 \cdot \exp(-t^2)$, with a maximum at 5 Gyr. In this case, the cutoff appears at higher luminosities (a compromise between the assumed age of the disk, 10.5 Gyr, and the the maximum in the stellar formation rate, 5 Gyr). All three cases were normalized to the value taken by the observed luminosity function at $\log(L/L_\odot) = -3$.

4. CONCLUSIONS

We have shown that the gravitational energy released by the settling of heavier species during the cooling of white dwarfs can delay the process during 1 to 3 Gyr, depending on the adopted phase diagram and on the abundance of the chemical species considered. We have also shown that the gravitational settling of minor species, in particular that of neon, is a potentially very important source of energy for white dwarfs and deserves further studies.

The precise form of the luminosity function as well as the position of the cutoff not only depend on the physics of white dwarfs, but also on the adopted SFR. The existence of a maximum in the stellar formation rate can distort the form of the luminosity function and provide severes underestimations of the age of the galactic disk.

Acknowledgements: This work has ben supported by the CICYT grant PB87-0304

5. REFERENCES

Barrat J.L., Hansen J.P., Mochkovitch R., (1988), *Astron. Astrophys.* **199**, L15.

D'Antonna F., Mazzitelli I. (1989) *Astrophys. J.* **347**, 934.

D'Antonna F., Mazzitelli I. (1990) *Ann. Rev. Astron. Astrophys.* in press.

García-Berro E., Hernanz M., Mochkovitch R., and Isern J., (1988a), *Astron. Astrophys,* **193**, 141.141.

García-Berro E., Hernanz M., Isern J., and Mochkovitch R. (1988b), *Nature* **333**, 644.

Hansen J.P., Torrie G.M., Vieillefosse J.P. (1977) *Phys. Rev.* **A16**, 2153.

Iben I. and Laughlin G. (1989) *Astrophys. J.* **341**, 312.

Ichimaru S., Iyetomi H., Ogata S. (1988), *Astrophys. J.* **334**, L17.

Isern J., Mochkovitch R., García-Berro E., Hernanz M. (1990) *Astron. Astrophys.* In press.

Lamb D.Q., Van Horn H.M. (1975) *Astrophys. J.* **200**, 306.

Liebert J., Dahn C.C. and Monet D.G. (1989), in "White Dwarfs". Ed. G.Wegner, *Lecture Notes in Physics* **328**, (Springer-Verlag) p.15.

Mochkovitch R., (1983), *Astron. and Astrophys* **122**, 212.

Mochkovitch R., García-Berro E., Hernanz M., Isern J., Panis J.F. (1990), *Astron. Astrophys.* **233**, 456.456.

Winget D.E., Hansen C.J., Liebert J., Van Horn H.M., Fontaine G., Nather R.E., Kepler S.O. and Lamb D.Q. (1987), *Astrophys. J. (Letters),* **315**, L77.

Wood M.A. and Winget D.M. (1989), in "White Dwarfs". Ed. G.Wegner, *Lecture Notes in Physics* **328**, (Springer-Verlag) p.282.

Terlevich: I will like to know what is the situation with white dwarfs in globular clusters?

Isern: Globular clusters are too far and white dwarfs are too faint to permit direct observations at present. Some candidates have been detected in ω Cen and M71.

The Challenge of Accurate Helium Abundances

G.J. FERLAND

Astronomy Department, The Ohio State University

1 INTRODUCTION

A precise, reliable, measurement of the primordial helium abundance can provide a test of big bang nucleosynthesis (Pagel 1989). The needed accuracy, much better than 5 percent, pushes nebular theory to new limits. The standard techniques and approximations used to measure He/H in nebulae are summarized by Osterbrock (1988). The two main issues (in addition to observational problems, Davidson and Kinman 1985) are a) do hydrogen and helium lines have their expected emissivity, and b) must a correction for (unobservable) neutral helium in the H^+ recombination zone be applied? These effects are central since they may introduce systematic errors into any abundance analysis.

Jack Baldwin, Peter Martin, and I recently completed a study of the Orion Nebula. A complete description of the work will appear in a future issue of the Astrophysical Journal (Baldwin et al. 1991). Orion was chosen for very careful study because it is a dusty HII region; the effects of dust on the ionization and thermal structure of the nebula and line emissivity should be more pronounced here than in low-z objects such as those used to determine the primordial helium abundance. The geometry, (a blister), ionization source (the O 6.5 star θ^1 Ori C), and dust properties (a large value of the ratio of total to selective extinction) are all well known. This is in sharp contrast to the case of the giant extragalactic HII regions — little is known in detail.

We obtained long-slit spectrophotometric observations with the primary aim of ensured linear spectrophotometric calibration; the slit extended from close to the Trapezium to a region nearly 5 minutes away. The well-known decrease in density and ionization (Osterbrock and Flather 1959; Peimbert and Torres-Peimbert 1977) was apparent in our data. We then computed photoionization models aimed at determining physical conditions and the helium ionization correction factor for the innermost, most highly ionized, regions of the nebula. The discussion which follows is based on this work.

2 THE HE IONIZATION CORRECTION FACTOR

One of the more important uses of photoionization model calculations is to calibrate indicators of the helium ionization correction factor (Stasinska 1982; Evans and
Dopita 1985; Mathis 1985; Osterbrock 1988). Previous calculations assumed spherical geometry and a dust-free HII region. Both assumptions tend to overestimate the
correction for neutral helium.

2.1 Blister vs Spherical Geometry

The Orion Nebula is the archetypical example of a "blister" geometry (Balick
et al. 1974; 1980). As the Trapezium stars move through the Orion Molecular
Cloud, a combination of winds and radiation pressure pushes back the surrounding
gas, forming a blister or cavity roughly centered on the stars. The situation is quite
complex; radial velocity data shows complicated flows whose pattern depends on the
species under study. It is thought that most hydrogen emission originates within
the flow from the ionization front more distant than the Trapezium; the intervening
ionized gas contributes some of the emission measure, and is also seen in absorption
against the Trapezium stars.

In our calculations we modeled the more distant flow just behind θ^1 Ori C as a
plane-parallel constant gas pressure slab, viewed perpendicular to the surface. The
distribution of ionization into this geometry is somewhat different from that across
a spherical nebula in which the ratio of outer to inner radii is large. In a spherical
geometry the r^{-2} dilution of radiation allows the gas to become fairly neutral in regions
where the photoelectric optical depth of the gas is still small. In a plane-parallel
geometry the gas only becomes neutral when significant photoelectric absorption has
taken place. It is difficult to compare results of the two geometries directly, but the
plane-parallel case does tend to have the higher level of ionization of helium for a
given [OII]/[OIII] spectrum.

2.2 Dust

A second major difference between our calculation and previous efforts is that
we explicitly included the effects of dust on the ionization and thermal structure of
the nebula. Dust in the Orion Nebula has been well studied, both in extinction along
the line of sight to the Trapezium (Bohlin and Savage 1981) and scattering (Mathis
et al. 1981). The ratio of total to selective extinction is anomalously large in Orion,
suggesting that the typical grain size is also larger than the "standard" interstellar
medium.

Grains can be heated by both absorption of resonance line radiation, especially

Lyα, and direct absorption of the incident stellar continuum. Which is the case in an individual object can be determined by examining the ratio of the total dust luminosity to the Hβ luminosity. If Lyα is the main dust heating source then L(IR)/L(Hβ) $\approx$ case B I(Lyα)/I(Hβ) $\sim$ 20–35, depending the fraction of 2s populations which undergo collisions to 2p before two-photon decay. In Orion the ratio is L(IR)/L(Hβ) $\sim$ 130, showing that absorption of the stellar continuum is the main dust heating source. Similarly, since the predominant fraction of the continuum from θ^1 Ori C occurs at ionizing energies, it follows that grain opacity is the dominant opacity within the HII region, and that grains are heated mainly by absorption of *ionizing* radiation.

This causes two major two differences between our calculations and dust-free ones. First, the opacity and resulting ionization structure of the nebula is quite different from what would have occurred had dust been ignored. In particular, grains selectively remove low-energy ionizing photons, resulting in a higher mean level of ionization (Aanestad 1989). This minimizes the correction for neutral helium. The second effect is on the thermal structure of the nebula; a significant fraction of grain absorptions of ionizing radiation result in grain photoionization (Spitzer 1948; Maciel and Pottasch 1982). The grains develop a electrostatic potential and reach a balance where they are self-shielded against the radiation field. They also deposit kinetic energy into the gas via the energy of freed photoelectron (this depends on both the energy of the incident photon and the grain potential) and remove energy from the gas by capturing electrons (with typical energy kT; Spitzer 1948; Draine 1978; Draine and Salpeter 1979). The second effect is the larger. We found that grain recombination cooling is the single most important gas coolant in the HII region. This process alters the entire thermal balance of the nebula, although the effects were most pronounced in the O^+– O^o region where the grain potential was smallest.

The photoionization calculations had as their aim first to reproduce the ultraviolet, optical, and infrared spectrum of the HII region near the Trapezium, and then to determine the correction for neutral helium. We found that no correction for neutral helium was needed for the innermost regions of Orion, a result that was not surprising considering the relatively early spectral type of the ionizing star, θ^1 Ori C (Osterbrock 1988).

3 H I AND HE I EMISSIVITIES

Once the correction for neutral helium in the ionized hydrogen zone is known, the next step is to convert the emission line intensity ratios into an abundance ratio. This can be done with a precision limited only by the photometric accuracy of the data and the knowledge of the line emissivities of the relevant ions.

3.1 Hydrogen Emissivities

Hydrogen lines are formed by radiative recombination in HII regions. The hydrogen line spectrum can be idealized as either of two limiting cases; case B, in which all higher Lyman lines scatter often enough to be converted into a Balmer line plus Lyα or the two-photon continuum, and case A, in which the Lyman lines are optically thin and freely escape. Case B is usually assumed in practice since Lyman line optical depths are generally large enough to ensure that Lyβ, γ, etc, do not escape. Case A is a mathematical fiction; if Lyman lines are optically thin and the gas photoionized then Lyman line continuum florescence will also be an important line excitation mechanism, since the integrated Lyman continuum oscillator strength is of order the oscillator strengths of individual Lyman lines.

Although case B is generally the best to assume, dust can alter the detailed emissivities in two ways. First, grains provide a background opacity which can destroy Lyman lines before they scatter often enough for case B conversion to occur (Capriotti 1966; Cox and Mathews 1969). This effect tends to increase as the level of ionization of the gas increases, since it depends on the ratio of the dust to line opacity (the latter depends on the gas neutral fraction). Line opacities decrease with increasing quantum number, causing very high levels to have near-case A populations under all conditions (Baldwin et al. 1991).

A second effect is also related to the presence of dust. Where no dust present, then the H^{+}–H^{o} ionization front would occur at a Lyman continuum optical depth of $\tau_{912} \sim 1$ if the continuum is relatively soft. The Lyα optical depth is $\sim 10^4$ times larger than this, with the result that a "typical" point in the HII region will see sufficient Lyman line optical depth for case B conversion to occur. This is in contrast with a dust-bounded geometry, in which the hydrogen ionization front occurs at a dust optical depth ~ 1, and a Lyman continuum optical depth $\tau_{912} \ll 1$. As a result Lyman line optical depths are also smaller, and these photons can escape from the illuminated face of the flow and go on to interact with gas and dust elsewhere in the blister. In the case of Orion, high resolution line observations reveal supersonic motions (Castañeda 1988), suggesting that, if the Doppler shift is not advantageous, Lyman lines escaping from one side of the blister will strike other portions of the nebula at a frequency where dust absorption is far more likely than line scattering. These processes further lower the emissivity of optical hydrogen lines, causing the hydrogen abundance to be underestimated if case B is assumed.

3.2 Helium Emissivities

The physics described above does not affect helium lines, since the strongest HeI lines arise from the triplets, which have no case A–case B distinction. However, collisional excitation from the metastable 2^3S level can increase the emissivity of certain lines in a way which can be computed if the density and temperature of the gas is known (Berrington and Kingston 1987; Clegg 1987). The effects are most pronounced for $\lambda10830$, but are still significant for $\lambda5876$ for conditions in Orion. Collisional excitation should be less important for $\lambda4471$ or $\lambda6678$.

The correction for collisional excitation can be made if the atomic processes populating the levels are understood. This has been questioned by work on planetary nebulae which discovered that the observed intensity of $\lambda10830$ relative to other helium lines is roughly half its expected value (see the discussion by Clegg 1987 and Clegg and Harrington 1989). Since $\lambda10830$ is predominantly collisionally excited, the inference is that either its collision strength is overestimated by a factor or two, or the metastable 2^3S level has half its expected population. The atomic processes affecting the level population have been extensively studied over the past twenty five years, and the general belief in the atomic physics community is that the data are accurate. The suggestion that charge transfer may depopulate the level (Clegg 1987) seems not to work; transfer with hydrogen would be needed if the process is to compete with electron exchange collisions, and charge transfer of hydrogen with metastable helium is known to be slow (Baldwin et al. 1991).

It seems likely that the solution to the too-weak $\lambda10830$ is that the line has been destroyed by dust within the nebula. For the well studied planetary IC 418 Clegg and Harrington (1989) quote a $\lambda10830$ optical depth through the nebula ~200. At a typical point within the nebula ($t_4=0.85$, $N_e = 1.45 \times 10^4$ cm^{-3}), the line cross section per proton (for He/H=0.1) in a typical component of the $\lambda10830$ multiplet is $\sim 10^{-18.5}$ cm^2, correspond to an escape probability of $\sim 10^{-2.5}$ (Hummer 1968). If dust within the nebula has ISM abundances then the dust cross section per proton is $\sim 1.7 \times 10^{-22}$ cm^2, corresponding to a destruction probability of $\beta F(\beta) \sim 10^{-2.5}$ (Hummer 1968). The result is, for an ISM dust abundance, *half of the 10830 photons will be destroyed*. Dust may well explain the discrepancy between observed and predicted 10830 intensities. If this is the solution to the paradox, then the population of the 2^3S level is indeed correct.

Whether or not dust weakens $\lambda10830$, there can be no doubt that the situation regarding excited state populations of triplet helium is presently unclear. Given this uncertainty the best course of action seems to be to avoid the problem as much as possible, i.e., determine abundances from lines will the smallest possible collisional contribution ($\lambda\lambda$ 4471 and 6678) while avoiding those lines which are expected to have a significant collisional contribution ($\lambda5876$).

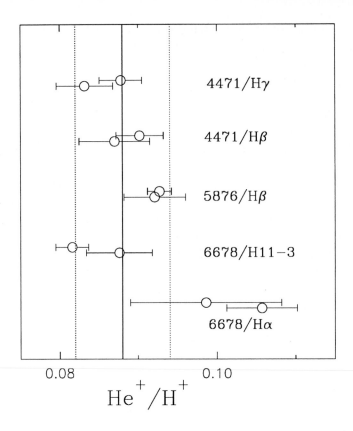

Figure 1: This figure shows several of the He$^+$/H$^+$ ionic abundance indicators. The long slit data set was divided into two groups, those nearer the Trapezium (the upper of each pair of points) and those further away. The presence of systematic error is obvious.

3.3 The Helium Abundance

The He$^+$/H$^+$ ionic abundance indicators we studied in Orion illustrate the various challenges to the determination of very accurate helium abundances (Figure 1). The data set was divided into two groups, the half nearest the Trapezium (where the nebula was brightest and the correction for both collisions and reddening were largest) and the outer half (where the nebula was fainter, statistical errors are important, but both the collisional and reddening corrections were smaller). The abundance ratios in the figure have been corrected for collisional excitation of helium lines, for not for deviations of hydrogen lines from case B emissivity.

The presence of systematic errors is obvious. We attributed these to two sources, uncertainties in the form of the reddening curve for Orion, and deviations from case B emissivity for the hydrogen lines. The 6678/Hα and 4471/Hγ abundance ratios have little dependence on the reddening correction. The helium abundance indicated by the line ratio with Hα is the highest in the data set, suggesting that the emissivity of Hα is below case B predictions. This is confirmed by the Balmer decrement; we found the Hα/Hβ ratio to be below case B.

The other abundance indicators are strongly affected by reddening; the λ6678/H11-3 ratio is only slightly affected by collisions or deviations from case B, but is most affected by reddening. The λ5876/Hβ ratio has a significant correction for both collisional excitation and reddening. Reddening corrections introduce errors in Orion for two reasons. First, the HII region is a strongly scattering extended atmosphere. The lines can really only be corrected for reddening by solving a radiative transfer problem in which the absorption and scattering parts of the extinction function are known separately. Unfortunately, the best data on Orion consist of *total* extinction studies of the Trapezium stars (Bohlin and Savage 1981). The absorption and scattering parts of the extinction curve are known only through theory (Baldwin et al. 1991). This introduces a systematic error which depends on details of the geometry and the grain albedo.

The second problem is that the form of the Orion extinction curve is not known with the precision required for accurate abundance determinations; the deduced extinction curves for each of the Trapezium stars differs from one another (Cardelli and Clayton 1988) by amounts which introduce $\sim$ 5 percent uncertainties in line ratios such as λ5876/Hβ. Worse yet, the reddening curve is not known over the baseline needed to correct the λ6678/H11-3 ratio.

The need for (uncertain) corrections for collisions, possible deviations from case B emissivity, and uncertainties in the form of the reddening curve, all introduce fundamental uncertainties in any abundance analysis of an HII region such as Orion. The helium abundance we quoted, He/H=0.088$\pm$0.006, reflected all of these uncertainties. The λ4471/Hγ ratio largely avoids these problems, and should be adopted as the preferred abundance indicator in future work. For Orion, the λ4471/Hγ data set alone indicates He/H = 0.086$\pm$0.004.

4 THE FUTURE; ADVANCES IN THE BASIC DATA

The next few years will witness a revolution in the calculation of model HII regions. There have long been problems in reconciling observed and model emission line spectra. Three sources of error can readily be identified. First, the effects of embedded grains on the ionization and thermal structure of the nebula are profound, and until recently have been largely ignored. A second major source of error lies in the stellar atmospheres. These have ionization edges coincident with the major opacity edges in the nebula, so the nebular ionization structure is sensitive to the stellar continuum. (We used plane-parallel line blanketed LTE atmospheres calculated by Kurucz.) A new generation of model atmospheres will soon be in hand (see, for example, the review by Kudritski and Hummer 1990) which will include the effects of winds.

The final major improvement will be in the basic atomic data, as a result of the Opacity Project (Seaton 1987). This effort should provide the definitive photoionization cross sections for atoms and ions of astrophysically abundant elements, including autoionization resonances. It will then be straightforward to use the Milne relation to deduce recombination coefficients which include the effects of low-lying resonances (see, for example, Nussbaumer and Storey 1983).

The case of sulphur illustrates how important these data are. Two abundance independent ionization indicators are present in spectra of HII regions; the [OIII] $\lambda 5007$ - [OII] $\lambda 3727$ ratio, and the [SIII] λ 9531 - [SII] λ 6720 ratio. The basic data for oxygen seem to be in good shape; photoionizaton cross sections, recombination coefficients (both radiative and dielectronic), and charge transfer rates all have been calculated using sophisticated methods. This is not the case for sulphur, however. For ionization of S^+ the two main references for photoionization cross sections are Chapman and Henry (1971) and Reilman and Manson (1979). These differ by nearly an order of magnitude. Our Orion calculations used preliminary Opacity Project cross sections which are intermediate between these two extremes. Unfortunately, no calculations of dielectronic recombination rate coefficients have been done for sulphur, despite the fact that this is expected to be the dominant recombination process. Rather than adopting a rate coefficient of zero, we used a rate coefficient which was the mean of those calculated for C^+, N^+, and O^+ (see Osterbrock 1988, Table 2.11). Judging from the scatter among the three, our adopted rate coefficient is uncertain by roughly 50 percent. This issue is important since the effects of this process on the predicted intensity of [SII] λ 6720 line are major; the intensity changes by nearly a factor of two when the rate coefficient is changed from zero to the mean for C, N. and O.

Uncertainties in the atomic data base should be nearly eliminated with the completion of the Opacity Project and its extensions to recombination and collisions. Opacity Project photoionization cross sections present a great deal of fine structure due to the many resonances present. Incorporating these data into existing ionization equilibria codes will present a major challenge because of the amount of data, the need to resolve resonances, and the likelihood that individual resonances will become optically thick. Meeting this challenge will be worth the effort since the Opacity Project data are likely to remain definitive for the next several decades.

5 SUMMARY AND QUESTIONS

This paper summarizes some of the challenges to very accurate helium abundances, based on work done on the Orion Nebula. Ionic abundances (i.e., the He^+/H^+ ratio) can best be obtained from the $\lambda 4471/H\gamma$ intensity ratio. This minimizes the correction for collisional excitation of helium lines and the deviation of hydrogen lines from case B emissivity. Drawbacks include the fact that the lines are weak, high resolution is required, and that the correction for underlying stellar absorption is most important for these lines. It is possible to check whether the hydrogen line spectrum is in case B by comparing helium abundances obtained from this ratio, and $\lambda 6678/H\alpha$. These disagreed by ~ 20 percent in Orion due to deviations from case B emissivity of $H\alpha$.

The next question is the correction for neutral helium in the ionized hydrogen zone. This is best determined from photoionization model calculations, which brings in questions concerning the model atmosphere, the basic atomic data, the geometry of the HII region, and the presence and abundance of grains. For work on low abundance objects, some sort of scaling law between the dust to gas ratio and the metalicity must be adopted–perhaps a constant dust-to-metals ratio is reasonable. The geometry of giant extragalactic HII regions is poorly known at present, although this has a major effect on the model calculations. Finally, it is vexing that, at this time, there are still fundamental questions concerning population and excitation mechanisms for helium lines (Clegg and Harrington 1989).

Many discussions of these and other matters with Jack Baldwin and Peter Martin, and the support of the National Science Foundation through grant AST 87-19607, are gratefully acknowledged.

6 REFERENCES

Aanestad, P.A., 1989, Ap.J., 338, 162.
Baldwin, J.A., Ferland, G.J., Martin, P.G., Corbin, M., Cota, S., Peterson, B., and Slettebak, A., 1991, Ap.J. in press.
Balick, B., Gammon, R.H., and Hjellming, R.M., 1974, Publ.A.S.P. 86, 616.
Balick, B., Gull, T.R., and Smith, M.G., 1980, Publ A.S.P. 92, 22.
Berrington, K.A., and Kingston, A.E., 1987, J. Phys. B. 20, 6631.
Bohlin, R.C., and Savage, B.D., 1981, Ap.J. 249, 109.
Capriotti, E.R., 1966, Ap.J. 146, 709.
Cardelli, J.A., and Clayton, G.C., 1988, A.J. 95, 516.
Castañeda, H.O., 1988, Ap.J. Suppl. 67, 1.
Chapman, R.D., and Henry, R.J.W., 1971, Ap.J. 168, 169.
Clegg, R.E.S., 1987, MNRAS 229, 31p.
Clegg, R.E.S., and Harrington, J.P., 1989, MNRAS 239, 869.

Cox, D.P. and Mathews, W.G., 1969, Ap.J. 155, 859.

Davidson, K., and Kinman, T.D., 1985, Ap.J. Suppl. 58, 321.

Draine, B.T., 1978, Ap.J. (Sup) 36, 595.

Draine, B.T., and Salpeter, E.E., 1979, Ap.J. 231, 77.

Evans, I.N., and Dopita, M.A., 1985, Ap. J. Sup 58, 125.

Hummer, D.G., 1968, MNRAS 138, 73.

Kudritski, R.P., and Hummer, D.G., 1990, Ann Rev Ast Ap 28, 303.

Maciel, W.J., and Pottasch, S.R., 1982, Ast Ap 86, 380.

Mathis, J.S., 1985, Ap.J. 291, 247.

Mathis, J.S., Perinotto, M., Patriarchi, P., and Schiffer, F.H., 1981, Ap.J. 249, 99.

Nussbaumer, H., and Storey, P.J., 1983, Ast Ap 126, 75.

Osterbrock, D.E., 1988, Astrophysics of Gaseous Nebulae and Active Galactic Nuclei,

Osterbrock, D.E., and Flather, E., 1959, Ap.J. 129, 26.

Pagel, B.E.J., 1989, *Evolutionary Phenomena in Galaxies*, p368, J.E. Beckman and
    B.E.J. Pagel, editors, Cambridge: Cambridge University Press).

Peimbert, M., and Torres-Peimbert, S., 1977, MNRAS 179, 217.

Reilman, R.F., and Manson, S.T., 1979, Ap.J. Sup 40, 815.

Seaton, M.J. 1987, J. Phys. B. 20, 6363.

Spitzer, L., Jr., 1948, Ap.J. 107, 6.

Stasinska, G., 1982, Ast Ap Sup 48, 299.

DISCUSSION

W.J. Maciel I have a comment regarding photoelectric heating by grains. I'd like to call attention to a paper on this subject by S. R. Pottasch and myself (Astron Astrophys 106, 1). I think this paper has been a bit overlooked in the literature, and the reason may be that due to a printer's mistake the title reads "...heating of HI regions", instead of "HII regions".

H.E. Smith I only have the comment that the giant HII regions used for abundance gradient work are generally very different from the Orion Nebula. In particular, they are very low density ($N \lesssim 10^2$ cm^{-3}) and are not strong infrared sources, suggesting that absorption/reemission of photoionizing radiation by grains may not be similarly important in these regions.

J.M. Vilchez Have you taken into account the possibility of very small grains in the outskirts of HII regions?

Reply: PAH's are observed at 3 and 5 microns in Orion, and contribute roughly one percent of the IR luminosity. They should not have much effect on the global energy budget of the nebula. Graphite and silicates were included in our calculations.

B. Pagel

1. I think your nitrogen abundances is about a factor of two too high, probably due to insufficient spectral resolution in your data.

2. I have argued for eight years that Orion should not be included in extrapolation to primordial helium and your comments on this are worthless. Furthermore, it is not clear that the dY/dZ slope is constant. Even if it is, you have got it wrong because your guess of the primordial value is too high. In Orion itself I am worried about the correction for neutral helium because the double ionization ratio $O^+ S^{++}/(O^{++} S^+)$ is so high that the correction is quite model-dependent.

Reply:

1. Our calculations reproduced the observed [NII] $\lambda 6548$ intensity. We have re-observed the nebula at very high resolution, and confirmed that the intensity we quoted in the paper is correct. N^+ is not a dominant stage of ionization of nitrogen, so infrared studies which detect N^{++} may indeed give more reliable measures of the nitrogen abundance (Simpson et al. ApJ 311, 895). Our observations were designed with helium in mind.

2. For the geometry and dust abundance we assumed, which are based on direct observations, the correction for neutral helium was not very model dependent, because it was so small. We successfully reproduced the observed O–S double ionization ratio. Previous calculations, which predicted a large correction for neutral helium, neglected dust and assumed a "planetary nebula" type of geometry. Neither is correct for Orion.

Nitrogen in H II Regions

DONALD R. GARNETT

Space Telescope Science Institute

1 INTRODUCTION

Measurements of the relative variation in abundances of different chemical elements are of great importance in chemical evolution studies. Such measurements constrain the relative yields of each element, testing models for nucleosynthesis in stars. The relative variation of two elements which are synthesized in stars of different masses can be used in principle to set constraints on variations in the stellar initial mass function (Scalo 1990). Comparison of the abundance of a secondary element to that of a primary element may provide information on the relative importance of gas inflow and outflow in a galaxy (Edmunds, this volume).

A great deal of data now exists on oxygen abundances in extragalactic H II regions, because the most important ions of oxygen have strong emission lines in the optical part of the spectrum. Most of the other elements seen in ionized nebulae, however, have at least one important ion which does not have strong emission lines in the optical. In those cases, one has to attempt UV or IR measurements (which have been limited in sensitivity) or rely upon photoionization models to estimate correction factors for unobserved ions (often dependent on uncertain input parameters). I have begun a series of studies of individual elements other than oxygen in extragalactic H II regions, using new detector technology to measure previously unobservable ions of a given element, and looking at possible systematic effects in the conversion from ionic abundances to elemental abundances using photoionization models. Here I describe an investigation of nitrogen in H II regions, primarily in metal-poor dwarf galaxies; the complete results appear in Garnett (1990).

Early studies of abundances in H II regions in the Magellanic Clouds (Peimbert and Torres-Peimbert 1974, 1976; Dufour 1975) determined that nitrogen was more depleted than oxygen compared to the Sun and suggested that perhaps the Magellanic Clouds were less evolved than the Galaxy. Numerous observational studies of spiral and irregular galaxies (summarized by Pagel 1985) confirmed the general trend that

N/O decreases as the oxygen abundance decreases. There was some hint that N/O leveled off to a constant value in the dwarf galaxies, but inspection of Figure 3 from Pagel (1985) suggests that the data could just as easily be interpreted by a continuous decrease in N/O.

At the same time, some questions arose regarding derived nitrogen abundances. Most abundance studies had used the approximation

$$\frac{N}{O} \approx \frac{N^+}{O^+}$$

(from Peimbert and Costero 1969) to derive nitrogen abundances from observations of [N II] and [O II]. However, far-infrared measurements of [N III] and [O III] fine-structure lines in Galactic H II regions (Lester *et al.* 1983, 1987) gave values of N^{+2}/O^{+2} which were much larger than the values of N^+/O^+ derived from optical spectra (Shaver *et al.* 1983); also, N^{+2}/O^{+2} showed a gradient across the disk of the Galaxy, unlike N^+/O^+. Furthermore, nitrogen abundances derived from observations of supernova remnants were systematically higher than those derived from optical observations of H II regions. These developments raised questions as to the accuracy of nitrogen abundance measurements derived from optical spectra.

2 BEHAVIOR OF NITROGEN IN H II REGIONS

The present study was undertaken to examine the following questions via a combination of observations and theoretical modeling:

(1) *Is N/O truly uncorrelated with O/H in the irregular galaxies, or does it continue to decrease with decreasing O/H?* This was addressed by obtaining the first measurement of the nitrogen abundance in the blue compact dwarf galaxy I Zw 18, which has the most metal-poor H II regions known. Dufour, Garnett, and Shields (1988) obtained $log(N/O) = -1.25$ in I Zw 18; further observations led to a slight downward revision to $log(N/O) = -1.36$ (Garnett 1990), but the basic conclusion remained the same: N/O does indeed appear to level off to a roughly constant value in the low metallicity galaxies. The observed N/O ratio in I Zw 18 appears to conflict with the hypothesis (Kunth and Sargent 1986) that this object is experiencing its first episode of star formation since nitrogen is produced mainly by intermediate mass stars.

(2) *Do photoionization models suggest an explanation for the discrepancy between the far-IR N^{+2}/O^{+2} measurements and the optical N^+/O^+ measurements?* A grid of photoionization models was constructed to explore the effects of variations of a number of parameters upon the ionization structure of an H II region: the effective temperatures of the ionizing stars, the heavy element abundances in the gas and in the

stellar atmospheres, the ionization parameter, and the choice of stellar atmosphere model. Some of the model results are shown in Figure 1. This figure shows how the ratio of the true N/O in the nebula to N^+/O^+ varies as a function of the O^+ fraction in the model nebula. Figure 1b shows model calculations for solar abundances, while Figure 1a shows calculations for one-tenth solar metallicity. The solid curves (labeled by K and T_{eff} in thousands of K) represent models employing ionizing fluxes from LTE atmospheres computed with ATLAS (Kurucz 1975) and having the appropriate metallicity; the dashed curves (labeled by M and T_{eff}) are models using Mihalas (1972) non-LTE atmospheres, which use only a crude treatment of the heavy element opacities.

Figure 1 shows that metallicity can have an enormous effect on the ionization structure, through modification of the stellar ionizing flux. At one-tenth solar metallicity (Fig. 1a), the differences between ionization models using LTE or non-LTE atmospheres (at a given T_{eff}) are relatively small, and the models suggest that the approximation N/O = N^+/O^+ is probably accurate to $\pm 20\%$. However, at solar metallicity (Fig. 1b), the models with the Kurucz atmospheres behave quite differently from those using the Mihalas atmospheres: N/O can be substantially larger than N^+/O^+ (in other words, N^{+2}/O^{+2} can be much larger than N^+/O^+) in the models with Kurucz atmospheres. These differences (which decrease with increasing effective temperature) occur mainly because of the different treatment of heavy element opacities in the stellar atmospheres; the opacity from bound-free transitions depletes the number of photons capable of ionizing O^+ more severely than those which can ionize N^+. To support this, I show in the figure a set of models computed using the 35,000 K non-LTE atmosphere of Anderson (1985), which includes a more realistic treatment of metal opacity than Mihalas. One finds that N^{+2}/O^{+2} can be much larger than N^+/O^+ in these models also; the models suggest, at least with regard to the ionization of nitrogen, that the LTE models behave similarly to non-LTE models having somewhat smaller T_{eff}. More recent non-LTE atmosphere computations by Kudritzki *et al.* (1990) also indicate that variations in metallicity affect ionizing fluxes.

The results shown in Figure 1 suggest that the effects of changing metallicity on nebular ionization structure are at least partly responsible for the discrepancy between the far-IR measurements and the optical measurements for nitrogen. The galactic radial abundance gradient would lead naturally to the N^{+2}/O^{+2} gradient observed by Lester *et al.* (1987) through its effect on stellar ionizing fluxes. It thus becomes critically important to measure both N^+ and N^{+2} abundances together in a sample of spiral disk H II regions. Unfortunately, the nebulae studied by the IR observers generally have not been the same ones studied in the optical. I conclude at present

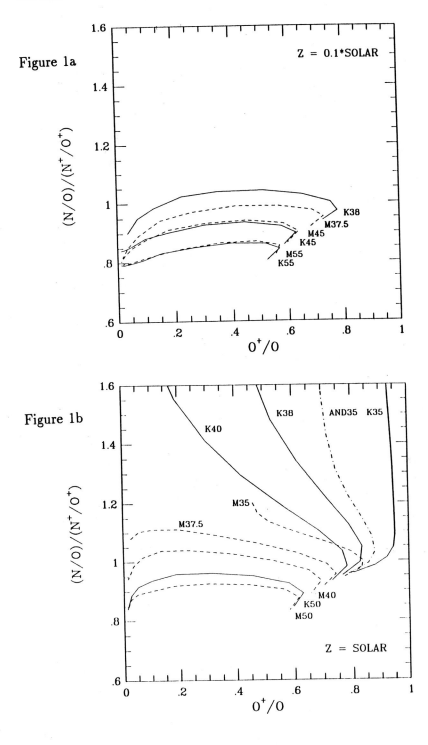

Figure 1a

Figure 1b

that nitrogen abundances in spiral disk H II regions are much more uncertain than is generally believed, and that we can not yet say whether or not radial gradients in N/O exist in spiral galaxies. A coordinated optical-IR study of H II regions in a nearby spiral (M33 is a good candidate) could resolve many of the uncertainties.

The situation appears less confused for low-abundance objects. The photoionization models indicate that the approximation $N/O = N^+/O^+$ is reasonably accurate for low-abundance nebulae. This conclusion is supported by far-IR measurements of [N III] in 30 Doradus by Lester *et al.* (1987), who find that $N^{+2}/O^{+2} \approx N^+/O^+$ in that object. It is therefore probably safe to use $N/O = N^+/O^+$ in nebulae where O/H is less than one-third the solar value. However, additional observations of [N III] in extragalactic H II regions are needed before any general conclusion can be made. It should be possible now, using HST, to detect the UV intercombination multiplet of N III] near 1750 Å in metal-poor H II regions. Another useful project would be to compare N^{+2} abundances derived from both UV and IR observations of planetary nebulae, where both the UV and IR transitions are relatively strong. Such a comparison would provide a valuable test of the accuracy of the atomic data for N^{+2}.

Keeping in mind the uncertainties with regard to nebular ionization for oxygen and nitrogen, I have plotted in Figure 2 an updated version of Pagel's (1985) figure showing the observed trend of N/O as a function of O/H in H II regions. Most of the data for emission line strengths have been taken from the literature (references are given in Garnett 1990), but the derived abundances have been updated to take into account changes in atomic coefficients. The data for blue compact dwarf and irregular galaxies are represented by filled circles and asterisks. The filled circles are those objects having more than one measurement of N/O, while the asterisks are objects having only one N/O measurement, but which have *log O/H* < −4.1; few objects with such low O/H have been observed more than once. The rest of the points represent data for spiral disk H II regions, shown for comparison. Because of the large uncertainties in the ionization corrections for nitrogen in metal-rich H II regions, only a few such objects are shown: the three Milky Way H II regions which have both optical *and* infrared data, and four H II regions in M81 from Garnett and Shields (1987), for which the abundances have been derived using detailed photoionization models of each object.

The mean value for log N/O in the 34 dwarf galaxies shown is *log N/O* = −1.46, with a 1σ statistical scatter of approximately ±0.12 dex. An interesting question is how much of the 0.5 dex spread in N/O at fixed O/H is real. Considering only those objects having more than one observation, an examination of the observational uncertainties combined with the theoretical uncertainty in the ionization correction gives a typical uncertainty in N/O between 0.1 and 0.15 dex for individual objects. With such uncertainties, the observed scatter in N/O is only significant at the $\pm 2\sigma$ level, although comparison of objects such as II Zw 40 and NGC 6822 which have been studied repeatedly suggests that real variations do exist. One particularly interesting case is that of VII Zw 403, which may have the lowest N/O ratio ever measured (Tully *et al.* 1981). A confirmation of the low N/O in this galaxy is needed; if the nitrogen abundance is truly low, VII Zw 403 may be the best candidate yet for a truly primordial galaxy.

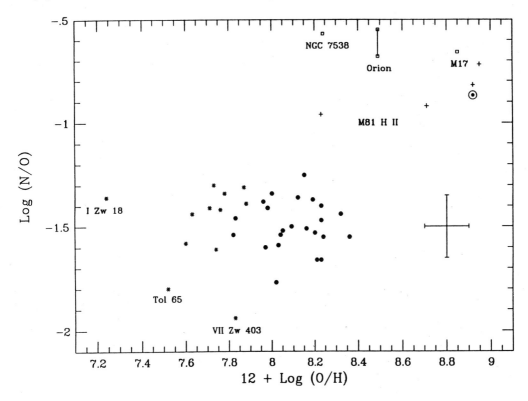

Figure 2

REFERENCES

Anderson, L. S. (1985), *Ap. J.* **298**, 848.

Dufour, R. J. (1975), *Ap. J.* **195**, 315.

Dufour, R. J., Garnett, D. R., and Shields, G. A. (1988), *Ap. J.* **332**, 752.

Garnett, D. R. (1990), *Ap. J.* **363**, 142.

Garnett, D. R., and Shields, G. A. (1987), *Ap. J.* **317**, 82.

Kudritzki, R.-P., Gabler, R., Kunze, D., Pauldrach, A., and Puls, J. (1990), in *Massive Stars in Starbursts*, ed. T. Heckman, C. Leitherer, C. Norman, and N. Walborn (Cambridge: Cambridge University Press), in press.

Kurucz, R. (1975), *SAO Special Report No. 309*.

Kunth, D., and Sargent, W. L. W. (1986), *Ap. J.* **300**, 496.

Lester, D. F., Dinerstein, H. L., Werner, M. W., Watson, D. M., and Genzel, R. L. (1983), *Ap. J.* **271**, 618.

Lester, D. F., Dinerstein, H. L., Werner, M. W., Watson, D. M., Genzel, R. L., and Storey, J. W. V. (1987), *Ap. J.* **320**, 573.

Mihalas, D. (1972), *Non-LTE Model Atmospheres for B and O Stars* (NCAR TN/STR 76).

Pagel, B. E. J. (1985), in *Production and Distribution of CNO Elements*, ed. I. J. Danziger, F. Matteucci, and K. Kjär (Garching: European Southern Observatory), p. 155.

Peimbert, M., and Costero, R. (1969), *Bol. Obs. Tonantzintla y Tacubaya* **5**, 3.

Peimbert, M., and Torres-Peimbert, S. (1974), *Ap. J.* **193**, 327.

Peimbert, M., and Torres-Peimbert, S. (1976), *Ap. J.* **203**, 581.

Scalo, J. M. (1990), in *Windows on Galaxies*, ed. A. Renzini, G. Fabbiano, and J. S. Gallagher (Dordrecht: Kluwer), in press.

Shaver, P. A., McGee, R. X., Newton, L. M., Danks, A. C., and Pottasch, S. R. (1983), *M. N. R. A. S.* **204**, 53.

Tully, R. B., Boesgaard, A. M., Dyck, H. M., and Schemmp, W. V. (1981), *Ap. J.* **246**, 38.

DISCUSSION

E. Skillman: Can the objects that have been observed optically be observed in the IR?

Garnett: Presumably.

Observed Nuclear Abundance Effects in Planetary Nebulae

MANUEL PEIMBERT

Instituto de Astronomía, Universidad Nacional Autónoma de México

ABSTRACT

The enrichment of He and heavier elements in PN shells is reviewed. PN are enriched in He, C and N. It is also possible that extreme population I PN also modify their O abundances. From the study of ten PN of extreme population II it is argued that some of them have modified their O and Ne abundances.

1 INTRODUCTION

The planetary nebula phase corresponds to a main stage in the stellar evolution of intermediate mass stars, those with masses in the $0.8 \leq M/M_\odot \leq 8M_\odot$ range, since most of them go through it (Peimbert 1990).

One of the most important aspects of the study of PN is the determination of their chemical abundances; the chemical composition of a PN is the result of the galactic chemical evolution prior to the formation of the progenitor star and of the modification of the initial abundances by nuclear reactions in their interior. Therefore the study of the chemical composition of PN can be used as a test of models of stellar evolution and of models of galactic chemical evolution. Some of these problems have been discussed by Maciel (1991).

It is possible to divide the elements in the PN shells in three groups: a) those that have been modified by the evolution of the central star like He, C and N, b) those that have been modified in some objects but not in others like O and Ne and c) those that have not been modified like S and Ar.

The evolution of the surface abundances of He, C, N and O for intermediate mass stars during their asymptotic giant branch phase has been predicted by stellar evolution models (*e.g.* Iben 1975, Iben and Truran 1978, Becker and Iben 1979,

1980, Renzini and Voli 1981, Iben and Renzini 1983, Renzini 1984, Wood and Faulkner 1986). The abundances predicted by the models have been compared with observations by many authors (*e.g.* Kaler *et al.* 1978, Aller 1983, Kaler 1983, Peimbert 1985, Clegg 1985, Kaler and Jacoby 1989, 1990, Henry 1989, Kaler *et al.* 1990).

In what follows I will review some of the main aspects related to the enrichment of the shells of PN. Recent discussions on the chemical abundances of PN and their relevance to several aspects of astrophysics are those by Aller (1990) and Peimbert (1991).

2 HELIUM, NITROGEN AND OXYGEN

There is a strong He-N correlation among disc PN (Torres-Peimbert and Peimbert 1977, Barker 1978, Kaler *et al.* 1978, Kaler 1978, 1979, Peimbert and Serrano 1980). This correlation is continuous and seems to comprise objects with main sequence masses of their progenitors in the 1.1 to 8 $M_\odot$ range, where there is a positive correlation between the mass of the progenitor star and the He-N enrichment.

The He-N correlation is due to at least three effects: a) the increase of the He/H, C/H and N/H ratios in the interstellar medium with time, which makes the initial He, C and N abundances of PN progenitor stars increase with their mass due to their lower ages, moreover the higher the C/H ratio the larger the N enrichment due to the first dredge-up, b) the presence of abundance gradients across the galactic disc which modifies the initial abundances (Peimbert and Serrano 1980) and c) the onset of the second dredge-up for stars with masses higher than about 2.5 $M_\odot$ that produces higher He/H and N/H ratios for progenitor stars with higher masses (Iben 1972, Iben and Truran 1978, Renzini and Voli 1981).

The He-N relationship was used to divide PN of type I from those of type II (Peimbert 1978), and to divide type II PN into types IIa and IIb (Faúndez-Abans and Maciel 1987). There are several observational arguments that indicate that indeed type I PN, those with $N(\text{He})/N(\text{H}) \geq 0.125$ or log $N(\text{N})/N(\text{O}) \geq -0.3$, come from more massive progenitors (*e.g.* Peimbert 1991 and references therein).

There is an O/H *versus* N/O anticorrelation for type I PN that seems to be real (Peimbert and Torres-Peimbert 1983, 1987, Henry 1989, 1990, Kaler *et al.* 1990) and which can not be explained by the available models. The O depletion reaches factors of 2-3 while the most favourable theoretical models produce depletions of a factor of 1.4 (Renzini and Voli 1981, Renzini 1984). There are at least two possible explanations for this result: a) part of the correlation could be due to an underestimate of the O/H abundances produced by shock wave enhancements of $\lambda 4363$ of [O III], b) the ratio of the mixing length to the pressure scale height might be larger than the largest values considered by the models increasing the efficiency of the ON cycling.

If the O/H *versus* N/O anticorrelation is due to ON cycling the O/Ne ratio should decrease with the increase of N/O, while if it is due to the use of an electron temperature higher than the real one the O/Ne ratio should *increase* with N/O because the Ne^{++} abundance is more sensitive to the electron temperature than the O^{++} abundance (see equation 1). Accurate O/Ne ratios are needed two choose between these two possibilities.

3 CARBON AND INHOMOGENEOUS PLANETARY NEBULAE

A30 and A78 show inner regions that are extremely H underabundant while the outer regions are normal (Jacoby 1979, Hazard *et al.* 1980, Jacoby and Ford 1983, Manchado *et al.* 1988). The abundances of the most contaminated inner regions of A30 and A78 are presented in table 1. Iben *et al.* (1983) have suggested that some of the central stars of PN experience a final thermal pulse after having achieved a white dwarf configuration, they suggest that A30 and A78 belong to this group and are in the postpulse, quiescent helium burning phase.

Table 1. Chemical abundances of A30 and A78, given in $12 + \log N(X)/N(H)$.

Object	He	C	N	O	Ne	References
Abell 30, 3	12.99	11.89	9.14	9.32	8.83	1
Abell 30	12.95	11.96	9.93	9.63	...	2
Abell 78	12.78	<11.73	9.19	9.89	9.52	1
Abell 78,4	12.43	...	11.09	11.09	10.41	3

References: (1) Jacoby and Ford (1983); (2) Peimbert (1983); (3) Manchado *et al.* (1988).

The C^{++} abundances derived from the C II $3d^2 D - 4f^2 F^0 \lambda 4267$ recombination line are in general higher than those derived from the [C III] $3s^2\ ^1S_0 - 3s3p^3 P_2 \lambda 1907 +$ C III] $3s^2\ ^1S_0 - 3s3p^3 P_1 \lambda 1909$ collisionally excited lines. The difference is typically of about a factor of four, but there are objects for which there is no difference and at least one object where the difference is about a factor of 50. General discussion of this problem have been given in the literature (*e.g.* Torres-Peimbert *et al.* 1980, Barker 1982, French 1983, Kaler 1986, Clegg 1989). The discrepancy could be due to spatial variations in the electron temperature produced by inhomogeneities in the C/H abundance ratio (Peimbert 1983); the regions with higher C/H ratios would be at lower temperatures and would contribute preferentially to λ 4267, while the regions with lower C/H ratios would be at higher temperatures and would contribute proferentially to $\lambda\lambda$ 1907+1909.

Torres-Peimbert *et al.* (1990) have computed a homogeneous model for NGC 4361 that explains the $\lambda\lambda$ (1907-1909) intensity but that predicts a λ 4267 intensity a factor of fifty smaller than observed. Torres-Peimbert *et al.* have also computed an inhomogeneous model with an inner shell with log C/H = -2.0 and an outer shell with log C/H = -3.9, in the inner shell T_e = 11300° K and in the outer shell T_e = 20100° K; this model can reconcile the C^{++} abundances derived from the λ 4267 and $\lambda\lambda$ 1907-1909 emission lines. The inhomogeneous model is in agreement with the very strong C IV emission lines shown by the central star of NGC 4361 (Méndez 1989).

Other examples of inhomogeneous PN are expected since according to Méndez *et al.* (1986) 35% of all the spectroscopically well studied PN central stars belong to the extreme helium rich class.

Another non homogeneous object is NGC 40 where the central star is a Wolf-Rayet with $N(He)/N(C) \approx 15$ and no H (Benvenuti *et al.* 1982); the inner layers of the nebula show very strong C IV λ 1550 lines, which apparently imply an overabundance of carbon, and the outer layers of the nebula show almost normal H, C, N and O abundances (Clegg *et al.* 1983). Moreover to explain the discrepancy between the temperature determined from the stellar nucleus and the much lower value inferred from the ionization of the nebula, Bianchi and Grewing (1987) have suggested the existence of a carbon curtain at the inner edge of the nebula.

Barker (1982, 1983, 1984, 1985, 1986), has found that the $\lambda\lambda$ 4267, 1907, 1909 discrepancy becomes largest closer to the central stars of NGC 6720, NGC 7009, NGC 6853, NGC 3242 and NGC 7662; a C/O ratio decreasing outwards in the parent star and a C rich wind ejected after the main ejection might also help to explain these results.

The C abundances derived from PN emission lines include only the gaseous component in the PN shell, and since part of the C might be locked up in grains inside the nebula, they correspond to lower limits only.

4 HALO PLANETARY NEBULAE

4.1 Oxygen and Neon

In table 2 solar neighborhood abundances are presented, where disk PN are those of type II, and an average population I set of abundances is defined to compare with those of halo PN (type IV). In table 3 the abundances of 10 halo PN relative to the average population I set are presented. In figures 1 and 2 the O/Ar *versus* Ar/H and the N/Ar *versus* Ar/H diagrams for these objects are presented.

Table 2. Chemical abundances of solar neighborhood objects, given in $12+\log N(X)/N(H)$

Object	He	C	N	O	Ne	S	Ar	References
⟨disk PN⟩	11.04	8.3-9.1	8.15	8.70	8.10	7.10	6.60	1,2,3
Orion Nebula	11.01	8.57	7.68	8.65	7.80	7.10	6.65	4,5,6
Sun	...	8.67	7.99	8.92	8.03	7.23	6.69	7,8,9
⟨Pop. I⟩	11.01	8.62	7.84	8.75	7.98	7.15	6.65	10

References: (1) Torres-Peimbert and Peimbert (1977); (2) Aller and Czyzak (1983); (3) Maciel (1991); (4) Peimbert and Torres-Peimbert (1977); (5) Torres-Peimbert *et al.* (1980); (6) Peimbert (1982); (7) Lambert (1978); (8) Lambert and Luck (1978); (9) Meyer (1985); (10) this paper.

The PN in the globular cluster M22 found by Gillett *et al.* (1986, 1989), GJJC-1, deserves further discussion. M22 is a very metal poor globular cluster with [Fe/H] ~ -1.8 (Cohen 1981, Alcaino and Liller 1983). The PN is very faint and only the emission lines λ 3869 of [Ne III] and λ 5007 of [O III] have been detected. From upper limits to the H line intensities Gillett *et al.* (1989) have found that the O/H and the Ne/H ratios are overabundant relative to those of the solar neighborhood by at least one and two orders of magnitude respectively, implying that a cloud rich in O and Ne was ejected by the central star.

Table 3. Chemical abundances of halo PN given in [X/H] = log (X/H) − log (X/H)$_{popI}$

Object	He	C	N	O	Ne	S	Ar	References
K648	−0.01	+0.08	−1.34	−1.05	−1.28	−2.00	−2.35	1,2,3
GJJC-1	...	...	...	>+1.25	>+2.05	...	...	4
BB-1	−0.03	+0.47	+0.50	−0.85	+0.02	−1.45::	−2.05	2,5
H4-1	−0.02	+0.69	−0.09	−0.35	−1.28	−1.95	−1.95	1,2
M2-29	...	...	−0.66	−1.40	−1.20	...	−1.30	6,7
NGC 4361	+0.01	+0.66	−0.48	−0.92	−0.41	...	−0.74	8
NGC 2242	−0.01	−0.23	−0.12	−0.72	−0.20	...	−0.76	8
DDDM-1	−0.07	<-1.52	−0.44	−0.65	−0.68	−0.65	−0.85	9,10
PRMG-1	−0.05	...	...	−0.65	−0.68	...	−0.85	11
PRTM-1	+0.03	<-1.02	<+0.16	−0.35	−0.08	...	−0.45	12

References: (1) Torres-Peimbert and Peimbert (1979); (2) Barker (1980, 1983) (3) Adams *et al.* (1984); (4) Gillett *et al.* (1989); (5) Torres-Peimbert *et al.* (1981); (6) Webster (1988); (7) Peña *et al.* (1991); (8) Torres-Peimbert *et al.* (1990); (9) Clegg *et al.* (1987); (10) Barker and Cudworth (1984); (11) Peña *et al.* (1989); (12) Peña *et al.* (1990).

Based on the high effective temperature of the ionizing star (50,000° K Cohen and Gillett 1989), a small O^+/O^{++} ratio is expected and consequently the effect of the charge exchange reaction $O^{+2} + H^0 \rightarrow O^+ + H^+$ (*e.g.* Pequignot *et al.* 1978, Pequignot 1980) on the O ionization structure can be neglected. Under these conditions $Ne^{++}/O^{++} = Ne/O$ and the Ne/O ratio is given by (Baluja *et al.* 1980, 1981, Pradhan 1974)

$$\frac{N(\text{Ne})}{N(\text{O})} = \frac{I(3869)}{I(5007)} 1.42 \exp (0.71 \times 10^4/T_e), \tag{1}$$

where the temperature term comes from the energy difference of the excited levels and the temperature dependence of the collision strengths.

Only an upper limit of 20000° K has been obtained for the PN in M22 (Gillett *et al.* 1989), therefore we have computed the Ne/O ratio based on equation (1) and for three different electron temperatures, see table 4; for $T_e < 20000°$ K it is found that [Ne/O]> +0.64, furthermore for reasonable T_e values the Ne/O ratio is comparable to that observed in BB-1.

Table 4. Ne to O ratios in halo PN, where [Ne/O] = log (Ne/O) - log (Ne/O)$_{popI}$

Object	$T_e(°K)$	log Ne/O	[Ne/O]
GJJC-1	5 000	+ 0.34	+ 1.11
GJJC-1	10 000	+ 0.03	+ 0.80
GJJC-1	20 000	− 0.13	+ 0.64
BB-1	...	+ 0.10	+ 0.87
H4-1	...	− 1.70	−0.93

From studies of K 648, H4-1 and BB-1, it has been found that Ar and S are underabundant by about two orders of magnitude, while O and Ne are underabundant by about one order of magnitude relative to the solar vicinity (see table 3). Two possibilities have been discussed in the literature to explain the different underabundances: a) that the enrichment of O and Ne in the interstellar medium has proceeded faster than that of Ar, and S and b) that the O and Ne excesses relative to Ar, and S were produced by the progenitors of the PN themselves (Peimbert 1973, 1981, Hawley and Miller 1978, Torres-Peimbert and Peimbert 1979, Barker 1980, 1983, Clegg *et al.* 1987, Clegg 1989, Torres- Peimbert *et al.* 1990). In what follows we will explore the second possibility.

From the study of main sequence stars it has been found that the ratio [el/Fe] in the halo is essentially identical, [el/Fe] $\simeq$ +0.4 for $-3 \leq$ [Fe/H] ≤ -1, for the wide range of α - elements from O through Mg, Si, S, Ca to Ti (Lambert 1989); this result implies that the ratio of two α elements remains unchanged for objects in that very low metallicity range. From the study of galactic and extragalactic H II regions covering a range $-1.5 \leq$ [O/H] ≤ 0, it has been found that the ratios of O, Ne, S and Ar have remained almost constant (*e.g.* Torres-Peimbert *et al.* 1989, Meyer 1989). These two results imply that the relative yields of these α elements did not change considerably during the lifetime of the Galaxy and that most of the abundances of these elements have been due to Type II and Ib supernovae.

From the previous considerations it follows that PN without self-enrichment should lie in a horizontal line with [O/Ar] and [Ne/Ar] equal to cero in figures 1 and 2. That is not the case for K 648, BB-1 and H4-1 in figures 1 and 2 nor for NGC 2242 and NGC 4361 in figure 2. Moreover K 648, BB-1, H4-1, NGC 2242 and NGC 4361 are C rich which supports the idea that these objects are more contaminated by freshly made O and Ne than DDDM-1, PRGM-1 and PRTM-1 that are C poor. There is no C/H determination for M2-29 but based on its position in figures 1 and 2 and on its high electron temperature I would expect this object to be C/H poor.

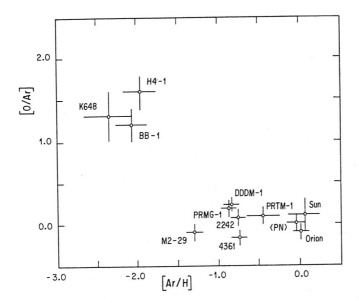

Figure 1. Plot of [O/Ar] *versus* [Ar/H] abundances for Halo PN, where [A/B] = log (A/B) − log (A/B)$_{popI}$. We have also plotted the abundances for the Sun, the Orion nebula and an average for type II PN.

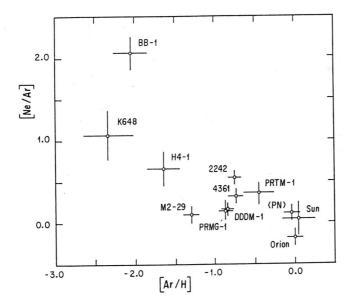

Figure 2. Same as figure 1 but for [Ne/Ar] *versus* [Ar/H].

The very large O/Ar and Ne/Ar overabundances in K 648, H4-1 and BB-1 support the idea that these objects present freshly made O and Ne. The Ne/O ratio varies by almost two orders of magnitude in halo PN, see table 4, this variation can be used to explore the mechanisms of enrichment of O and Ne in stellar evolution models.

4.2 Carbon

Of the seven halo PN with C/O abundance determinations five of them show C/O > 1 (see table 3). These values are difficult to explain because AGB carbon stars are not found in globular clusters nor are they predicted by current AGB models with core mass typical of halo PN, $M_c < 0.6 M_\odot$ (*e.g.* Renzini and Voli 1981, Iben and Renzini 1984). To explain C/O > 1 values Renzini (1989) has proposed that these objects eject their envelope during one thermal pulse. Thermal pulses, also known as helium shell flashes, have often been regarded as possible triggers for the envelope ejection. The mass that could be eyected during one pulse peak for an object with $M_c = 0.54 M_\odot$ is about $0.1 M_\odot$ a value considerably higher than the envelope masses of NGC 4361, NGC 2242, K 648 and the PN in M22 that amount to about 0.074, 0.020, 0.018 and 0.0006 $M_\odot$ respectively (Torres-Peimbert *et al.* 1990, Peimbert 1973, Gillett *et al.* 1989).

4.3 Pregalactic Helium Abundance

Peimbert (1983) and Clegg (1989) have derived the pregalactic or primordial helium abundance Y_p based on the halo PN. To determine Y_p from the observed Y value of a given PN it is necessary to estimate two effects: the helium enrichment of matter previous to the formation of the progenitor star due to galactic evlution, ΔY_{GE}, and the helium enrichment due to the stellar evolution of its progenitor, ΔY_{SE}, *i.e.*

$$Y = Y_p + \Delta Y_{GE} + \Delta Y_{SE} = Y_i + Y_{SE}, \tag{2}$$

where ΔY_{GE} can be approximated by $Z \, \Delta Y / \Delta Z$, Y_i is the initial helium abundance of the star and ΔY_{SE} is given by

$$\Delta Y_{SE} = \Delta Y^1 + \Delta Y^2 + \Delta Y^3, \tag{3}$$

where the terms on the right hand side are due to the three dredge-up episodes

(Renzini and Voli 1981).

DDDM-1 is the best halo PN to determine Y_p because it has the lowest O/H, Ne/H and Ar/H abundance ratios of the two objects without C enrichment. From the line intensities for DDDM-1 by Clegg *et al.* (1987) and assuming that the depopulation rate of the level 2^3S of He^0 is due to radiative transitions to the 1^1S state and to triplet-singlet exchange collisions (Clegg 1987, Peimbert and Torres-Peimbert 1987) it is found that N(He)/N(H) = 0.087, which corresponds to $Y = 0.258$ for $Z = 0.0025$ (see below). From the values of table 3 and the assumptions that O constitutes 60% of the heavy elements with which the progenitor star was formed and that the galactic chemical evolution is characterized by $\Delta Y/\Delta Z = 3.6$ (Torres-Peimbert *et al.* 1989) it follows that $Z = 0.0025$ and that $\Delta Y_{GE} = 0.009$. From the work by Renzini and Voli (1981), the mass of $\sim 0.9 M_\odot$ for the progenitor star of DDDM-1 and the very low C/H ratio it follows that $\Delta Y^1 = 0.02$, $\Delta Y^2 = 0$ and $\Delta Y^3 = 0$. Therefore from equations (2) and (3) it follows that $Y_p = 0.229 \pm 0.012$, where the error includes uncertainties in: ΔY^1, the collisional excitation correction and the $\Delta Y/\Delta Z$ ratio.

The Y_p value derived from DDDM-1 is in excellent agreement with the results of 0.230 ± 0.006 by Torres-Peimbert *et al.* (1989) and of 0.229 ± 0.004 by Pagel and Simonson (1989) derived from extragalactic H II regions.

Due to the high C abundances of the other halo PN it is difficult to estimate ΔY_{SE}, either due to ΔY^3 or to other possible enrichment processes. From a very good value of Y_p it is possible to invert the procedure and to predict the ΔY_{SE} value needed to explain the observed Y values in halo PN, this result would become a strong restriction for stellar evolution models. To derive accurate Y values for H4-1, K 648 and BB-1 the line intensities have to be corrected by non-linearity effects in the detectors used, furthermore the He I line intensities have to be corrected by collisional excitation effects from the 2^3S He^0 level.

REFERENCES

Adams, S., Seaton, M.J., Howarth, I.D., Aurriere, M. and Walsh, J.R. 1984, *Monthly Notices Roy. Astron. Soc.*, **207**, 471.

Alcaino, G. and Liller, W., 1983, *Astron. J.*, **88**, 1330.

Aller, L.H. 1983, in *IAU Symp. 103 Planetary Nebulae*, ed. D.R. Flower (Dordrecht: Reidel) p. 1

Aller, L.H. 1990, *Publ. Astron. Soc. Pacific*, **102**, 1097.

Aller, L.H. and Czyzak, S.J. 1983, *Astrophys. J. Suppl.*, **51**, 211.

Baluja, K.L., Burke, P.G. and Kingston, A.E. 1980, *J. Phys. B*, **13**, 829.

Baluja, K.L., Burke, P.G. and Kingston, A.E. 1981, *J. Phys. B*, **14**, 119

Barker, T. 1978, *Astrophys. J.*, **220**, 193

Barker, T. 1980, *Astrophys. J.*, **237**, 482.

Barker, T. 1982, *Astrophys. J.*, **253**, 167.

Barker, T. 1983, *Astrophys. J.*, **270**, 641.

Barker, T. 1984, *Astrophys. J.*, **284**, 589.

Barker, T. 1985, *Astrophys. J.*, **294**, 193.

Barker, T. 1986, *Astrophys. J.*, **308**, 314

Barker, T. and Cudworth, K.M. 1984, *Astrophys. J.*, **278**, 610.

Becker, S.A. and Iben, I., Jr. 1979, *Astrophys. J.*, **232**, 831

Becker, S.A. and Iben, I., Jr. 1980, *Astrophys. J.*, **237**, 111

Benvenuti, P., Perinotto, M. and Willis, A.J. 1982, in *IAU Symp. 99 Wolf-Rayet Stars*, eds. C. de Loore and A.J. Willis (Dordrecht: Reidel), p. 453.

Bianchi, L. and Grewing, M. 1987, *Astron. Astrophys.*, **181**, 85.

Clegg, R.E.S. 1985, in *Production and Distribution of C,N,O Elements*, eds. I.J. Danziger, F. Matteucci and K. Kjar (Garching: ESO), p. 261

Clegg, R.E.S. 1987, *Monthly Notices Roy. Astron. Soc.*, **229**, 31p.

Clegg, R.E.S. 1989, in *IAU Symp. 131 Planetary Nebulae*, ed. S. Torres-Peimbert (Dordrecht: Kluwer), p. 139.

Clegg, R.E.S., Seaton, M.J., Peimbert, M. and Torres-Peimbert, S. 1983, *Monthly Notices Roy. Astron. Soc.*, **205**, 417.

Clegg, R.E.S., Peimbert, M. and Torres-Peimbert, S. 1987, *Monthly Notices Roy. Astron. Soc.*, **228**, 59.

Cohen, J.G. 1981, *Astrophys. J.*, **247**, 869.

Cohen, J.G. and Gillett, F.C. 1989, *Astrophys. J.*, **346**, 803.

Cohen, J.G. and Liller, W. 1983, *Astron. J.*, **88**, 1330.

Faúndez-Abans, M. and Maciel, W.J. 1987, *Astron. Astrophys.*, **183**, 324

French, H.B. 1983, *Astrophys. J.*, **273**, 214

Gillett, F.C., Jacoby, G.H., Joyce, R.R., Cohen, J.G., Neugebauer, G., Soifer, B.T., Nakajima, T. and Matthews, K. 1989, *Astrophys. J.*, **338**, 862.

Gillett, F.C., Neugebauer, G., Emerson, J.P. and Rice, W.L. 1986, *Astrophys. J.*, **300**, 722.

Hawley, S.A. and Miller, J.S. 1978, *Astrophys. J.*, **220**, 609.

Hazard, C., Terlevich, R., Morton, D.C., Sargent, W.L.W. and Ferland, G. 1980, *Nature*, **285**, 463.

Henry, R.B.C. 1989, *Monthly Notices Roy. Astron. Soc.*, **241**, 453.

Henry, R.B.C. 1990, *Astrophys. J.*, **356**, 229.

Iben, I. Jr. 1972, *Astrophys. J.*, **178**, 433

Iben, I., Jr. 1975, *Astrophys. J.*, **196**, 525.

Iben, I. Jr., Kaler, J.B., Truran, J.W. and Renzini, A. 1983, *Astrophys. J.*, **264**, 605.

Iben, I. Jr. and Renzini, A. 1983, *Ann. Rev. Astron. Astrophys.*, **21**, 271.

Iben, I. Jr. and Renzini, A. 1984, *Phys. Rep.*, **105**, 329.

Iben, I., and Truran, J.W. 1978, *Astrophys. J.*, **220**, 980.

Jacoby, G.H. 1979, *Publ. Astron. Soc. Pacific*, **91**, 574.

Jacoby, G.H. and Ford, H.C. 1983, *Astrophys. J.*, **266**, 298.

Kaler, J.B. 1978, *Astrophys. J.*, **226**, 947.

Kaler, J.B. 1979, *Astrophys. J.*, **228**, 163.

Kaler, J.B. 1983, *Astrophys. J.*, **271**, 188.

Kaler, J.B. 1986, *Astrophys. J.*, **308**, 337.

Kaler, J.B. 1988, *Astrophys. J.*, **226**, 947.

Kaler, J.B., Iben, I. Jr. and Becker, S.A. 1978, *Astrophys. J.*, **224**, L63.

Kaler, J.B. and Jacoby, G.H. 1989, *Astrophys. J.*, **345**, 871.

Kaler, J.B. and Jacoby, G.H. 1990, *Astrophys. J.*, **362**, 491.

Kaler, J.B., Shaw, R.A. and Kwitter, K.B. 1990, *Astrophys. J.*, **359**, 392.

Lambert, D.L. 1978, *Monthly Notices Roy. Astron. Soc.*, **182**, 249.

Lambert, D.L. 1989, in *Cosmic Abundances of Matter*, ed. C.J. Waddington (New York: Amer. Inst. Phys.), p. 168.

Lambert, D.L. and Luck, R.E. 1978, *Monthly Notices Roy. Astron. Soc.*, **183**, 79.

Maciel, W.J. 1991, these proceedings.

Manchado, A., Pottasch, S.R. and Mampaso, A. 1988, *Astron. Astrophys.*, **191**, 128.

Méndez, R.H. 1989, private communication.

Méndez, R.H., Miguel, C.H., Heber, U. and Kudritzki, R.P. 1986, in *IAU Coll. 87 Hydrogen Deficient Stars and Related Objects*, eds. K. Hunger, D. Schönberner, and N. Kamesware (Dordrecht: Reidel), p. 323.

Meyer, J.-P. 1985, *Astrophys. J. Suppl.*, **57**, 151.

Meyer, J.-P. 1989, in *Cosmic Abundances of Matter*, ed. C.J. Waddington (New York: Amer. Inst. Phys.), p. 245.

Pagel, B.E.J. and Simonson, E.A. 1989, *Rev. Mexicana Astron. Astrofis.*, **18**, 153.

Peimbert, M. 1973, *Mem. Soc. R. Sci. Liege Ser 6*, **5**, 307.

Peimbert, M. 1978, in *IAU Symp. 76 Planetary Nebulae*, ed. Y. Terzian (Dordrecht: Reidel), p. 215.

Peimbert, M. 1981, in *Physical Processes in Red Giants*, ed. I. Iben, Jr. and A. Renzini (Dordrecht: Reidel) p. 409.

Peimbert, M. 1982, *Ann. N.Y. Acad. Sci.*, **295**, 24.

Peimbert, M. 1983, in *Primordial Helium*, ed. P.A. Shaver, D. Kunth and K. Kjar (Garching: European Southern Observatory), p. 267.

Peimbert, M. 1990, *Rev. Mexicana Astron. Astrofis.*, **20**, 119.

Peimbert, M. 1991, *Rep. Prog. Phys.*, **54**, in press.

Peimbert, M. and Serrano, A. 1980, *Rev. Mexicana Astron. Astrof.*, **5**, 9.

Peimbert, M. and Torres-Peimbert, S. 1977, *Monthly Notices Roy. Astron. Soc.*, **179**, 217.

Peimbert, M. and Torres-Peimbert, S. 1983, in *IAU Symp. 103 Planetary Nebulae*, ed. D.R. Flower (Dordrecht: Reidel), p. 233.

Peimbert, M. and Torres-Peimbert, S. 1987, *Rev. Mexicana Astron. Astrofis.*, **14**, 540.

Peña, M., Ruiz, M.T., Maza, J. and González, L.E. 1989, *Rev. Mexicana Astron. Astrofis.*, **17**, 25.

Peña, M., Ruiz, M.T., Torres-Peimbert, S. and Maza, J. 1990, *Astron. Astrofis.*, **237**, 454.

Peña, M., Torres-Peimbert, S. and Ruiz, M.T. 1991, in preparation.

Pequignot, D. 1980, *Astron. Astrophys.*, **81**, 356.

Pequignot, D., Aldrovandi, S.M.V. and Stasinska, G. 1978, *Astron. Astrophys.*, **120**, 249.

Pradhan, A.K. 1974, *J. Phys. B*, **7**, L503.

Renzini, A. 1984, in *Stellar Nucleosynthesis*, ed. C. Chiosi and A. Renzini (Dordrecht: Reidel), p. 99.

Renzini, A. 1989, in *IAU Symp. 131 Planetary Nebulae*, ed. S. Torres-Peimbert (Dordrecht: Kluwer), p. 391.

Renzini, A. and Voli, M. 1981, *Astron. Astrophys.*, **94**, 175.

Torres-Peimbert, S. and Peimbert, M. 1977, *Rev. Mexicana Astron. Astrofis.*, **2**, 181.

Torres-Peimbert, S. and Peimbert, M. 1979, *Rev. Mexicana Astron. Astrofis.*, **4**, 341.

Torres-Peimbert, S., Peimbert, M. and Daltabuit, E. 1980, *Astrophys. J.*, **238**, 133.

Torres-Peimbert, S., Peimbert, M. and Fierro, J. 1989, *Astrophys. J.*, **345**, 186.

Torres-Peimbert, S., Peimbert, M. and Peña, M. 1990, *Astron. Astrophys.*, **233**, 540.

Torres-Peimbert, S., Rayo, J.F. and Peimbert, M. 1981, *Rev. Mexicana Astron. Astrofis.*, **6**, 315.

Webster, B.L. 1988, *Monthly Notices Roy. Astron. Soc.*, **230**, 377.

Wood, P.R. and Faulkner, D.J. 1986, *Astrophys. J.*, **307**, 659.

Nebular abundances and the chemical evolution of the Galaxy

W. J. MACIEL

IAG/USP - Caixa Postal 30.627 - CEP 01051 São Paulo SP, Brazil

1 INTRODUCTION

Planetary nebulae (PN) form a subsystem particularly suitable to study the structure and chemical evolution of the Galaxy. As remnants of intermediate mass stars (IMS), PN reproduce some characteristics of a range of interstellar conditions, and are also probes of the inner workings of these stars.

In this work, evidences based on the distribution, kinematics, and chemical composition of PN are taken into account in order to determine average abundances, distance independent correlations, and radial gradients.

2 CLASSIFICATION OF GALACTIC PN

We adopt Peimbert (1978) classification scheme with modifications. The PN are divided into four main types, namely Type I, II, III, and IV. Bulge, or Type V PN (Maciel, 1989) are not treated here.

PN with high mass progenitors (2.4 - 8 $M_\odot$) These are Peimbert's Type I PN. They are the youngest PN, associated with the galactic thin disk. Estimated heights from the galactic plane are $\mid z \mid \ll 1$ kpc, and their peculiar radial velocities are $\mid \Delta v \mid < 60$ km/s (for a discussion on the kinematic properties of PN see Dutra and Maciel, 1990). Their progenitor stars are massive enough to have contributed a significant amount of He and N, so that He/H ≥ 0.125, and log N/O > -0.30. Therefore, they are important in the enrichment of the interstellar medium.

PN with intermediate mass progenitors (1.2 - 2.4 $M_\odot$) These are Peimbert's Type II PN. Still associated with the thin disk ($\mid z \mid < 1$ kpc, and $\mid \Delta v \mid < 60$ km/s), their progenitors have intermediate masses, so that their chemical enrichment must have been small. Therefore, they are important in the study of galactic gradients.

Type II PN may be further divided into Types IIa and IIb (Faúndez-Abans and Maciel, 1987a). Type IIa have been formed from progenitor stars near the high mass bracket, so that they may present slightly enriched abundances of O, S, Ne, and Ar, and some definite enrichment in He and N. Type IIb PN originate from less massive stars, and may be relatively underabundant in the main heavy elements (O, Ne, S), presenting a very small contamination of He and N. Therefore, this subtype is useful in the determination of the pregalactic helium abundance.

PN with low mass progenitors (1 - 1.2 M$_\odot$) These are basically Peimbert's Type III PN. They are generally older, having been formed at a time when the interstellar medium was poorer in heavy elements. Therefore, they may present underabundances of elements such as O, S, and Ne. Moreover, their low mass prevent strong He or N enrichments. They have spatial distribution and kinematic properties associated with the thick disk, $\mid z \mid \leq 1$ kpc and $\mid \Delta v \mid \geq 60$ km/s.

Halo PN These are PN ejected from *very* low mass halo stars (0.8 - 1 M$_\odot$). Only a few objects belong to this class, with low heavy element abundances and large deviations from disk kinematics. They are important in the study of halo-disk abundance variations (not treated here; see Faúndez-Abans and Maciel, 1988) and in the determination of the pregalactic He abundance, although their small number obviously limits their application in statistical studies.

3 THE DATA

The present sample consists of 151 PN, namely: 53 of Type I, 60 of Type II (32 Type IIa and 28 Type IIb), 33 of Type III and 5 of Type IV. The abundances have been collected from the recent literature, with additional data by the IAG/USP group (see references in Faúndez-Abans and Maciel, 1986, 1987b, and also Aller and Keyes, 1987 and Freitas Pacheco et al., 1989; 1990). The adopted abundances are accurately derived averages, with uncertainties of 0.01 for He/H and 0.1–0.2 dex for the heavy elements. The electron temperatures are from the [OIII] lines $\lambda 4959$, 5007 A, with a typical uncertainty of 1000 K.

Distances are the main problem, as usual. We have generally used statistical distances of Maciel (1984), which is the largest available scale. However, whenever a discrepancy is observed between these and existing astrophysical distances (extinction/spectroscopic), the latter have been adopted.

4 AVERAGE ABUNDANCES OF GALACTIC PN

Table 1 shows average abundances of galactic PN. Each type is characterized by a single value, but the observed scatter may produce some overlap. In particular, disk PN are affected by abundance gradients (section 6), and only 5 halo nebulae are included, so that any averages must be viewed with caution.

The average behaviour of the PN types is better seen in Figure 1, where the PN abundances are given relative to galactic H II regions (Peimbert and Torres-Peimbert, 1977; see also Henry, 1990). Some general conclusions are:

(i) The He enrichment in PN is small, except for Type I (and perhaps Type IIa) objects. This is strengthened if a small correction of ~ 0.01 due to collisional excitation is applied (cf. Maciel and Leite, 1990).

Table 1 - Average abundances of planetary nebulae

ratio	Type I	Type IIa	Type IIb	Type III	Type IV
He/H	0.143	0.107	0.106	0.104	0.102
log O/H + 12	8.66	8.70	8.57	8.39	8.02
log N/H + 12	8.61	8.19	7.79	7.78	7.41
log S/H + 12	7.00	6.95	6.84	6.69	5.64
log C/H + 12	8.77	8.81	8.75	8.51	8.54
log Ne/H + 12	8.02	8.07	7.91	7.72	7.14
log Ar/H + 12	6.63	6.46	6.30	6.12	5.03
log Cl/H + 12	5.43	5.31	5.03	5.06	
log N/O	-0.07	-0.50	-0.76	-0.63	-0.59
log S/O	-1.69	-1.77	-1.74	-1.77	-2.36
log C/O	0.12	0.12	0.19	0.03	0.54
log Ne/O	-0.64	-0.63	-0.66	-0.68	-0.88
log Ar/O	-2.07	-2.24	-2.27	-2.27	-2.99
log Cl/O	-3.20	-3.41	-3.55	-3.39	

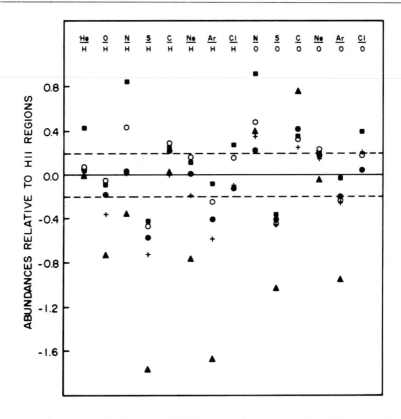

Figure 1 – Abundances of PN relative to H II regions. All differences are given as the logarithm of the abundances except for He/H, where the straight number ratio is plotted. Symbols are as follows: Type I: squares; Type IIa: empty circles; Type IIb: filled circles; Type III: crosses; Type IV: filled triangles.

(ii) N is enriched in PN, especially for Types I and IIa; for Types IIb and III, N abundances are essentially the same as those of H II regions.

(iii) C is somewhat enriched, especially for Types I-II. The enrichment is lower than that of N, probably due to C-N conversion. However, carbon abundances are not accurately known (cf. Peimbert, 1990; this conference).

(iv) Oxygen is "normal" or slightly depleted in PN of Types I-II. For Types III-IV the depletion is stronger, probably reflecting the poorer interstellar conditions at the time of formation of the central stars.

(v) From Table 1, we have log $(C+N)/H + 12 = 8.57, 8.76$, and 9.00 for halo, disk (Types II-III) and Type I PN, respectively. Adopting the value 8.59 for H II regions (Peimbert and Torres-Peimbert, 1977) we see that for Type IV and disk PN the total C + N agree with the interstellar values within the errors, assuming that the chemical evolution has been slow since the formation of the PN central stars. For planetary nebulae of Type I, a larger discrepancy is obtained, which can be only partially explained by uncertainties in the C abundances. On the other hand, log $(C+N+O)/H + 12 = 8.68, 8.97$, and 9.16 for halo, disk and Type I PN, comparable to the H II region value (8.98), within the uncertainties. Therefore, the use of C +N +O abundances suggests that some O is converted into N in PN.

(vi) Ne follows closely the O abundances, especially for Types I-III.

(vii) Ar and S generally behave like O, with an important difference: their underabundance is stronger, suggesting that the interstellar enrichment of O and Ne has proceeded faster than that of Ar and S (cf. Torres-Peimbert, 1984).

(viii) Cl seems to be also similar to O and Ne in opposition to S and Ar, although the Cl abundances are scarce and less accurate.

5 DISTANCE INDEPENDENT CORRELATIONS

(a) Elements synthesized by large mass stars: Ne, S, Ar, and Cl

According to stellar evolution and nucleosynthesis calculations (cf. Renzini and Voli, 1981; Wood and Faulkner, 1986), IMS are not expected to produce significant amounts of oxygen and of S, Ne, Ar and Cl. The present analysis indicates that the ratio O/H is well correlated with the corresponding ratios of these elements, as exemplified by Figure 2 for Ar/H. The slopes are essentially equal to one, so that these elements can be considered as tracers of interstellar abundances, as is oxygen. Secondary components are probably ruled out, since the ratios relative to oxygen are essentially constant. This allows some investigation on the Initial Mass Function and chemical yields of IMS to be done, based on results derived for PN.

(b) Elements synthesized by intermediate mass stars: He, N, C

The synthesis of He, N and C is partially made by IMS, which can be seen from Figure 1 and Figure 3, where N/O is given as a function of He/H. There

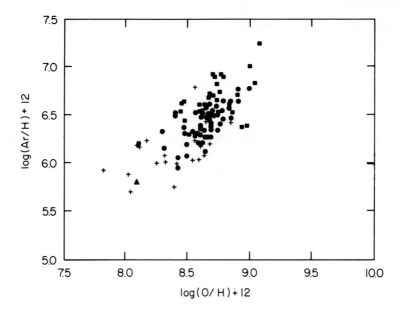

Figure 2 – log (Ar/H) + 12 as a function of log (O/H) + 12 (symbols as in Figure 1).

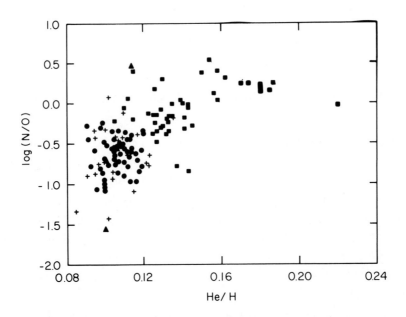

Figure 3 – log (N/O) as a function of He/H (symbols as in Figure 1).

is an agreement with predictions of stellar evolution, in the sense that PN with the highest He/H and N/O ratios have massive progenitors. In this figure, the separation of the types is clear: Type I PN on the top, Type II in the middle, with Type IIa on top of IIb, and Types III and IV on the bottom left.

Figure 4 shows N/O x O/H, which is essentially a scatter diagram if all PN are taken together. For Type II a positive correlation is obtained, as expected by simple models of chemical evolution, if N were essentially secondary. Here again Type IIa can be distinguished, lying on top of Type IIb nebulae. For Type I (and also for Type III !) there is some tendency for an anticorrelation, as proposed by several authors (see for example, Peimbert, 1984; Henry, 1990). The anticorrelation can be expected if N increases in these objects partially via the ON cycle instead of the CN cycle. This is confirmed by a very good correlation between N/O x N/H, with some Type I PN deviating from the general trend.

A plot of C/O x He/H confirms some carbon production. Central stars in the low mass range (Types II-IV) agree with theoretical predictions of Renzini and Voli (1981), showing a higher spread in C/O than in He/H. Type I PN have varying C/O ratios, and halo PN generally present a rather large C abundance, suggesting that the third dredge-up is efficient even for low mass stars (Peimbert, this conference).

(c) Pregalactic He abundance and enrichment ratio

Maciel (1988) and Maciel and Leite (1990) have shown that Type IIb planetary nebulae can be used in addition to H II regions in order to make a better estimate of the pregalactic helium abundance Y_p and of the helium to metals enrichment ratio $\Delta Y/\Delta Z$. Recent results (Maciel and Leite, 1990) show that $Y_p = 0.230 \pm 0.004$ (1σ) and $\Delta Y/\Delta Z = 3.4 \pm 0.4$, in agreement with independent determinations (Pagel and Simonson, 1989; Peimbert, 1990; this conference; Reeves, this conference).

6 GALACTIC GRADIENTS

Electron temperature gradients are known for Type II PN (Maciel and Faúndez-Abans, 1985). The present sample gives $dT_e/dR = 530$ K kpc^{-1} for 57 PN, similar to the observed value for H II regions. Type I objects follow the same trend, but several PN have higher temperatures than expected for their position, which reflects the higher temperatures of their central stars and the influence of shock waves on temperatures derived from the [O III] $\lambda 4363/5007$ A ratio (Peimbert, 1990). The lower heavy element abundances of Type III PN produce high temperature objects, especially if their central stars are also hot.

The *temperature* gradient reflects the *abundance* gradients observed. Radial gradients from the main heavy elements have been studied by Faúndez-Abans and Maciel (1986, 1987b), and the results presented here leave no doubt as to their

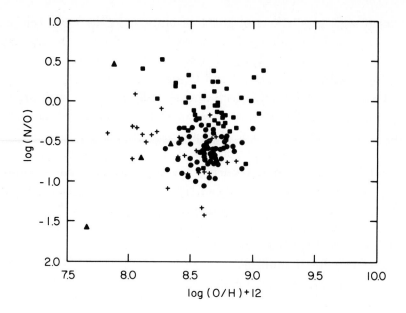

Figure 4 – log (N/O) as a function of log (O/H) + 12 (symbols as in Figure 1).

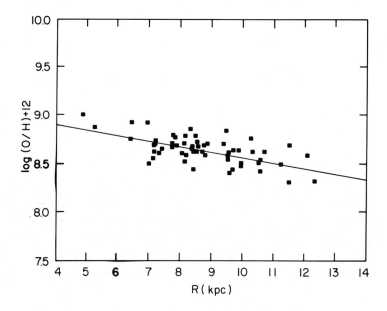

Figure 5 – O/H radial gradient ($R_0 = 8.5$ kpc) for Type II PN.

existence. As an example, Figure 5 shows the O/H gradient (with $R_0 = 8.5$ kpc) for Type II nebulae, where $d\log(O/H)/dR = -0.06$ kpc^{-1}, in agreement with earlier results by Faúndez-Abans and Maciel (1986).

The elements S (cf. Edmunds, this conference), Ne, Ar and Cl present very similar gradients for samples of 40-60 nebulae. For N, the same gradient is obtained for Type IIb PN (Faúndez-Abans and Maciel, 1987a). In view of the previous discussion on carbon, it can be expected that no C gradients can be derived. Plots of C/H x R for Type II PN are essentially flat, so that the C abundances are changed during the stellar evolution in a more complex way. On the other hand, the He/H gradient (Faúndez-Abans and Maciel, 1986) is also confirmed, amounting to dHe/dH $= -0.0025$ kpc^{-1}, somewhat lower than estimated earlier.

Acknowledgements – I thank J. A. Freitas Pacheco, C. C. M. Leite and C. M. Dutra for several discussions. This work was partly supported by CNPq and FAPESP.

References

Aller, L.H., Keyes, C.D., 1987, *Ap. J. S.* **65**, 405

Dutra, C.M., Maciel, W.J., 1990, *Rev. Mex. Astron. Astrof.* (in press)

Faúndez-Abans, M., Maciel, W.J., 1986, *A. Ap.* **158**, 228

Faúndez-Abans, M., Maciel, W.J., 1987a, *A. Ap.* **183**, 324

Faúndez-Abans, M., Maciel, W.J., 1987b, *Ap. S. S.* **129**, 353

Faúndez-Abans, M., Maciel, W.J., 1988, *Rev. Mex. Astron. Astrof.* **16**, 105

Freitas Pacheco, J.A., Costa, R.D.D., Maciel, W.J., Codina-Landaberry, S.J., 1989, *An. Acad. Bras. Ci.* **61**, 389

Freitas Pacheco, J.A., Maciel, W.J., Costa, R.D.D., Barbuy, B., 1990, (preprint)

Henry, R.B.C., 1990, *Ap. J.* (in press)

Maciel, W.J., 1984, *A. Ap. S.* **55**, 253

Maciel, W.J., 1988, *A. Ap.* **200**, 178

Maciel, W.J., 1989, *IAU Symp. 131*, ed. S. Torres-Peimbert, Kluwer

Maciel, W.J., Faúndez-Abans, M. 1985, *A. Ap.* **149**, 365

Maciel, W.J., Leite, C.C.M., 1990, *Rev. Mex. Astron. Astrof.* (in press)

Pagel, B.E.J., Simonson, E.A., 1989, *Rev. Mex. Astron. Astrof.* (in press)

Peimbert, M., 1978, *IAU Symp. 76*, ed. Y. Terzian, Reidel

Peimbert, M., 1984, *Rev. Mex. Astron. Astrof.* **10**, 125

Peimbert, M., 1990, *Rep. Prog. Phys.* (in press)

Peimbert, M., Torres-Peimbert, S., 1977, *M. N. R. A. S.* **179**, 217

Renzini, A., Voli, M., 1981, *A. Ap.* **94**, 175

Torres-Peimbert, S., 1984, *Stellar Nucleosynthesis*, ed. C. Chiosi, A. Renzini, Reidel

Wood, P.R., Faulkner, D.J., 1986, *Ap. J.* **307**, 659

Abundance ratios in Planetary Nebulae and the Mass of the Central Stars

G. STASIŃSKA[1]), R.TYLENDA[2])

[1]) DAEC, Observatoire de Paris-Meudon, F-92195 Meudon Cedex, France
[2]) Copernicus Astronomical Center, Chopina 12/18, PL-87100 Toruń, Poland

Abstract: By comparing observational data to theoretical evolutionary tracks, we have determined the masses of the central stars for a sample of about 90 planetary nebulae. We have then studied the relation between these masses and the abundance ratios in the nebulae. We found that N/O is definitely correlated with the central star mass, while Ne/O is not. There is also a marginal correlation of the star mass with O/H.

1 INTRODUCTION

From theoretical considerations, the nitrogen to oxygen ratio in planetary nebulae is expected to be related to the progenitor mass (Renzini and Voli, 1981). The mass of the central star is also believed to be somehow linked to the progenitor mass (see e.g. Weidemann, 1990). Therefore, one expects a correlation between the nitrogen to oxygen ratio in the planetary nebulae and the masses of the central stars. A number of authors have claimed that the position of nitrogen rich planetary nebulae in the HR diagram indeed supports this view (e.g. Kaler, 1983, Gathier and Pottasch, 1989, Kaler and Jacoby, 1989). However, the interpretation of the observed HR diagram of planetary nebulae nuclei in term of masses is not straightforward, and the effect of systematic errors has to be taken into account .

We have reexamined this problem by considering all the planetary nebulae for which there were published data allowing to derive simultaneously the abundances with reasonable acuracy and the parameters needed to derive the central star masses. We have estimated the star masses using several diagrams - i.e. not only the HR diagram - , and in a manner which eliminates the effects of systematic errors.

2 THE NEBULAR ABUNDANCES AND THE CENTRAL STAR PARAMETERS

The references to the observational data can be found in Stasińska and Tylenda (1990, herinafter ST90) .

The abundances were derived from optical spectra of planetary nebulae using standard empirical methods (e.g. Torres-Peimbert and Peimbert, 1977). The line emissivities were calculated with the electron temperatures deduced from the [O III] ratios and the electron densities deduced from the [S II] ratios. In order to deal with a sample where the uncertainties on the abundances are not too important, we have eliminated objects with a density greater than $2\ 10^4\ cm^{-3}$ and/or a temperature greater than $2\ 10^4 K$.

The stellar parameters were derived as explained in ST90: the effective temperature T_{eff} was obtained with the Zanstra method for He II or H I; the luminosity L, the absolute visual magnitude M_V and the age were computed in a straightforward manner, assuming Shklovsky (1956) distances.

3 DETERMINATION OF THE CENTRAL STAR MASSES

The fact that the theoretical evolution of the nuclei of planetary nebulae is so strongly mass dependent leads to the idea that the comparison of observations with theoretical evolutionary tracks may provide an estimate of this mass. But the observational determination of the stellar parameters suffers from systematical biases due to the presence of the surrounding nebula (the stellar luminosity and age are overestimated for ionization-bounded nebulae, whereas the effective temperature, and consequently also the luminosity are underestimated for density-bounded nebulae). Therefore, we have compared the observations not directly to the theoretical tracks followed by the central stars, but to tracks which we have constructed by deriving the parameters of the model central stars from the properties of a surrounding model nebula in exactly the same way as this was done for observed objects.

The stars were assumed to evolve according to the theoretical models of post AGB stars computed by Schönberner (1981, 1983). The standard model nebula was a simple geometrical model consisting of an uniform density shell of mass $0.2M_\odot$, having a constant thickness $\Delta R/R=0.3$ with an outer boundary expanding at 20km s^{-1}. Note that the properties of this standard model nebula are compatible with the observations of planetary nebulae in the galactic bulge (Stasinska et al., 1991).

We have thus determined the apparent theoretical tracks in the (log L, log T_{eff}), (M_V, log t_{exp}) and (log T_{eff}, log f) planes (f being a distance-independent parameter, see ST90). By placing the observational points in the same diagrams, we were then able to derive three different estimates of the central star masses for each planetary nebula of our sample. A careful study of these results allowed us to attribute for each star a "best mass estimate", M*.

A more detailed description and discussion of the whole procedure appears in ST90 and Tylenda et al. (1991).

4 ABUNDANCE RATIOS AND THE MASSES OF THE CENTRAL STARS

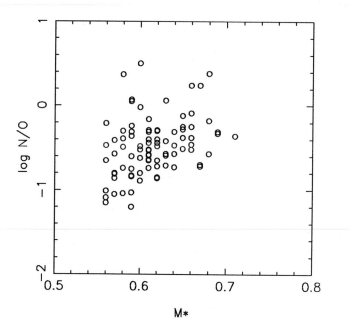

Fig. 1: N/O versus M*

Figure 1 shows the values of the N/O ratios as a function of M*. We see that these two parameters are correlated. Statistical tests show this correlation to be significant above the 99.9% level. Considering that real objects may deviate substantially from our adopted standard model for the evolution of planetary nebulae, which may introduce relatively large uncertainties in our mass determinations, the true correlation between N/O and M* is certainly much stronger than seen in this figure. Note, by the way, that we do not see any clearcut separation between the type I planetary nebulae as defined by Peimbert (1978) and the other nebulae of the sample.

If, qualitatively, the trend between N/O and M* is in agreement with what was expected, there remains a quantitative problem. Our sample of planetary nebulae contains only objects with central star masses smaller than $0.7M_\odot$, and in about 40% of them, the N/O ratio is larger than 0.4. The theoretical dredge-up models of Renzini and Voli (1981) do not predict such large values for core masses smaller than $0.7M_\odot$. One might of course question the N/O abundance ratios derived empirically from optical spectra of planatary nebulae. From a study of HII regions, Rubin et al. (1988) have found that the N/O ratios derived from the far infra-red lines tend to be substantially larger than the ones derived from optical studies. If this were confirmed, and if it were proved that optical data give too small N/O ratios, the problem would be even worse.

Note, in passing, that the upper limit of the star masses in our sample is simply due to the fact that, by June 1990, there were no published observations allowing a good determination of N/O in nebulae with massive nuclei. In a more recent study, Kaler et al. (1990) exhibit an HR diagram for large planetary nebulae with derived N/O abundances in which about 20 objects seem to have nuclei more massive than 0.7M☉.

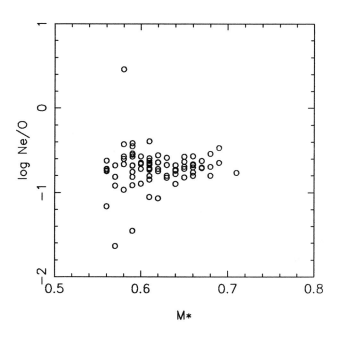

Fig. 2: Ne/O versus M*

Figure 2 shows the values of the Ne/O ratios as a function of M*. No correlation is visible there. This means that, if the relative abundances of Ne and O were to be significantly altered by nucleosynthesis in intermediate mass stars (see Clegg, 1989 and Henry 1989), the efficiency of the process(es) involved would not be particularly related to the final core mass.

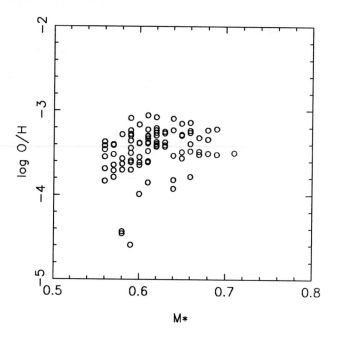

Fig. 3: O/H versus M*

Finally, Figure 3 displays the values of the O/H ratio versus M*. There seems to be a tendency for this ratio to increase with M*, and this tendency subsists at the 90% level of significance even when one disregards the 3 points with the lowest values of O/H. This can be interpreted as due to the fact that more massive nuclei originate from more massive progenitors which, being younger than less massive ones, have formed out of a medium richer in oxygen. Part of the dispersion observed in Figure 3 would be due to the fact that the planetary nebulae of our sample have different locations in the Galaxy.

References

Clegg.R.E.S.:1989, *"Planetary Nebulae"*, IAU Symp. 131, ed Torres-Peimbert (Reidel)

Gathier,R., Pottasch, S.R., 1989, *Astron. Astrophys.* **209**, 369

Henry, R.B.C.:1989, *Mon. Not. R. astr. Soc.* **241**, 453

Kaler,J.B.: 1983 *Astrophys. J.* **271**, 188

Kaler,J.B., Jacoby,G.H.: 1989, *Astrophys. J.* **345**,871

Kaler,J.B., Shaw, R.A., Kwitter,K.B.: 1990, *Astrophys. J.* **359**,392

Peimbert, M.: 1978,*"Planetary Nebulae"*, IAU Symp. 76, ed. Terzian (Reidel)

Renzini,A., Voli,M.: 1981, *Astron. Astrophys.* **94**, 175

Rubin, R.H., Simpson, J.P., Erickson, E.F.,Haas, M.R:1988, *Astrophys. J.* **327**, 377

Schönberner, D.: 1981, *Astron. Astrophys.* **103**, 119

Schönberner, D.: 1983, *Astrophys. J.* **272**, 708

Shklovsky,I.S.: 1956, *Astr. Zh.* **33**, 222 and 315

Stasińska,G., Tylenda, R.:1990 , *Astron. Astrophys.* in press (ST90)

Stasińska,G., Tylenda, R., Acker, A., Stenholm, B.: 1991 , *Astron. Astrophys.*, subm.

Torres-Peimbert,S., Peimbert, M.: 1977, *Rev. Mex. Astron. Astrophys.* **2**,181

Tylenda, R., Stasińska,G., Acker, A., Stenholm, B.: 1991 , *Astron. Astrophys.*, subm.

Weidemann,V.:1990, *Ann.Rev. Astron.Astrophys.* **28**, 103

DISCUSSION

D.Garnett: Harriet Dinerstein and Mike Werner have measured N^{+2}/O^{+2} from the far infra-red lines in several planetary nebulae. A preliminary analysis of NGC2440 shows that N^{+2}/O^{+2} from the infra-red lines agrees fairly well with N^{+2}/O^{+2} from UV measurements, and also with N^{+}/O^{+} from optical measurements, so there may not be a problem with N/O determinations in planetaries

Understanding Composite Stellar Populations

D. ALLOIN

Observatoire de Paris, Section de Meudon

1. INTRODUCTION

I am particularly happy to present this review for B. Pagel who has been, over the past five years, such a careful reader of some pieces of the work I shall describe hereafter.

Rather than to provide a complete review on stellar populations in galaxies, I shall restrict the scope of this presentation to the analysis of composite populations in galaxies where individual stars cannot be resolved. I shall first discuss the current methods used to analyze composite stellar populations. Then I shall describe more extensively a new approach which has proven to be a powerful tool to understand composite stellar populations resulting from a high star formation rate in the past. The cosmological impact of this work is obvious : as it brings us a better knowledge of giant elliptical galaxies in our local environment, any confrontation with their high-redshift counterparts may reveal intrinsic cosmological effects in addition to the passive evolution of galaxies.

Our main interests in analyzing composite stellar populations are indeed to recover the chemical evolution and the star formation history of galaxies : some kind of an observational point of view on galaxy formation and evolution processes.

A composite stellar population is the present-day result of successive stellar generations which have occured in a given object : this succession is represented by the star formation rate SFR(t). For any stellar generation, the duration of the star forming episode is assumed to be small with respect to its age. And because we are looking today at the evolved products of these stellar generations, *age is the parameter to consider first*. The stellar generation formed at a given epoch in the life of the galaxy, arises from gas with some metallicity enrichment Z which increases from one generation to the next. In places where the SFR has been quite large, Z can reach values substantially above that in the solar neighbourhood. Then, one expects the star formation processes and stellar evolution of the corresponding stellar generation to be modified according to the value of Z. Therefore, *the metallicity Z is a second parameter* to be taken into account in the analysis of astrophysical objets where the SFR has been large.

Let us provide a few observational facts in support of the latter argument :

(i) the spectrum of massive elliptical galaxies exhibits huge metallic features which cannot be reproduced by any spectral mixture from stellar libraries where Z is only solar or sub-solar. This led Spinrad and Taylor, as early as 1971, to invoke the existence of super metal-rich stars, (ii) the analysis of HII regions across the disc of spiral galaxies indicates the presence of an abundance gradient of the heavy elements. Values of the metallicity extrapolated to the centre can be as large as four times the solar one (Belley and Roy, 1991) as shown on Fig.1, (iii), M giant stars observed in the bulge of our Galaxy appear to be substantially more metal-rich than in the solar neighborhood (Frogel, 1988 ; Rich, 1988).

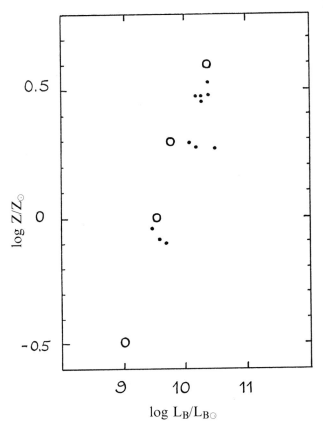

Fig. 1 : The metallicity vs luminosity relationship, for the central regions of galaxies. The luminosity L_B refers to the entire galaxy and is provided on the $H_0 = 100$ km s^{-1} Mpc^{-1} distance scale. Black dots correspond to the metallicity values for a sample of spiral galaxies, extrapolated from the abundance gradient obtained through HII regions analysis (Belley and Roy, 1991). Open circles represent the maximum metallicity enrichment in a typical 1 kpc central region, met by a sequence of early type galaxies from giant to dwarf. These values have been derived from stellar population analysis (Bica, 1988).

2. CURRENT METHODS IN THE ANALYSIS OF COMPOSITE STELLAR POPULATIONS

Two main directions have been followed up to now and are described below :

2.1. Evolutionary Synthesis

This method consists in building a model which starts with the mass of material, M, available to form the galaxy. Stars are produced at a rate SFR(t), with a certain initial mass function (IMF) characterized by its slope and its low and high stellar mass cutoffs ; then, the stars are let to evolve according to stellar evolution theory. It is possible to follow, step by step, the chemical evolution of the galaxy and to track the amount of gas available for subsequent star formation. The main parameter to adjust is SFR(t). At any epoch, one can build the composite H-R diagram which will contain all the evolved products of the previous successive stellar generations. If, for each component in this H-R diagram {characterized by its mass, age and Z} some common *observable* quantity can be collected, such as broad band photometry or spectral information, one will be able by summing up all these components to predict the to-day state of the observable from the composite stellar population modeled. From a confrontation between the prediction and the observation, one adjusts SFR(t). This is a schematic description of what is called evolutionary synthesis, developed since the early 70's (Tinsley and Gunn, 1972 ; Rocca-Volmerange et al., 1981 ; Bruzual, 1983 ; Arimoto and Yoshii, 1986) and available through large computer codes. This kind of approach is appealing for its consistency. Yet, the values of many parameters have to be assumed and the observables used to constrain the model may not always be stringent enough. For example, it is well established now that in a very simple case of composite stellar population such as in a star cluster, broad-band photometry does not allow to separate age from metallicity effects. In general in these codes, the impact of metallicity changes with time is not fully taken into account : in particular, its incidence on the IMF is not considered. Finally, when the point is reached in the evolutionary synthesis, of predicting the characteristics of the light from the composite population, - the observables - , some components in the composite H-R diagram cannot be taken into account because they are not observed individually. This is the case of stars with a metallicity above solar, with the exception of M giants.

The apparent uniqueness of the solution in evolutionary synthesis reflects the way the model is built, from M to a composite luminosity L. In fact, along the computation, an error-bar should be attached to the value of every parameter. The influence of the uncertainty on each parameter should be examined on the final composite light prediction (for a same SFR(t)). But also the combined effect of all the error-bars on the parameter values should be considered, in order to really test the ability of the observables commonly used to constrain the SFR.

Evolutionary synthesis techniques benefit continuously from any progress in the field of stellar evolution or star formation, as this reduces the error-bars on parameters used in the computation. The problem of missing representations for some components in the composite H-R diagram, such as for stars with a large metal content, could be solved by introducing modeled spectra or colors from stellar atmosphere computations. This has been attempted already at some spectral features like the CaII infrared triplet (Smith and Drake, 1990 ; Erdelyi-Mendes and Barbuy, 1990), and for the spectral energy distribution (Barbaro and Olivi, 1986). This solution looks promising.

2.2. Stellar Population Synthesis

In stellar population synthesis, the aim is to reproduce the present-day light, L, from a composite stellar system by the mixture of individual stars from a star library. In a sense, this can be seen as a static approach because evolutionary aspects of the problem are not taken into account in a systematic way.

Although their degree of complexity has increased with time, stellar population syntheses directly stem from the pioneering work of Whipple (1935), Stebbins and Whitford (1948) and de Vaucouleurs and de Vaucouleurs (1959). Many stellar population syntheses have been performed over the past twenty years (Mc Clure, 1969 ; Spinrad and Taylor, 1971 ; Alloin et al., 1971 ; Faber, 1972 ; Williams, 1976 ; O'Connell, 1976 ; Turnrose, 1976 ; Rose, 1985 ; Pickles, 1985). Differences among these studies are related to the star library content, the choice of the observables to be fitted, either broad-band photometry or spectral energy distribution or spectral features, and the algorithm used to find a solution.

The main difficulty in this approach is to input astrophysical constraints to populate the composite H-R diagram in a way which is consistent with stellar evolution and chemical evolution within a galaxy. The solution derived through stellar population synthesis is strongly dependent on the completeness of the star library. In addition, as many kinds of stars enter the composite system, much more than the number of observables, the uniqueness of the solution can be severely questionned. One way to improve this situation is certainly to add astrophysical constraints which will link groups of stars together, and to increase the number of observables to be fitted. Yet, ideally one would like to use a complete star library including all spectral types and luminosities over an extended range of Z, hence increasing the number of components to be combined. It seems difficult in this method to reduce much the number of free parameters. Therefore, population synthesis from a star library should be used with caution, or in a situation for which the star library is complete and astrophysical constraints are easy to apply.

3. AN ALTERNATIVE : THE STAR CLUSTER METHOD

3.1. Outline of the Method

Facing the limitations of the methods just described, we developed a new approach with the aim of :

(i) retaining the evolutionary aspect of the problem,

(ii) reducing the number of free parameters,

(iii) taking into account the metallicity as a parameter.

This method starts with the characteristics of the composite light, L, to be reproduced and is carried out to the derivation of SFR(t). Instead of using a star library, we use a star cluster library parametrized by age and Z. Each of the successive stellar generations will be represented by the integrated properties of a star cluster at the corresponding age and metallicity. This allows to retain astrophysical constraints of the problem : indeed, the star cluster is formed with an IMF which, although not explicit, does not have to be assumed in any manner and for which the Z dependence is included. To characterize the integrated light from the star cluster, we consider spectral signatures rather than broad-band photometry (Hartwick, 1980), as spectral features are better discriminators of age and metallicity which are the main parameters in this approach.

3.2. The Star Cluster Base and the Grid of Integrated Spectral Properties

We have collected at the European Southern Observatory (La Silla) integrated spectra from 3700 to 9600Å, for a sample of 63 star clusters, at a 12Å spectral resolution matching the velocity dispersion observed in central regions of massive galaxies. This sample is made of globular clusters and a few, well-concentrated, rich open clusters in our Galaxy ; as well, we have considered clusters from the Magellanic Clouds. They span a range in age from 2×10^6 yr up to 1.6×10^{10} yr and a range in metallicity $-2 \leq [Z/Z_\odot] \leq 0.1$ (Bica and Alloin,

1986a, 1987a), thanks to the inclusion of clusters towards the bulge of our Galaxy, like NGC 6528, which exhibit a high metal content (may be even larger their 0.1). The age and metallicity attached to each cluster are taken from the literature and their determination depend on stellar evolution theory. Would the reference points of the age or metallicity scale be modified, these changes could be implemented easily, producing a change in SFR(t), while the relative contributions from all components in our synthesis would remain the same.

We have collected in the same conditions, spectra from the central regions (~ 1kpc) of a 164 galaxy sample from the Shapley-Ames catalogue (Bica, Alloin, 1987b). The observed metallic features for massive galaxies (30% for the galaxy sample) call for a metal content possibly as large as $[Z/Z_\odot] \sim 0.6$. Therefore, the actual star cluster base is not fully suitable to represent such populations. We have overcome this difficulty in the following way : we have analyzed the behaviour with age and Z, of 88 consecutive windows across the spectrum, which isolate atomic, molecular and continuum features. From their behaviour, at every age, over the metallicity range $-2 \leq [Z/Z_\odot] \leq 0.1$, we have extrapolated their value at $[Z/Z_\odot] \sim 0.6$. This procedure led us to build a *grid* of the star cluster spectral properties, extended to a high metallicity, from which elements are extracted at age and metallicity bins, to be used in the population synthesis (Bica and Alloin, 1986b).

3.3. Synthesis Using Elements from the Grid of Star Cluster Spectral Properties

The information stored in the full set of spectral windows is highly redundant and must be sorted out. Through the analysis of the equivalent width W = f (age, Z) for all spectral windows, we have been able to recognize the features which are primarily sensitive to one only of the two parameters Z or age : they are the best to constrain the population synthesis. Also, strong features which can be measured with precision should be preferred, in order to better focus on a solution. And finally, the selected windows should be well distributed across the wavelength range covered in the analysis. The set of spectral features is made of CaII K 3933Å, CN 4200Å, CH 4300Å, MgI+MgH 5175Å, CaII 8542-8662Å and Balmer lines (if they are free of emission). The equivalent widths of these features, observed in the composite population, must be reproduced in the population synthesis. The difference $|W_{obs} - W_{synth}|$ will be minimized for the chosen set of spectral features. The various steps of the method are summarized on Fig.2 and extensively described in Bica (1988).

Fig. 2. : Successive steps of the star cluster method

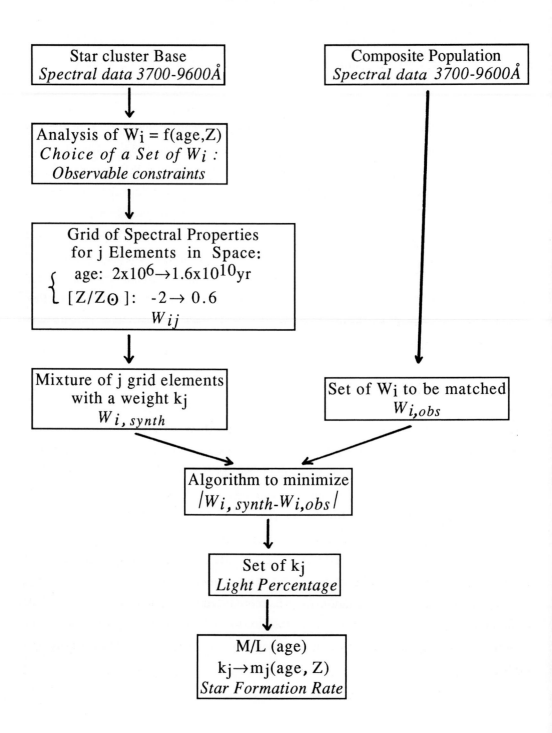

An example is shown on Fig.3, following the population synthesis performed in this way, for an elliptical galaxy of moderate mass.

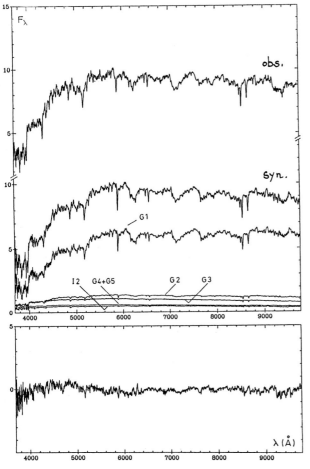

Fig.3 : Graphic representation of the synthesis. Upper frame, the observed spectrum and the synthesized one. The contributing elements from the grid are represented by star cluster groups G1, G2, G3, G4, G5 and I2 (see Bica, 1988 for further details). The lower frame shows the difference spectrum between the observed and synthetic ones. After Bica (1988).

3.4. Uniqueness of the Solution

The algorithms used to derive the mixture of grid elements which reproduces at best the characteristics of the composite object light, the $W_{i, obs}$, can be either direct combinations (Bica, 1988) or a multi-minimization procedure (Schmidt et al., 1989). Their respective merits and drawbacks are discussed in the paper just mentionned.

However, another question is that of the uniqueness of the solution, i.e., of the set of $\{k_j\}$, which has been addressed in a recent work (Schmidt et al., 1991). Firstly, the grid of $W_{i,j}$ used in these techniques to represent the star cluster spectral properties has been analyzed with respect to its principal components and reduced to minimize degenerative effects due to internal correlation. Secondly, a series of tests have been performed by building simulated composite spectra from the $W_{i,j}$ grid, and attempting to recover without subjective interference the original k_j proportions with population synthesis techniques. The results show that, although a given population synthesis problem can be difficult to solve due to its degenerate nature or due to the lack of pertinent constraints, the input proportions are reproduced in general within an accuracy from 5 to 10%. An illustration of these tests is given in Fig.4.

Fig.4a : Some examples of the simulated spectra (a), (b), (c), (d) (Schmidt et al., 1991).

Fig.4b : Success of the population synthesis in recovering the initial $\{k_j\}$ for each of the previous simulated spectra (a), (b), (c) and (d), after Schmidt et al. (1991).

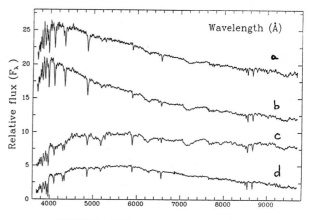

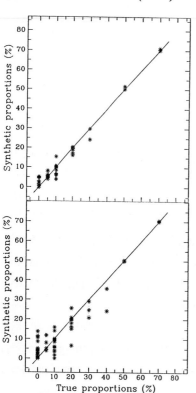

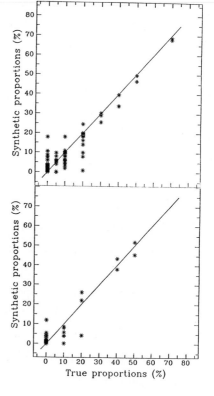

3.5. Derivation of the Star Formation Rate

The populations synthesis aims at reproducing the characteristics of the light, L, from the composite population. However, we wish to access as well the mass contribution related to each of the successive stellar generations represented by the $\{k_j\}$, in order to recover the star formation history in the object. This further step requires the knowledge of the M/L evolution of a star cluster. It has derived by Bica et al. (1988), using the evolutionary synthesis code by Yoshii and Arimoto (1987). We illustrate this procedure on Fig.5, for the case of a dwarf elliptical galaxy having suffered a recent burst of star formation.

It should be noted that the use of a star cluster base imposes a logarithmic time scale. The age bins displayed on Fig.5 for example, provide a temporal resolution which is suitable at ages from 10^7 to 10^9 yr (recent or intermediate-age star formation). For older stellar generations, any information regarding the star formation rate is lost, while that on the metallicity enrichment is preserved. But then, evolutionary synthesis techniques can relay the population synthesis. This fact points towards the conclusion that a *combined use* of the two techniques will provide the best approach to understand composite stellar populations.

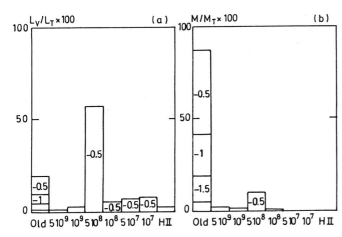

Fig.5 : The V flux luminosity fractions, $L_V/L_{V,T}$ in percentage (a) and the corresponding mass fractions M/M_T in percentage (b), for the successive stellar generations in a dwarf elliptical galaxy having suffered a recent burst of star formation. The metallicity is indicated, $[Z/Z_\odot]$, whenever possible. Notice that the recent star formation event dominates the light while it represents very little of the galaxy mass.

4. CONCLUSION AND PROSPECTIVE

The star cluster population synthesis, interfaced with the evolutionary synthesis, constitutes a powerful tool. It can be used to date star formation bursts, to recover the metallicity enrichment achieved in a composite population and the star formation rate. It allows to study the composite stellar populations in galaxies at large redshift and such a work is in progress. The method can be improved in several ways :

(i) by implementing in the star cluster base, clusters with a high metallic content, such as those found in M31.

(ii) by extending the wavelength coverage from 3700Å down to 1200Å, or from 9600Å upwards to 2.5μ.This will increase the number of non-redundant observables and therefore, of constraints.

These two steps have already been started and we are confident that they will bring a significant advance.

REFERENCES

Alloin, D., Andrillat, Y., Souffrin, S., 1971, Astron. Astrophys. **10**,401

Arimoto, N., Yoshii, Y., 1986, Astron. Astrophys. **164**, 260

Barbaro, G., Olivi, F., 1986, in Spectral Evolution of Galaxies, eds. C. Chiosi, A. Renzini, Reidel p.283

Belley, J., Roy, J.-R., 1991, ApJ. in press

Bica, E., 1988, Astron. Astrophys. **195**, 176

Bica, E., Alloin, D., 1986a, Astron. Astrophys. **162**, 21

Bica, E., Alloin, D., 1986b, Astron. Astrophys. Sup. **66**, 171

Bica, E., Alloin, D., 1987a, Astron. Astrophys. **186**, 49

Bica, E., Alloin, D., 1987b, Astron. Astrophys. Sup. **70**, 281

Bica, E., Arimoto, N., Alloin, D., 1988, Astron. Astrophys. **202**, 8

Bruzual, G., 1983, ApJ. **273**, 105

Erdelyi-Mendes, M., Barbuy, B., 1990, Astron. Astrophys. submitted

Faber, S., 1972, Astron. Astrophys. **28**, 109

Frogel, J., 1988, Ann. Rev. Astron. Astrophys. **26**, 51

Hartwick, F., 1980, ApJ. **236**, 754

Mc Clure, R., 1969, A.J. **74**, 50

O'Connell, R., 1976, ApJ. **206**, 370

Pickles, A., 1985, ApJ. **296**, 340

Rich, R.M., 1988, A.J. **95**, 828

Rocca-Volmerange, B., Lequeux, J., Maucherat-Joubert, M., 1981, Astron. Astrophys. **104**, 177

Rose, J., 1985, A.J. **90**, 1927

Schmidt, A., Bica, E., Dottori, H., 1989, M.N.R.A.S. **298**, 925

Schmidt, A., Copetti, M., Alloin, D., Jablonka, P., 1991, M.N.R.A.S. in press

Smith, G., Drake, J., 1990, Astron. Astrophys. in press

Spinrad, H., Taylor, B., 1971, ApJ. Sup. **22**, 445

Stebbins, J., Whitford, A., 1948, ApJ. **108**, 403

Tinsley, B., Gunn, J., 1976, ApJ. **203**, 52

Turnrose, B., 1976, ApJ. **210**, 33

De Vaucoulours, G., De Vaucouleurs, A., 1959, P.A.S.P. **71**, 83

Whipple, F., 1935, Harvard College Obs. Circular 404

Williams, T., 1976, ApJ. **209**, 716

Yoshii, Y., Arimoto, N., 1987, Astron. Astrophys. **188**, 13

Discussion

V. Rubin : What is the spatial resolution of the galaxy observations

D. Alloin : As a mean, we observe the composite light within a central 1 kpc region.

B. Pagel : I think it would be desirable to have atlases of stellar spectra with non-solar abundances, low as well as high.

D. Alloin : Yes indeed. This would allow to test evolutionary synthesis codes on the very simple case of a star cluster. This has been done for a star cluster with solar abundance, but it is more critical to perform it on metal-rich and metal poor star clusters.

V. Trimble : What do you do about matching the continuum of very metal-rich populations ?

D. Alloin : Only the equivalent widths are synthesized to obtain the population, in order to avoid problems with an intrinsic reddening. But we can then see what the continuum looks like and the results are very interesting for instance in the case of face-on versus edge-on spirals.

M. Peimbert : Have you compared the synthetic M/L ratios with those derived from dynamical arguments ?

D. Alloin : Not really because the synthetic M/L ratios refer to the central region of galaxies, while dynamical arguments will point towards the whole galaxy properties.

ABUNDANCE GRADIENTS IN SPIRAL DISCS

Angeles I. Díaz

Depto. de Física Teórica, C-XI, Univ. Autónoma de Madrid, 28049 Madrid, Spain.

Abstract

The problem of determining abundances in regions of low excitation and therefore relatively high metallicity is addressed.

From published data on HII regions in spiral galaxies it is shown that the oxygen radial distribution is steeper in the central parts of galactic discs and that there seems to be an inverse relation between the oxygen gradient and the effective radius of galaxies. More and better abundance determinations in HII regions in inner galactic discs are needed in order to confirm these facts.

1 INTRODUCTION

Observational work on galaxies provides information about observable quantities at the present time. In order to deduce something about those quantities at earlier epochs the history of the evolution of galaxies has to be known. This is very important, for instance, when we want to get information about abundances at the time of galaxy formation.

The discs of spiral galaxies represent probably the last components of galaxies that have been formed. Their initial chemical composition was probably very close to the primordial one and it has changed with time, getting enriched, by processes of star formation and evolution, with the elements synthesized in the stellar interiors. The present day distribution of abundances over spiral discs are then very important to know since they constitute the most powerful constraint for models in which the chemical evolution of galaxies is tried to be reconstructed.

2 ABUNDANCE DETERMINATION IN LOW EXCITATION HII REGIONS

It is not difficult (or so we thought) to determine present day abundances for the most common elements in the insterstellar medium of galaxies. Giant HII regions can provide this information through the analysis of their emission line spectra.

Thanks to this kind of analysis it is nowadays wellestablished that spiral discs show abundance gradients with O/H decreasing with radial distance from the galactic centre (see for example Pagel & Edmunds 1981). However, the information is by no means complete. Most spectroscopic studies concern regions of relatively high excitation and are therefore restricted to the outer parts of a given disc. The partial radial distributions of abundances thus obtained are then extrapolated to the galaxy centres.

The reason why we do not have good information about abundances in the inner parts of galaxies is not only that those regions are more difficult to observe but that, once you observe them, the required abundance analysis is more difficult to perform. The measurement of temperature indicator lines like those of [OIII] λ 4363 Å or [NII] λ 5575 Å is needed in order to calculate abundances with a certain degree of confidence, but these lines are intrinsically weak and therefore are not observable in low excitation or low surface brightness regions. To overcome this problem, empirical or semiempirical calibrations of strong, easily observable, lines against electron temperature or abundance have been proposed. Maybe the most widely used of these calibrations is the one proposed by Pagel *et al.* (1979) in which the oxygen abundance is calibrated against the ratio R_{23} of the optical lines $\log[OII]+\log[OIII]/H_\beta$. The calibration was originally made using HII regions of high excitation for which the [OIII] λ 4363 Å was observed and measured and a theoretical model of the giant low excitation HII region of M 101, S 5, made by Searle & Sargent (1978).

The success of the empirical calibration method led several groups to construct photo-ionization model sequences governed by a single parameter (McCall, Rybski & Shields 1985; Dopita & Evans 1986). These studies showed the calibration to be properly defined at the low metallicity end where it is empiricaly determined, but at high metallicities where HII region models have to be used, different model sequences give different results depending on the asumptions underlying the models.

What actually happens is that R_{23} works very well as an abundance indicator at metallicities at which the cooling mechanisms are dominated by the oxygen abundance through the optical lines used for its defintition. However, in regions of very high metallicity (higher than solar) where electron temperatures are very low (< 6000 K), the lines in the infrared become important and R_{23} is no longer sensitive to the oxygen abundance (see the figure presented by Edmunds in this volume). Besides that other heavy elements like C, N and S start to contribute substantially to the cooling complicating the picture.

Therefore, one parameter sequences are not a good representation of the true nature of HII regions over the whole range of abundances.

A two-parameter approach has been proposed by Vílchez & Pagel (1988) introducing a second parameter, $\eta = \frac{O^+/O^{++}}{S^+/S^{++}}$ which reflects the characteristics of the ionizing radiation. An equivalent observational parameter, $\eta' = \frac{[OII]/[OIII]}{[SII]/[SIII]}$, can be used in order to avoid the calculation of ionic fractions which would require an estimation of the electron temperature. The use of the method requires the observation of the IR [SIII] lines at $\lambda\lambda$ 9069, 9532 Å, which have the advantage of being easily observable even in low excitation regions (see Díaz et al. 1990). Assuming that η is independent of abundance, density and ionization parameter it is possible, with the help of adecuate photo-ionization models, to determine simultaneously the temperature of the ionizing stars and the oxygen abundance of the gas.

In reality, things are not so simple; the emission line espectrum of an HII region depends on three parameters: the shape of the ionizing continuum, the ionization parameter and the chemical composition. For giant HII regions, ionized by a young star cluster, these three parameters reflect the effective temperature of the stars that dominate the ionization, the number of ionizing photons that they provide and the physical conditions (density and abundance) of the gas. The two first parameters depend on the Initial Mass Function and stellar evolution which in turn may be affected by metal content in different ways.

Figure 1 shows the effect of each of these parameters on R_{23} and η' according to simple photo-ionization models. They have been computed using the photo-ionization code "Cloudy" (Ferland 1990) and represent nebulae ionized by single stars of different effective temperatures, from 35000 K to 55000 K whose spectral energy distributions have been taken from the NLTE atmosphere models of Mihalas (1972). The models assume spherical symmetry and uniform density and have been computed for different values of the ionization parameter and metallicity. We can see from the figure that, as expected, η' depends mainly on the effective temperature of the ionizing stars, but it depends on ionization parameter as well, being almost independent of metal abundance. On the other hand, R_{23} is seen to depend mainly on chemical composition and effective temperature of the ionizing stars and is almost independent of ionization parameter. Therefore an estimation of the ionization parameter, that can be made from the S^+/S^{++} ratio (Mathis 1985), is needed in order to simultaneously determine both effective temperature and chemical abundance in a given HII region. Yet, both R_{23} and η' depend to some extent on the relative abundances of S/O and N/O but we can take this to be a second order effect.

Figure 2 shows the application of the method outlined above to HII regions in the spiral galaxy M 33 (data from Vílchez et al. 1988). Models with metallicities 1/2 solar and solar are shown. Line codes are as in Figure 1. HII region data (solid dots) are labelled by their ionization parameter. The method allowed us to derive the metal abundance of the inner HII regions in M 33 which implied an oxygen

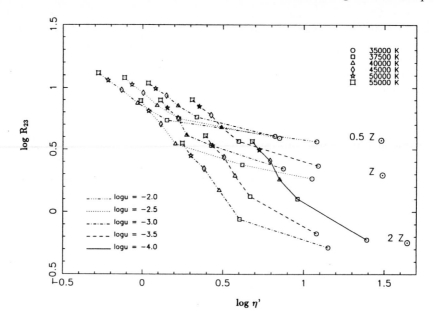

Figure 1: *The relation between R_{23} and η' derived from simple photo-ionization models*

abundance gradient steeper in the inner part of the galactic disc than expected from simple extrapolation of that derived for the outer part of the galaxy. If this effect is characteristic of M 33 or common to other galaxies is difficult to know given the scarcity of good abundance data for HII regions in the inner parts of galaxies. It seems to be present also in M 81 for which Garnet & Shields (1987) derived abundances in a good number of regions.

3 *ABUNDANCE GRADIENTS IN GALACTIC DISCS*

A comparison of O/H gradients in different galaxies is difficult to make due to their different scales and therefore it is custommary to represent abundances against galactocentric distance normalized to a characteristic radius of the galaxy as, for instance, the de Vaucouleurs photometric radius or the effective radius of the galactic disc. Even when represented in this way, gradients give the impression of being widely different from galaxy to galaxy (see for example Fig. 5 of Edmunds 1989), although a global trend of diminishing abundance with galactic radius is apparent (see Fig. 1 of Díaz 1989).

A relation between gas metal abundance and surface mass density over galactic discs has been suggested by McCall (1982) and evidence for its existence has been presented by Edmunds & Pagel (1984) and Garnett & Shields (1987). Figure 3 shows the relation between oxygen abundance and surface mass density for HII regions in a

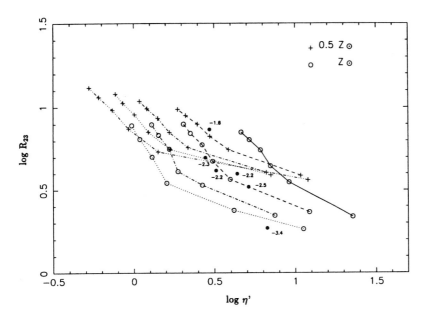

Figure 2: *The R_{23} vs η' relation for the HII regions of M 33 (data from Vílchez et al. (1988).*

number of spiral galaxies. The trend of increasing abundance with increasing surface mass density can be seen, but the scatter is very large.

There seems to be no obvious relation between the magnitude of the abundance gradient of a given galaxy and its luminosity, mass or morphological type. However, there are reasons to expect that some relation should exist between these latter properties and the average metal content of a galaxy since there are well known correlations between these quantities in elliptical galaxies. Therefore we should allow for the fact that different spiral galaxies will have, in general, different average metallicities depending on total mass and luminosity and abundances in a galactic disc should be referred to that average metallicity. In this way the abundance gradient shows up as a purely differential effect and the scatter is greately reduced. This can be seen in Figures 4 and 5 where the oxygen abundance, relative to that at the effective radius of each galaxy, is shown as a function of normalized radius and surface mass density respectively (note that in Figure 5 the surface mass density increases to the left in order to provide a representation similar to that of the radial abundance gradient). A more uniform picture seems to emerge. With a certain amount of scatter, which is expected in view of the large abundance error attached to individual HII regions (typically between 0.1 and 0.2 dex), all observed regions seem to follow the same relation. Three things are apparent from this figure: 1) there are not large differences in the relative abundance distributions of different galaxies, except for

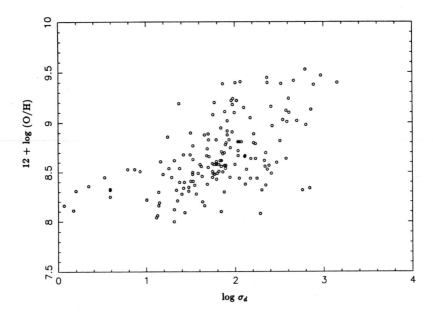

Figure 3: *Oxygen abundance vs surface mass density for HII regions in a number of spiral galaxies. References to abundaces can be found in Díaz (1989). Surface mass density data have been taken from McCall (1982).*

barred galaxies (solid dots) in which the oxygen abundance remains almost constant; 2) the distributions seem to be steeper in the inner part of galactic discs ($R \leq 1.75$ R_{eff}); and 3) there is very little gradient in the outer parts of discs ($R > 1.75$ R_{eff}). If we fit a straight line to the data in Fig. 4 for $R \leq 1.75$ R_{eff} we get the relation:

$$(O/H)_R = (O/H)_{R_{eff}} \times 10^{-a[(R/R_{eff})-1]}$$

with $a = 0.39 \pm 0.03$. This means that there is an inverse correlation between the slope of the abundance gradient and the effective radius of a galaxy: galaxies with smaller effective radii develop steeper abundance gradients.

This fact sems to be reproduced by relatively simple chemical evolution models. Mollá *et al.* (1990) have studied the time evolution of abundance gradients in spiral galaxies. For M 101 and NGC 2403, two galaxies with the same morphological type (SABcd) but different effective radii (6.91 and 2.62 Kpc respectively), the oxygen gradient is seen to develop earlier and become steeper in NGC 2403, the galaxy with the smaller effective radius, than in M 101 where most of the gradient develops in the last 2 Gyr of the life of the disc. However, it is important to realize that the abundance gradient is a differential effect and therefore provides information about the different chemical evolution in the inner and outer parts of galactic discs. Therefore good determinations of abundances are needed both in the inner and the outer parts of

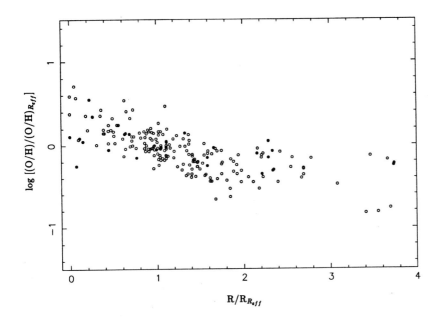

Figure 4: *Abundance distribution of HII regions as a function of normalized radius. Oxygen abundances have also been normalized, to that at the effective radius of each galaxies. References for the data can be found in Díaz (1989).*

galactic discs before any systematics in the behaviour of abundance gradients can be found. Extrapolations of gradients determined by straight line fits to several points restricted to a relatively small part of the disc are probably not good representations of radial abundance distributions.

REFERENCES

Díaz, A.I. , 1989. *In "Evolutionary Phenomena in Galaxies", Eds. J.E. Beckman & B.E.J. Pagel, Cambridge University Press, p377.*

Díaz, A.I. Terlevich, E., Pagel, B.E.J., Vílchez, J.M. & Edmunds, M.G., 1991. *rmx*, in press

Dopita, M.A. & Evans, I.N., 1986. *Astrophys. J.* , **307**, 431.

Edmunds, M.G., 1989. *In "Evolutionary Phenomena in Galaxies", Eds. J.E. Beckman & B.E.J. Pagel, Cambridge University Press, p356.*

Edmunds, M.G. & Pagel, B.E.J., 1984. *Mon. Not. R. astr. Soc.* , **211**, 107.

Ferland, G., 1990. *Hazy: an Introduction to Cloudy.*

Garnett, D.R. & Shields, G.A., 1987. *Astrophys. J.* , **319**, 662.

Mathis, J.S., 1985. *Astrophys. J.* , **291**, 247.

McCall, M.L., 1982. *PhD Thesis, University of Texas, Austin, USA.*

McCall, M.L., Rybski, P.M. & Shields, G.A., 1985. *Astrophys. J. Suppl.* , **57**, 1.

Mihalas, D., 1972. *"Non-LTE Model Atmospheres for B and O Stars", NCAR-TN/STR-76.*

Mollá, M., Díaz, A.I. & Tosi, M, 1990. *In "Dynamical and Chemical Evolution of Galaxies", Eds. J.J. Franco, F. Matteucci, F. Ferrini.*

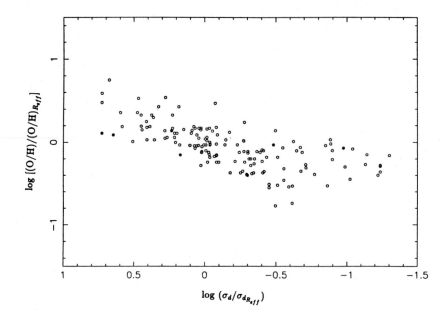

Figure 5: *Oxygen abundances as a function of surface mass density normalized to the effective radius of each galaxy.*

Pagel, B.E.J. & Edmunds, M.G., 1981. *Ann. Rev. Astr. Astrophys.* , **19**, 77.
Pagel, B.E.J., Edmunds, M.G., Blackwell, D.E., Chun, M.S. & Smith, G., 1979. *Mon. Not. R. astr. Soc.* , **189**, 95.
Searle, L. & Sargent, W., 1978. *Astrophys. J.* , **222**, 821.
Vílchez, J.M. & Pagel, B.E.J., 1988. *Mon. Not. R. astr. Soc.* , **231**, 257.
Vílchez, J.M., Pagel, B.E.J., Díaz, A.I., Terlevich, E. & Edmunds, M.G., 1988. *Mon. Not. R. astr. Soc.* , **235**, 633.

DISCUSSION

P. François: Concerning the models of chemical evolution of galaxy discs, what prescription did you take for the star formation rate as a function of radius?

A. Díaz: In our models the galaxy disc is divided in rings 1 Kpc wide. In each of the rings the SFR is exponentially decreasing with time. The initial SFR in each ring is derived from the present day surface gas and mass distributions.

E. Skillman: Have you looked at the classification scheme of Zaritsky *et al.* into low and high excitation HII regions?

A. Díaz: Not for all the regions presented here. But I guess most of the regions inside one effective radius will be of low excitation according to their classification scheme. All HII regions of Zaritsky *et al.* sample are at $R \geq 2\,R_{eff}$, so none of them is in the "inner" part of the discs.

M. Peimbert: Have you computed the "effective" yield (that derived assuming the simple model) and the true yield as a function of galactocentric distance for the different models?

A. Díaz: No, I have not. But $\log \left[(O/H)/(O/H)_{R_{eff}} \right]$ can easily be expressed as a function of the ratio of the corresponding "effective" yields.

The Mass-Metallicity Relationship for Dwarf Irregular Galaxies

EVAN D. SKILLMAN

Astronomy Department, University of Minnesota

1 HISTORY AND OBSERVATIONS

1.1 History

Since the nebular abundance studies of the Magellanic Clouds in the mid 1970's (Peimbert & Torres-Peimbert 1974, 1976; Dufour 1975; Dufour & Harlow 1977; Pagel *et al.* 1978), it has been suggested that there might be a correlation between galaxian mass and the metallicity of the interstellar medium. Surveys of the abundances in a number of H II regions in irregular galaxies by Lequeux *et al.* (1979), Talent (1980), and Kinman & Davidson (1981) all produced a clear correlation of oxygen abundance with both galaxy mass and luminosity.

1.2 The Observations

In the last few years, my collaborators and I have been using this relationship to find very low metallicity objects for the study of relative abundances (Skillman *et al.* 1988, Skillman, Terlevich & Melnick 1989, Skillman, Kennicutt & Hodge 1989). Figure 1 (next page) is a plot of oxygen abundance versus luminosity for dwarf irregular galaxies. Under the assumption of comparable mass/light ratios, the correlation between metallicity and luminosity would also produce a mass metallicity relationship. Here I have plotted oxygen abundance versus luminosity instead of versus mass because luminosities are generally available, but reliable masses are more difficult to obtain. For those galaxies with reliable mass estimates, a correlation is also found between abundance and mass.

Not all dwarf galaxies comply with the metallicity-luminosity relationship shown in Figure 1. For example, many blue compact dwarf galaxies (BCDGs), which derive a significant fraction of their total luminosity from their high surface brightness, active star forming regions, lie to the left in Figure 1. If the underlying correlation is between mass and metallicity, then this would be expected since the BCDGs are likely to have lower mass/light ratios. While the correlation in Figure 1 is clear, it is also clear that there is significant scatter at any given absolute magnitude. The uncertainty in the oxygen abundance is usually less than 0.1 dex, and distance uncertainties may produce uncertainties in the luminosity as large as a magnitude (or, in the case of IC 5152, larger). Given a reasonable scatter in the mass/light ratios, a tighter correlation between mass and metallicity may underlie the relationship in Figure 1.

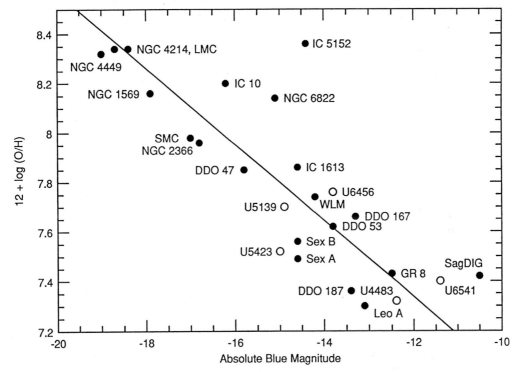

Figure 1. Oxygen abundance plotted versus galaxy luminosity for dwarf irregular galaxies (from Skillman, Kennicutt, and Hodge 1989). New observations of dwarf galaxies in the M 81 group (from Skillman, Terlevich, Kennicutt, and Garnett, in prep.) have been added as open circles.

On the other hand, the fundamental relationship may not be between mass and metallicity, but between surface density and metallicity. Mould, Kristian, and DaCosta (1983) first discovered that dwarf elliptical galaxies also show a strong correlation between metallicity and luminosity, and Aaronson (1986) showed that the metallicity *vs.* luminosity relationships for the two classes of galaxies are roughly identical in both slope and zero-point. Since low luminosity, high surface brightness, and relatively metal rich dwarf ellipticals have been observed, Bothun and Mould (1988) have suggested that surface density may be the fundamental parameter determining metallicity (see also Edmunds & Phillips 1989). The same may hold true for the dwarf irregulars, and is worth investigating observationally.

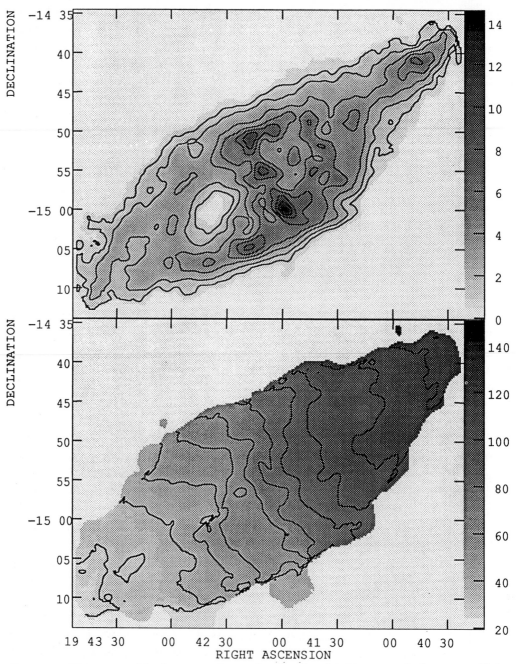

Figure 2. The neutral hydrogen column density (multiply greyscale by 2×10^{20} for atoms cm^{-2}) and the velocity field (greyscale in $-$ km s^{-1}) for the Local Group galaxy NGC 6822 (Skillman, van Woerden & Albinson, in prep).

Additional measurements of dynamical masses and mass surface densities will hopefully clarify the situation. There is a shortage of reliable mass estimates for dwarf irregulars, and one of the reasons is demonstrated in Figure 2. This figure shows the HI distribution and velocity field for the Local Group dwarf irregular NGC 6822. Optically, NGC 6822 is aligned roughly N-S, about 20 arc minutes in size, and with an axis ratio of about 2 (Hodge 1977). Note that the optical and HI morphologies are completely different. Note also that the velocity field is well ordered, meaning that a reliable mass distribution can be obtained, but only by studying the complete HI distribution derived through radio interferometry. Unlike the case for most spiral galaxies, a single dish line profile cannot be converted into a reliable total mass estimate because estimates of the appropriate inclination and radius from optical images can be specious. Since reliable mass estimates must be based on the more time consuming interferometric observations, results are lagging behind abundance measurements, but there is progress.

2 THEORIES OF THE PHYSICAL BASIS

I will now consider the circumstances that would give rise to a mass-metallicity relationship. In the simple model of chemical evolution plus infall/outflow, one can vary four parameters to achieve a given present day ISM metallicity. Those parameters are the yield, the star formation history, the infall, and the outflow. Here there is immediate trouble because we can only measure the *effective* yield; none of the four important parameters are directly observable.

2.1 The Yield

Studies of the effective yields in dwarf irregular galaxies have shown that almost all are consistent with values of the effective yield between about 0.002 and 0.003 (Pagel & Edmunds 1981, Pagel 1986), which is lower than the range of values normally derived for spiral galaxies of ≈ 0.01 (McCall 1982). Note that most studies of the ionizing clusters of H II regions result in flatter IMF slopes or higher mass cut-offs, or both, with lower abundance (Terlevich 1986, Viallefond 1986), which results in a larger true yield at lower abundance. Thus, one must take care not to interpret the effective yields as real yields, and it is unlikely that dwarf galaxies have low metallicities due to lower real yields.

2.2 The Star Formation History

It is possible to obtain a mass-metallicity relationship if one starts with a relationship between mass and age. In the extreme, some have suggested that the least chemically enriched galaxies (*e.g.*, I Zw 18) are just forming at the present epoch. Huchra (1977) first argued against the presence of young galaxies based on a comparison of model results with UBV colors. Currently, the strongest evidence against the young galaxy hypothesis is the study of relative abundances. Edmunds & Pagel (1978) introduced the use of the N/O ratio as an age estimator for the bulk of the star formation in a galaxy (based on the difference in masses of the stars primarily responsible for each element). Since most low metallicity galaxies have N/O ratios in a relatively narrow range (Pagel 1985, Garnett 1990), it is unlikely that we have caught galaxies in formation (see also the discussion of I Zw 18 in Dufour, Garnett, and Shields 1988).

The essence of the star formation history is the average star formation rate and the age of the galaxy. In the stochastic self-propagating star formation model of dwarf galaxies (Gerola, Seiden, & Schulman 1980), the average rate of star formation decreases with decreasing galaxy size. This is because for smaller galaxies a larger fraction of the galaxy participates in a single star forming event, which leads to a higher ratio of spontaneous/stimulated star formation events. This results in a mass-metallicity relationship. Unfortunately, the single physical process that may determine the ability of a disk to recover from a burst and proceed with the next generation of star formation (gravity) is not included in these models, and determining the appropriate coefficients for stimulated and spontaneous star formation may not be possible observationally.

A star formation threshold is another interesting possibility. Evidence for a massive star formation threshold in irregular galaxies has been presented by Skillman (1987). If a disk must rise above a certain gas surface density before star formation can occur, then, in the simplest picture of galaxy evolution, lower mass galaxies with lower mass surface densities will halt star formation at relatively higher values of gas mass fraction. Phillips, Edmunds, and Davies (1990) have shown how a star formation threshold will result in a mass-metallicity relationship, and have also presented a list of complications to this simple scheme.

2.3 Infall
If infall were increasingly more important for lower mass galaxies, then this might produce a mass-metallicity relationship. Two lines of evidence argue against infall as the primary cause of the mass-metallicity relationship: 1) There is no correlation of the presence of an extended H I halo with the abundance relative to the trend in Figure 1 (Skillman, Kennicutt, and Hodge 1989), *i.e.,* those galaxies in Figure 1 with large H I halos are not metal-poor for their luminosity; and 2) There is no correlation of effective yield with galaxy mass, *i.e.,* if infall were more important for low mass galaxies, they should have correspondingly lower effective yields.

2.4 Outflow
Outflow has been a very popular method for producing the mass-metallicity relation-

ship in dwarf galaxies. Outflow is more efficient than infall at reducing the effective yield, so the relatively low effective yield of dwarf irregulars is suggestive. Dekel and Silk (1986) propose that dwarf irregulars have lost most of their mass through winds. A real problem for this theory is the He/O ratio. If most of the gas is lost during the initial burst of star formation as the consequence of the evolution of massive stars, then the subsequent production and replenishment of He due to low mass stars will result in a large He/O ratio. In fact, He/H correlates well with O/H (Pagel and Simonson 1989). If the lowest mass galaxies are most efficient at shedding their interstellar gas, then δHe/δO should decrease with increasing O/H. Given the factor of 4 difference between the effective yields of dwarfs (where winds might be important) and spirals (which should retain most of their metals), one might expect a change in δHe/δO of a factor of 4. There is no evidence for this (Torres-Peimbert, Peimbert & Fierro 1989).

3 SUMMARY

Of the four parameters capable of producing a mass-metalicity relationship, I think that the yield, infall, and outflow are unlikely candidates. This leaves the star formation history as the remaining candidate. For this parameter, at least three different possibilities remain. 1) Dwarf galaxies may be younger than more massive galaxies; 2) the average star formation rate may depend on the size/mass of the galaxy; and 3) a star formation threshold may reduce the average star formation rate for dwarf galaxies. While the first of these three options seems the least likely, unfortunately, deciding between these three alternatives is observationally a very difficult problem. Enlarging the sample of dwarf galaxies with accurate measurements of metallicity (O/H), relative abundances (He/O, N/O, C/O), dynamical mass, gas mass, surface mass densities, and surface brightness at ultraviolet, optical, and infrared wavelengths will aid in further constraining the speculative possibilities.

4 REFERENCES

Aaronson, M. (1986) in it Stellar Populations, eds. C.A. Norman, A. Renzini, & M. Tosi, Cambridge University Press, Cambridge, p45.

Bothun, G.D., & Mould, J.R. (1988) ApJ, 324, 123.

Dekel, A., & Silk, J. (1986) ApJ, 303, 39.

Dufour, R.J. (1975) ApJ, 195, 315.

Dufour, R.J., Garnett, D.R., & Shields, G.A. (1988) ApJ, 332, 752.

Dufour, R.J., & Harlow, W.V. (1977) ApJ, 216, 706.

Edmunds, M.G., & Pagel, B.E.J. (1978) MNRAS, 185, 77p.

Edmunds, M.G., & Phillipps, S. (1989) MNRAS, 241, 9p.

Garnett, D.R. (1990) ApJ, 363, 142.

Gerola, H., Seiden, P.E., & Schulman, L.S. (1980) ApJ, 242, 517.

Hodge, P.W. (1977) ApJ Suppl, 33, 69.

Huchra, J.P. (1977) ApJ, 217, 928.

Kinman, T.D., & Davidson, K. (1981) ApJ, 243, 127.

Lequeux, J., Peimbert, M., Rayo, J.F., Serrano, A., & Torres-Peimbert, S. (1979) A&A, 91, 269.

McCall, M.L. (1982) Ph.D. thesis, University of Texas.

Mould, J.R., Kristian, J., & DaCosta, G.S. (1983) ApJ, 270, 471.

Pagel, B.E.J. (1985) in *Production and Distribution of C, N, O Elements*, eds. I.J Danziger, F. Matteucci, & K. Kjär, ESO, Munich, p155.

Pagel, B.E.J. (1986) Highlights Astr., 7, 551.

Pagel, B.E.J., & Edmunds, M.G. (1981) ARAA, 19, 77.

Pagel, B.E.J., Edmunds, M.G., Fosbury, R.A.E., & Webster, B.L. (1978) MNRAS, 184, 569.

Pagel, B.E.J., & Simonson, E. (1989) Rev. Mex. A.&A., 18, 153.

Peimbert, M., & Torres-Peimbert, S. (1974) ApJ, 193, 327.

Peimbert, M., & Torres-Peimbert, S. (1976) ApJ, 203, 581.

Phillips, S., Edmunds, M.G., & Davies, J.I. (1990) MNRAS, 244, 168.

Skillman, E.D. (1987) in *Star Formation in Galaxies*, ed. C.J. Lonsdale Persson, NASA, Washington, p263.

Skillman, E.D., Kennicutt, R.C., & Hodge, P.W. (1989) ApJ, 347, 875.

Skillman, E.D., Melnick, J., Terlevich, R., & Moles, M. (1988) A&A, 196, 31.

Skillman, E.D., Terlevich, R., & Melnick, J. (1989) MNRAS, 240, 563.

Talent, D.L. (1980) Ph.D. thesis, Rice University.

Terlevich, R. (1986) in *Star Forming Dwarf Galaxies*, ed. D. Kunth, T.X. Thuan, & J. Tran Thanh Van, Editions Frontieres, Gif sur Yvette, p395.

Torres-Peimbert, S., Peimbert, M., & Fierro, J. (1989) ApJ, 345, 186.

Viallefond, F. (1986) in *Star Forming Dwarf Galaxies*, ed. D. Kunth, T.X. Thuan, & J. Tran Thanh Van, Editions Frontieres, Gif sur Yvette, p207.

5 QUESTIONS

G. Ferland: High metallicity H II regions emit only infrared fine structure lines and Balmer lines. They would be hard to detect optically.

E. Skillman: Since we survey dwarf galaxies in Hα, and then take spectra of the H II regions to determine oxygen abundance, it is unlikely that we are selecting against high metallicity H II regions in low luminosity galaxies.

Abundances, Star-Formation and Nuclear Activity in Galaxies

R. CID FERNANDES[1], H. A. DOTTORI[2], S. M. VIEGAS[1] and R. B. GRUENWALD[1]

[1]IAG/USP - Caixa Postal 30.627 - CEP 01051 São Paulo, Brasil
[2]IF/UFRGS - Caixa Postal 15051 - CEP 90069 Porto Alegre, Brasil

1 INTRODUCTION

The question of the energy source in active galactic nuclei (AGN) is a long standing one. Up to the present day the most widely accepted paradigm is that of an accretion disk surrounding a super massive black hole (the so called "monster") lying in the nucleus of the galaxy. Starbursts were early discarded as a possible alternative model due to the fact that no stars were known to be hot enough to produce an ionizing spectrum with the required hardness. Terlevich and Melnick (1985, hereinafter TM) reacessed the Starburst hypothesis for the activity phenomenum. They have followed the evolution of the emission line spectrum of an HII region powered by a young star cluster containing WARMERS—massive stars which reach T_{eff}'s of $\approx$ 150000 K as a result of the strong mass loss. Their results have shown that such a system evolves from an initial normal phase (HII galaxy) to an active one (Seyfert 2 and/or LINER). It is our purpose in this contribution to investigate in detail the Starburst-WARMERS model for narrow line AGN, exploring mainly the metallicity and the chemical enrichment (by stellar winds) effects.

2 WARMERS and the Role of the Metallicity

Virtually all evolutionary calculations for massive stars which include mass loss agree in that stars above a certain initial mass go through an extremely hot and luminous phase near the He ZAMS. Because of the computed surface abundances, these hot (over 10^5 K) stages are usually associated to WR stars. Nevertheless, it is generally thought that the dense wind of these stars might decrease the surface temperature and alter the output spectra as a whole. A quantitative approach to this point, however, must await for future progress on WR atmospheres modeling. There are, anyway, some indications that at least some WR stars may be much hotter than commonly believed, as the ones reported by Barlow and Hummer

(1982), Davidson and Kinman (1982) and specially the one analysed by Dopita et al. (1990). WARMERS are supposed to be "extreme WR stars", of the WC or WO types (TM). It is clear that the whole WARMERS picture for AGN relies upon the controversial T_{eff} question, since the key novelty introduced by WARMERS is the UV emission harder than OB stars, which distinguishes active from HII-like ionizing continuum.

In the theory of radiatively driven winds mass loss rates are expected to grow with metal content (Z) because greater opacities increase the efficiency of line acceleration by radiation pressure. In the context of stellar evolution, greater $\dot{M}$ enhances the peeling of the outer layers, thus favouring the onset of WR/WARMERS phases. For instance, the minimum initial mass for a star to reach such stages is significantly lower the higher Z is. It is therefore clear that high Z star forming regions will naturally favour WR/WARMERS occurrence (see Azzopardi et al. 1988 for observational evidences). In AGN, several evidences indicate abundances in the 1–2 $Z_\odot$ range (e. g., Bonatto et al. 1989). HII nuclei, on the contrary, tend to be metal poor. Terlevich (1990), for example, report a remarkable lack of highly metallic HII galaxies on a survey of 450 emission-line objects. The Starburst-WARMERS scenario provides a natural interpretation for this dichotomy, since the low Z on HII-galaxies makes them inappropriate *locci* for a WARMERS "culture".

3   THE IONIZING SPECTRUM

To obtain the total spectrum of a theoretical star cluster one needs: *a)* a set of evolutionary tracks, *b)* an initial mass function (IMF) and *c)* model atmospheres. We developed a code which calculates the spectrum of each star from a 2-D interpolation of model atmosphere spectra in the $\log g \times T_{eff}$ plane, then sums up the contribution of each star (weighted by the IMF) to the total luminosity (L_ν). The stellar evolutionary tracks used are the Maeder and Meynet (1988—MM) ones. The IMF was taken as $\phi(M) \propto M^{-(1+x)}$; the normalization constant is proportional to the total cluster mass (M_T), i. e. , to the mass of gas converted into stars in the burst, which was supposed to be instantaneous. The adopted mass limits are $M_{up} = 120$ and $M_{low} = 3\,M_\odot$. Model atmospheres were taken from Kurucz (1979) for stars in the 5500–50000 K range, Clegg and Middlemass (1987) for the 50000–70000 K range and Wesemael (1981) pure He models for $T_{eff} > 70000$ K. (The critical point here is the use of standard static, plane-parallel LTE models for the presumably extended, non-LTE envelopes of WARMERS. This illustrates the above stressed fact that proper WR/WARMERS model atmospheres are not available yet.)

The evolution of the ionizing spectrum is plotted in Figure 1 together with a corresponding emission line spectra, and comprises two main periods:

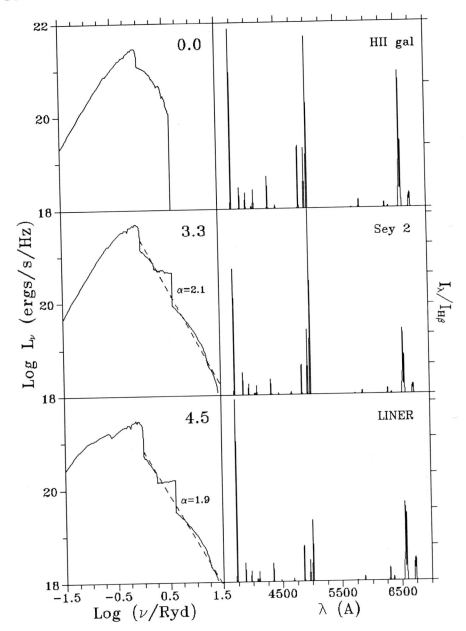

Figure 1: Evolution of the ionizing spectrum and the corresponding nebular emission lines. L_ν is normalized to $M_T = 1\ M_\odot$. The top, right labels in each figure are the age in Myrs and the spectral type of the emitting nucleus. Dashed lines show the fitted power-laws. The emission line spectra correspond to a $U = 0.87\ 10^{-3}, 10^{-3}$ and $0.33\ 10^{-3}$ ($t = 0$, 3.3 and 4.5 Myrs respectively), $n = 10^3$ cm^{-3} and $Z = Z_\odot$ model sequence. [OII]3727,3729, [NeIII]3868,3967, H$_\epsilon$ 3970, [SII]4070,4077, H$_\delta$ 4101, H$_\gamma$ 4340, [OIII]4363, HeII 4686, H$_\beta$ 4861, [OIII]4959,5007, [NII]5755, HeI 5876, [OI]6300,6363, [NII]6548,6584, H$_\alpha$ 6563 and [SII]6717,6731 Å lines have been included in the plot. Ticks in the intensity scale are multiples of H$_\beta$ intensity.

HII PHASE At age 0 the UV spectrum is dominated by O and B ZAMS stars and is typical of a high ionizing temperature HII region.

ACTIVE PHASE At 3.3 Myrs, the UV emission is drastically increased and extended to higher energies by the high T_{eff} photons emerging from the first WARMERS ($\approx$150000 K). At 4.5 Myrs this high energy emission is weaker, since stars with initial masses $> 55\ M_\odot$ have already died (type Ib supernovae). The IR enhancement reflects the presence RSGs in the cluster.

After $\approx$ 6 Myrs all massive stars have evolved and a weak "inactive" source remains—see, however, Terlevich 1990 for the role of type II SNe, which are ignited from this age on. Note that the overall shape of the UV spectra in the active phase is similar to a power law. A $L_\nu \propto \nu^{-\alpha}$ fit to the spectrum above 13.6 eV yields $\alpha = 2 \pm 0.2$, where $\pm$ 0.2 accounts for different cluster ages and IMF slopes. Most photoionization models for Seyferts 2 and LINERs use this kind of ionizing spectrum with $\alpha \approx$ 1.5–2 (Ferland and Netzer 1983, Binette 1985, Viegas Aldrovandi and Gruenwald, 1988—VAG). Thus, from our fitted α's, we expect photoionized clouds to have AGN-like emission line spectra prior to any photoionization calculations! Here, however, this shape is a natural consequence of the cluster evolution rather than an *ad hoc* assumption. Moreover, the noticeable contribution of RSGs to the IR emission from 4.5 Myrs on is a possible key to understand the observational evidence for the presence of such stars in AGN (Terlevich et al., 1990). From the obtained emission rate above the Lyman limit (Q_H) and the observed H$_\alpha$ luminosity we estimate the total cluster mass (M_T) to roughly range between 5 10^3 and 5 10^7 $M_\odot$. The (type Ib) SN rate is $\approx$ 2 10^{-9} SNe per year per unit $M_T/M_\odot$ (χ=1.35) during the active phase.

4   PHOTOIONIZATION MODELS

The ionizing spectra obtained were then used as an input to the photoionization code AANGABA (Gruenwald and Viegas 1990) to calculate the output emission line spectra. The cloud input parameters are the hydrogen density (n) and chemical abundances. Single density models with $\log n/\mathrm{cm}^{-3} = 2, 3, 4, 5, 6$ were computed and then used to calculate integrated models as in VAG. A power-law distribution function for the densities of the clouds is assumed, $f(n) \propto n^{-\beta}$, with β in the 0–3 range (see VAG). Models were computed for $Z = 1/10, 1/2, 1, 2$ and 3 $Z_\odot$. Several values of the ionization parameter ($U = Q_H/4\pi c\, r^2 n$) were tested—note that in a starburst scenario (U) evolves according to $Q_H(t, IMF)$, and is the essential quantity governing the emission lines changes during the active phase.

The transition of the emission line spectrum from an initial HII to an active (Seyfert 2 and/or LINER) phase on classical diagnostic diagrams by the time first

Cid Fernandes et al.

WARMERS appear is in very good agreement with TM models (see Fig. 1 for a pictorial view). Instead of emphasizing the evolutionary aspects, well studied by TM, we concentrate on the effect of varying model parameters in order to verify the abrangence of the Starburst-WARMERS scenario. In Figure 2 we compare our results to observations in the [NII]/H$_\alpha$ × [OIII]/H$_\beta$ diagram. The fundamental parameters in this plot are U, β and Z. A general conclusion drawn from the study of this and several other diagrams is that our models apply to a good deal of observed objects, but certainly not to all of them, since AGN areas are not totally covered (a more detailed analysis of these models will be published elsewhere). This is clearly seen in Figure 2 where no models with solar scaled Z reproduce [NII]/H$_\alpha$>1, a remarkable feature of many real objects, particularly LINERs. We can see no alternative, within the context of pure photoionization calculations, other than a Nitrogen overabundance to explain such cases (see Fig. 2). Oxygen being by far the major coolant, overabundances of Nitrogen leave the ionization structure essentially unchanged, while strongly enhancing its own emission lines. As we will see, this chemical "anomaly" may be understood in terms of the contamination from WR/WARMERS winds.

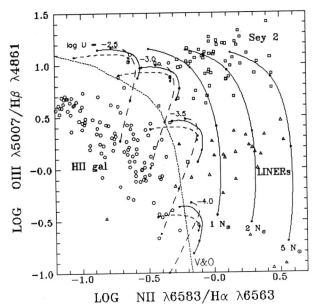

Figure 2: Integrated models. The four sets of curves correspond to the labeled values of the ionization parameter; solid, small and long dashed lines correspond to $Z = 1$, 1/2 and 2 $Z_\odot$. The arrow shows the direction of increasing β (points are plotted for $\beta = 0$, 1.5 and 3). The curves labeled 1, 2 and 5 N$_\odot$ are $n = 10^4$ cm^{-3} homogeneous models with solar abundances except for nitrogen. The curve labeled V&O is the "dividing line" of Veilleux and Osterbrock (1987).

From the point of view of photoionization alone, we conclude that the "optimum" metallicity for WARMERS-AGN models range between 1/2 and 2 $Z_\odot$, with $Z_\odot$ being strongly favoured. The existence of an upper limit for Z poses an interesting point in the context of the Starburst-WARMERS hypothesis. Since the ocurrence of such stars is expected to be favoured in high Z enviroments (and inhibited for metal-poor media) one would guess that the higher the metal-content the "more active" the nucleus. At $Z > 2\,Z_\odot$, however, cooling from forbidden lines prevents the matching of line ratios observed on Seyferts 2 and LINERs. Conversely, lower metallicities would not drive enough mass loss to throw massive stars into WARMER stages.

5 CHEMICAL ENRICHMENT BY STELLAR WINDS

The MM tables provide the detailed chemical evolution of the surface abundances of H, He, C, N and O. These data (together with $\dot{M}$) allow us to easily calculate the total returned mass on any of the above elements by an instantaneous burst with a given IMF. To calculate the relative importance of this source of ISM chemical enrichment the system cloud + cluster + stellar-wind-ejecta is considered to be a closed system in the sense that the gaseous mass at each instant is the sum of an original mass of gas plus the mass input by winds, which is a reasonable assumption for the time scales of interest. The evolution of the gas mass and chemical abundances is then followed by solving the standard set of equations for the chemical evolution of galaxies (namely, the conservation of mass and the production rate for each element). The solution for the evolution of the number abundance of an element i relative to H (A_i) in this "simple model" is

$$A_i(t) = \frac{A_i(0)\gamma + \mu_i^\star(t,\chi)/a_i X_H^0}{\gamma + \mu_H^\star(t,\chi)/X_H^0} \tag{1}$$

where $\gamma \equiv (M_0 - M_T)/M_0$, M_0 is the existing mass of gas before the burst; $\mu_i^\star(t,\chi)$ is the total wind returned mass on i per unit $M_T/M_\odot$, up to a time t and for an IMF slope χ; X_H^0 is the initial hydrogen mass fraction (≈ 0.7) and a_i is mass number of i. The mathematical role of γ is to quantify the proportion of pre-existing to the input gas, the lower γ the more significant this chemical "pollution" will be and vice-versa. The pattern of chemical changes in the cloud is shown in Figure 3. Two main features are evident in this picture. First, lower γ's increase the pollution, as expected. Second, nitrogen is the first element to become noticeably overabundant (since N is expelled before than C and O).

What can one expect from all these pollution calculations concerning the starburst-WARMERS scenario for AGN? First of all, we have shown that such effects can be very significant, depending mainly on γ, i. e. , on the amount of gas to be enriched. Second, and more important, the fact that nitrogen is the first species to become overabundant may be the key to understand the N overabundances claimed for some nuclei, specially LINERs (Binette 1985, Sthorchi-Bergmann and Pastoriza 1989). From the point of view of photoionization models, calculations with overabundant N do bring much improvement (Fig. 2). It is vital to understand, however, that if oxygen is also enhanced by the same factor this will no longer be true, i. e., no improving would occur due to the cooling role of O. Figure 3 shows that such a situation (high N an low O) does hold for ≈ 1 Myr. It must be also recalled that the wind does not mix itself instantaneously to the cloud. A typical "diffusion" time scale would be ≈ 0.5–1 Myr (the time required for a 10^3 Km s^{-1} wind to travel some hundreds of pc), which corresponds to a horizontal shift in the direction of increasing time in Fig. 3. Since we expect to see emission lines only while WARMERS are alive to heat the clouds, we conclude that no matter what strong abundance anomalies are introduced during the late stages of the cluster evolution, they would not be observed! Moreover, the peak of N pollution occurs between 3.5 and 4.5 Myrs (τ_{dif}=0.5 Myrs), coinciding with the more luminous phase of the ionizing cluster, making the whole nucleus more easily observable (conversely, older systems, therefore more C and O polluted, should be rarely observed). The fact that such anomalies are more frequently reported in LINERs is consistent with the picture that this nuclei have less ionized gas, what make them more susceptible to pollution effects.

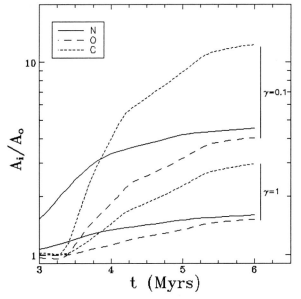

Figure 3: Evolution of the C, N and O abundances with respect to their initial values for two values of γ. Initial abundances are solar and χ=1.35.

6   CONCLUSIONS

In summary, our results confirm that, as long as optical line-ratios are concerned, the Starburst-WARMERS model for narrow line AGN do work well for a great fraction of objects (but not all). Metallicities, however, are constrained to a fairly narrow range around $Z_\odot$. Our study of the chemical pollution by the strongly processed material ejected by WR/WARMERS turned out to explain the frequently suggested nitrogen overabundance on active nuclei in a very natural and straight-forward way.

Acknoledgements: We thank Dr. Thaisa Sthorchi-Bergmann and Dr. Roberto Terlevich for making available their data banks and Dr. J. A. F. Pacheco for very fruitfull discussions. Most of this work was supported by the Brazilian institutions CNPq and CAPES.

References

Azzopardi, M.; Lequeux, J.; Maeder, A., 1988, *Astron. & Astrophys.*, **189**, 34.

Barlow, M. J.; Hummer, D. G., 1982, in "Wolf-Rayet Stars: Observations, Physics, Evolution", IAU Symp. 99, p. 387, eds. C. W. H. de Loore and A. J. de Willis

Binette, L., 1985, *Astron. & Astrophys.*, **143**, 334.

Bonatto, C.; Bica, E.; Alloin, D., 1989, *Astron. & Astrophys.*, **226**, 23.

Clegg, R. E. S. & Middlemass, D., 1987, *Mon. Not. R. ast. Soc.*, **228**, 759.

Davidson, K. & Kinman, T. D., 1982, *Publ. Ast. Soc. Pac.*, **94**, 634.

Dopita, M. A.; Lozinskaya, T. A.; McGregor, P. J.; Rawlings, S. J., 1990, *Astrophys. J.*, **351**, 563.

Ferland, G. J. & Netzer, H., 1983, *Astrophys. J.*, **264**, 105.

Gruenwald, R. B. & Viegas, S. M., 1990, *Astrophys. J. Suppl. Ser.*, submitted.

Kurucz, R. L., 1979, *Astrophys. J. Suppl. Ser.*, **40**, 1.

Maeder, A. & Meynet, G. 1988, *Astron. & Astrophys. Suppl. Ser.*, **76**, 411.

Maeder, A., 1990, *Astron. & Astrophys. Suppl. Ser.*, in press.

Storchi-Bergmann, T. S. & Pastoriza, M. G., 1989, *Astrophys. J.*, **347**, 195.

Terlevich, R. & Melnick, J., 1985, *Mon. Not. R. ast. Soc.*, **213**, 841.

Terlevich, E.; Díaz, A. I.; Terlevich, R., 1990, *Mon. Not. R. ast. Soc.*, **242**, 271.

Terlevich, R., 1990, in "Structure and Dynamics of the Interstellar Medium", IAU Colloquium 120, G. Tenorio-Tagle, M. Moles and J. Melnick (Eds.), p. 343.

Veilleux, S. & Osterbrock, D., 1987; *Astrophys. J. Suppl. Ser.*, **463**, 295.

Viegas-Aldrovandi, S.M. & Gruenwald, R. B., 1988, *Astrophys. J.*, **324**, 683.

Wesemael, F., 1981, *Astrophys. J. Suppl. Ser.*, **45**, 177.

QUESTIONS

B. Pagel: Do you also expect an enrichment in Helium?

R. Cid Fernandes: Yes, of $\approx$ 10–60% for γ between 1 and 0.1.

J. Vilchez: I have two comments. First, perhaps you are aware of the new Maeder's tracks for metallicities between 0.1 and 2 $Z_{\odot}$.

R. Cid Fernandes: Yes, we are. We have redone the pollution calculations for these new "yields" and, although significant changes are introduced, the qualitative picture we outlined here remains valid. Actually, the much larger mass loss rates in these new models make the overall enrichment still more efficient for $Z = 1$ and 2 $Z_{\odot}$. Of course, for smaller Z's the lower $\dot{M}$ works in the opposite sense.

Vilchez: The second point is that I wouldn't be so confident to say that this or that is the exact ammount of mass needed to have N pollution. What we do observe is that sometimes you have enrichment and sometimes not.

R. Cid Fernandes: I fully agree with you, and that is the role of γ, to regulate the effectiveness of pollution. Anyway, this is a simple model—although it certainly holds for the system cluster + cloud + stellar-ejecta considered as a whole. A detailed picture for the real wind-enrichment should go through the hydrodynamics of the wind-ISM interaction.

The Diffuse Ionized Medium in NGC 1068

J. BLAND-HAWTHORN

Dept. of Space Physics and Astronomy, Rice University

ABSTRACT   Spectrophotometric observations with the Hawaii Imaging Fabry-Perot Interferometer (HIFI) of the Hα and [NII]λ6548,λ6584 lines has detected a *diffuse ionized medium* (DIM) pervading the inner $\sim$10 kpc ($\sim$135$''$) of the nearby, luminous Seyfert galaxy NGC 1068. The physics and dynamics of this energetic gas are distinct from the population of adjacent disk HII regions.

1 INTRODUCTION

Most spiral galaxies appear to have moderately active nuclei (Keel 1983), yet the impact of this activity on the large-scale, interstellar medium (ISM) of the surrounding disk remains largely unexplored. Recent studies of the nearby Seyfert galaxy NGC 1068 (Baldwin, Wilson and Whittle 1987; Pogge 1988; Bergeron, Petitjean and Durret 1989) find a wide variety of ion species confined to the radio jet axis ($\sim$45°; Wilson and Ulvestad 1983) at large radii (>20$''$), implying a connection with the active nucleus. From the direct nuclear spectrum alone, the similar widths of the permitted and forbidden lines covering a wide range of ionization classify NGC 1068 as a type 2 Seyfert galaxy (Khachikian and Weedman 1974). However, the polarized nuclear spectrum exhibits broad permitted lines on a featureless continuum (Antonucci and Miller 1985), and in the ultraviolet, broad FeII lines are also observed (Snijders, Netzer and Boksenberg 1986). These observations provide strong evidence for a "buried" type 1 Seyfert nucleus in NGC 1068 obscured from direct view, whose radiation escapes anisotropically before being scattered into our line-of-sight. Recently, we have demonstrated that the brightest components of the narrow line region envelope the radio jet as a conic outflow with opening angle $\sim$80° inclined by $\sim$45° to the disk plane (Cecil, Bland and Tully 1990). While the reviewed observations demonstrate that a Seyfert nucleus can energize the ISM over great distances along its radio axis, the Fabry-Perot observations furnish evidence for a much wider influence throughout the disk, that is, both along and perpendicular to the radio axis. We discuss the characteristics of this gas, its source of ionization, and its implications for the extended ISM in NGC 1068.

2 OBSERVATIONS AND RESULTS

The Hα and [NII]λ6548,λ6584 observations were made on the nights of December

5 and 6, 1985 with the HIFI system at the f/8 Cassegrain focus of the CFH 3.6m telescope (Bland and Tully 1989). Eighty data frames were obtained over a 77Å range at a spectral resolution of 65 km s^{-1} FWHM in an exposure time of 15 800 s. The data were parametrized with ~60 000 triple Gaussian fits to obtain emission-line velocities, dispersions and total fluxes (Bland-Hawthorn, Sokolowski and Cecil 1991). In Fig. 1(a), the brightest HII regions in Hα are confined to an elliptic ring with diameter ~ 3 kpc and major axis ~45° that is centered on the Seyfert nucleus. This same region exhibits high concentrations of molecular gas, high rates of star formation, and an oval, stellar bar aligned with the 3 kpc ring (q.v. Scoville *et al.* 1988). In contrast, the [NII]λ6584 flux distribution shown in Fig. 1(b) is characterized by diffuse, filamentary emission that extends out to ~5 kpc radius. The [NII]λ6584/Hα map reveals that the discrete HII region population is characterized by line ratios <0.5, attaining values ~0.3 in the regions of highest line intensity. In between the HII regions, a diffuse ionized medium exhibits [NII]λ6584/Hα ratios that are closer to unity, much higher than observed in HII regions with solar abundance. This difference is emphasized further in the line dispersions where HII region complexes have typically ~100 km s^{-1} FWHM while the diffuse gas has 150–250 km s^{-1} FWHM, corrected for instrumental broadening.

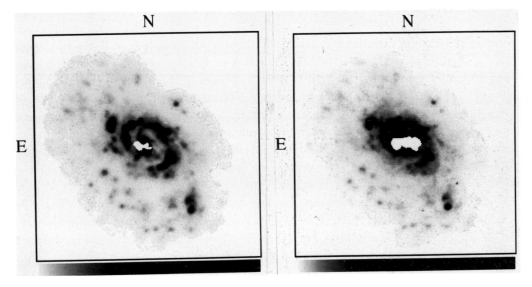

Fig. 1. (a) integrated Hα emission for a 129″×129″ field of view sampled at 0.43″ per pixel; (b) integrated [NII]λ6584 emission.

The spectral collage in Fig. 2 illustrates the dramatic increase in the [NII] line intensity relative to Hα as the Hα surface brightness decreases. In between the HII regions, the [NII] intensity profile declines as ≈ r^{-2}, with no perceptible gradient in the [NII]λ6584/Hα ratios, across roughly elliptic isophotes that are aligned with the stellar bar. The DIM is characterized by strong [NII]$\lambda6584$ emission (3.0×10^{41} erg s^{-1}) when compared to Hα (1.2×10^{41} erg s^{-1}), with most of the [NII] luminosity arising from within the elliptic ring and from two diametric, narrow arcs along the radio axis where [NII]$\lambda6584$/H$\alpha \approx 2$–3. Away from the radio axis, the DIM is characterized by [NII]$\lambda6584$/H$\alpha \approx 1.1\pm0.2$. Evans and Dopita (1987) recognized a non-stellar component of excitation superimposed on their HII region spectra that we now ascribe to the DIM. From their published line strengths, we subtract the contribution from the HII regions to put constraints on some important line diagnostics. We find that [OII]($\lambda3726+\lambda3729$)/Hα ($\approx1.0\pm0.5$) and [SII]($\lambda6716+\lambda6731$)/Hα ($\approx0.4\pm0.1$) are also significantly higher while [OIII]($\lambda4959+\lambda5007$)/Hα ($\approx0.4\pm0.2$) is much lower than observed in solar-abundance HII regions. In the next section, we deduce a thickness of ~1 kpc for the emitting volume in the disk. The diffuse Hα and [NII] luminosities provide a constraint on the filling factor ϕ and the average electron density n_e, such that $\phi n_e^2 \approx 0.01$, or equivalently, $n_e \approx 0.1\phi^{-0.5}$. Presently, there are no reliable limits on ϕ which represents the biggest uncertainty on the equation of state for the DIM.

3 PHYSICAL STATE OF THE DIM

The Burbidges (1962) first noted [NII]$\lambda6584$/Hα ratios in the nuclear regions of external galaxies that are significantly larger than those found in solar-abundance HII regions. The origin of these peculiar ratios has been controversial, with explanations invoking excitation variations or enhanced nitrogen abundance (q.v. Rose and Searle 1982). Historically, high [NII]$\lambda6584$/Hα ratios in HII regions, supernova remnants and planetary nebulae are thought to reflect changes in the N/H or N/O abundance (Pagel 1989). Recently, an overabundance of N/O, and possibly S/O, has been found over the inner 2–5 kpc in a sample of Seyfert 2 and LINER galaxies (Bergmann, Bica and Pastoriza 1990). However, we emphasize that the work of Evans and Dopita (1987) shows that the high [NII]$\lambda6584$/Hα ratios in NGC 1068 *do not reflect a substantial overabundance of nitrogen*. In fact, these authors find solar N/O and S/O abundances (although O/H is enhanced by $<30\%$), for 13 HII regions at radii $20''$ to $60''$. Moreover, we stress that Hα line emission is not substantially diluted by underlying Balmer absorption (FWZI $\sim80\text{\AA}$) which arises from pressure broadening in the atmospheres of early-type stars. The Hα absorption produces a 7% modulation that is indistinguishable from a slowly varying continuum over the HIFI bandpass. After correcting for the broad Balmer feature, if all the remaining continuum arises from stars later than spectral type A, the Doppler-broadened absorption decreases [NII]$\lambda6584$/Hα in the general disk by $<20\%$ and negligibly along the radio axis. It remains to enhance [NII]$\lambda6584$/Hα with either a dilute radiation field (e.g. Halpern and Steiner 1983) or a higher electron temperature, without further ionizing the N$^+$ ions (e.g. Ferland and Mushotzsky 1984). Thus, we conclude that the excitation conditions in the DIM are quite different from conditions within the HII regions.

The measured line widths observed by HIFI throughout the DIM imply an average dispersion of $\sigma \approx 85$ km s^{-1}, a value that is much larger than the thermal value for the ionized gas ($T_e \approx 10^4$K). This suggests that the DIM is supported predominantly by turbulent pressure although the contribution from magnetic fields or cosmic rays is presently unknown. We can approximate the scale height of the diffuse gas if we make the assumption that the DIM is generally in hydrostatic equilibrium supported only by its internal pressure. Then we may write

$$\frac{d(\rho \sigma_z^2)}{dz} = -\rho g_z$$

where ρ is the total mass density at a height $|z|$ above the plane. The properties σ_z and g_z are the components of velocity dispersion and gravitational acceleration perpendicular to the disk respectively, where g_z is given by Poisson's equation. If we make the assumption that $g_z \sim 4\pi G\mu$, for a midplane density ρ_0, the vertical mass density distribution has the form $\rho(z) = \rho_0 e^{-|z|/h}$ where the scale height is given by $h = \sigma_z^2 (4\pi G\mu)^{-1}$. We obtain a lower limit on the surface density μ from the HIFI kinematic data by characterizing the properties of the DIM at a radius of 3 kpc ($\sim 10^{22}$ cm). Here, for an assumed inclination angle of $\sim 40°$, the rotation curve is flat with a deprojected velocity, $v_c \sim 175$ km s^{-1}. The circular velocity is somewhat uncertain in that the *entire* inner disk shows evidence for strong elliptic streaming which does not always align with the 2μm stellar bar. We approximate the local surface density using $\mu \sim v_c^2 (2\pi Gr)^{-1}$. The estimated surface density is ~ 400 M$_\odot$pc^{-2} which indicates a vertical scale height for the DIM of $h \sim 400$ pc. In the starburst ring, the CO line dispersion spans the range 13–50 km s^{-1} (Planesas, Scoville and Myers 1990). The low dispersion when compared with the DIM and the high surface density indicate a molecular scale height of ~ 50 pc.

4 HEATING BY HOT OR EVOLVED STELLAR POPULATIONS

The bolometric luminosity of NGC 1068 ($\sim 3 \times 10^{11}$L$_\odot$) is shared by the Seyfert nucleus and the extended starburst region (Telesco *et al.* 1984). Regions of vigorous star formation are thought to produce large numbers of massive stars that give rise to energetic, gaseous outflows (Fabbiano and Trinchieri 1984). Evans and Dopita (1986) see a broad feature at $\lambda 4660$ in a single knot that they interpret as the permitted CIII + CIV blend powered by WC stars, Snijders, Briggs and Boksenberg (1982) find strong P-Cygni MgII, SiIV, CIV and NV line profiles, indicating wind speeds ~ 1000 km s^{-1} from Of stars. While hot winds can collisionally ionize and heat the gas, published shock models are stretched to produce [NII] lines as strong as Hα (Shull and McKee 1979). However, the inferred [OIII]/[OII] and [SII]/[OII] ratios are typical of shock velocities greater than 100 km s^{-1}. For a wide range of shock conditions, the strengths of [NII] and [SII] are linked fairly closely (Cox and Raymond 1985). The models produce [NII]/[SII]~ 1 as compared with [NII]/[SII]~ 3 inferred from the long-slit observations. A stronger argument against shocks is that most of the star formation appears confined to the 3 kpc ring while the DIM surface brightness peaks near the nucleus and continues to decline well beyond the starburst ring.

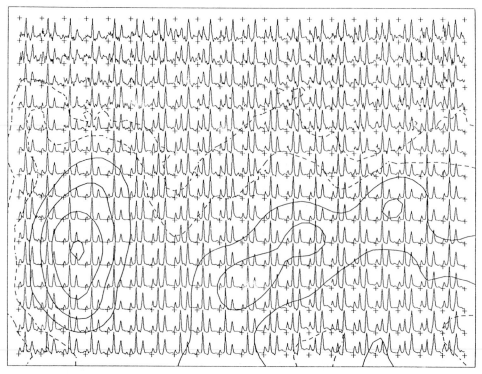

Fig. 2. A grid of 20×15 self-scaled HIFI spectra for a 26″×19″ field of view NE of the nucleus with superimposed contours that track the Hα surface brightness.

Aperture photometry indicates that NGC 1068 is bluer than expected for its Sb classification (Smith, Weedman and Spinrad 1972), which raises the possibility that the DIM is energized by an unrecognized population of massive, young stars that track the mass distribution. Alternatively, an old stellar population (e.g. hot white dwarves, planetary nebulae) with a characteristically larger velocity dispersion could produce a diffuse, ionizing radiation field. However, Wynn-Williams, Becklin and Scoville (1985) find that their upper limit on the free-free radiation and the observed colors of the inner disk are both consistent with an extended population characterized by B5 stars. Moreover, for the derived nebular abundance, photoionization by early-type stars produce [NII]λ6584/Hα ratios much lower than those observed. Thus, stars do not seem to be capable of producing the observed excitation patterns.

5 HEATING BY THE ACTIVE GALACTIC NUCLEUS

In a model developed fully elsewhere (Sokolowski, Bland-Hawthorn and Cecil 1991), we argue that the different spatial distributions of high and low ionization lines in NGC 1068 reflect a difference in the ionization parameter as seen by the gas along and away from the radio axis. We show that the DIM is excited by scattered radiation from the nucleus, in contrast to the highly ionized gas that sees the nuclear continuum directly. The ionization parameter is defined by $U = L(4\pi r^2 c n_e)^{-1}$ where $L/4\pi r^2$

is the ionizing flux incident on a filament with source distance r. Our calculations show that over a wide range of power-law indices, the [OIII]λ5007, [NeIII]λ3868 and [NeV]λ3426 to Hα ratios maximize for $U \geq 10^{-2}$, while the [OII]λ3727, [NII]λ6584 and [SII]λ6716 ratios to Hα maximize for $U < 10^{-2}$. This is to be expected for a gas in thermal equilibrium because larger U and flatter power-law slopes induce higher excitation. We deduce that the spatial confinement of high ionization species to the radio axis is characteristic of $U \approx 10^{-2}$, and the ubiquitous distribution of low ionization species throughout the DIM results from a more dilute ionizing continuum, $U \approx 5\times10^{-4}$. The higher value of U arises from direct, anisotropic radiation of the nuclear continuum along the radio axis, as argued by others (Baldwin, Wilson and Whittle 1987; Pogge 1988). A fraction L_S of the intrinsic nuclear luminosity L_I is scattered into the disk, such that $L_S = (1 - e^{-\tau_S})\left(\frac{\Delta\Omega}{4\pi}\right)L_I$ where $\Delta\Omega/4\pi$ is the opening angle through which the ionizing radiation escapes. In principle, we do not distinguish between Thomson and Rayleigh scattering because we require only that the scattering optical depth be $\tau_S \sim 0.1$, which is not contradicted by the observations for either mechanism (Antonucci and Miller 1985; Bailey *et al.* 1988). From UV and X-ray observations of the nucleus (Monier and Halpern 1987; Koyama *et al.* 1989), we infer the intrinsic ionizing luminosity to be about one quarter of the nuclear IR luminosity of $\sim$1.5$\times10^{11}$L$_\odot$ (Telesco *et al.* 1984). If $\Delta\Omega/4\pi \sim 0.25$ is implied by the [NeV] distribution (Bergeron *et al.* 1989), there is sufficient scattered nuclear flux for even moderate dust extinction ($<$0.5 mag kpc^{-1}) to ionize the DIM. Begelman (1985) has derived the equation of state for an ISM photoionized by an AGN and finds that runaway X-ray heating can push gas up high into the galactic halo. The kinetic ionization parameter Ξ, which describes the thermal and the ionization equilibrium in a photoionized gas, is given by $\Xi = L(r^2 c k n_e T_e)^{-1}$ which leads to $\Xi \sim 10\phi^{0.5}L$ at a radius of 3 kpc. Along the radio axis, Ξ may be high enough to cause filaments to become thermally unstable through runaway X-ray heating. In our model of the DIM, Ξ is a factor of $\sim$40 smaller than anticipated by Begelman as the circumnuclear, molecular torus blocks most of the direct radiation from the nucleus. If we approximate the scattered ionizing luminosity by $L_S \sim 5\times10^{42}$ erg s^{-1}, for filling factors in the range $10^{-4} < \phi < 1$, this leads to $0.1< \Xi <10$. It is noteworthy that, if we assume $T_e \approx 10^4$K, *thermally* stable phases do exist for the DIM filaments when photoionized by a hard X-ray source (Lepp *et al.* 1985), although this is sensitive to the assumed cooling processes. These filaments would tend to dissipate on a sound crossing time unless confined by an external medium (ram pressure, thermal pressure, etc.). While scattered radiation from the nucleus can control the ionization balance, photo-excitation cannot inject sufficient energy into the DIM to account for the observed bulk motions. In the picture of the 3-phase ISM proposed by McKee and Ostriker (1977), the supernovae and their remnants regulate the energy and mass balance of the ISM. Wang and Cowie (1988) have suggested that vigorous star formation can keep all phases of the ISM turbulent. This may provide a natural explanation for much of the turbulent motion inferred from the [NII] line dispersion.

ACKNOWLEDGMENTS   This work is funded in part through NSF grant AST 88-18900 that supports the HIFI "Nearby Active Galaxies" program. I wish to thank Jim Sokolowski and Gerald Cecil for stimulating discussions and their involvement with much of this research.

Antonucci, R.R.J., and Miller, J.S., 1985, *Ap. J.,* **297,** 621.
Bailey, J.A., Axon, D.J., Hough, J.H. *et al.,* 1988, *M.N.R.A.S.,* **234,** 899.
Baldwin, J.A., Wilson, A.S., and Whittle, M., 1987, *Ap. J.,* **319,** 84.
Begelman, M.C., 1985, *Ap. J.,* **297,** 492.
Bergeron, J., Petitjean, P., and Durret, F., 1989, *Astr. Ap.,* **213,** 61.
Bergmann, T.S., Bica, E., and Pastoriza, M.G., 1990, *M.N.R.A.S.,* **245,** 749.
Bland, J., and Tully, R.B., 1989, *A. J.,* **98,** 723.
Bland-Hawthorn, J., Sokolowsi, J., and Cecil, G.N. 1991, *Ap. J.,* accepted.
Cecil, G., Bland, J., and Tully, R.B., 1990, *Ap. J.,* **355,** 70.
Cox, D.P., and Raymond, J.C., 1985, *Ap. J.,* **298,** 651.
Evans, I.N., and Dopita, M.A., 1986, *Ap. J. (Letters),* **310,** 15.
Evans, I.N., and Dopita, M.A., 1987, *Ap. J.,* **319,** 662.
Fabbiano, G., and Trinchieri, G., 1984, *Ap. J.,* **286,** 491.
Ferland, G. J., and Mushotzsky, R. F., 1984, *Ap. J.,* **286,** 42.
Halpern, J. P., and Steiner, J. E., 1983, *Ap. J. (Letters),* **269,** L37.
Keel, W.C., 1983, *Ap. J.,* **269,** 466.
Khachikian, E.Y., and Weedman, D.W. 1974, *Ap. J.,* **192,** 581.
Koyama, K., Inoue, H., Tanaka, Y. *et al.,* 1989, *Pub. A.S. Jap.,* **41,** 731.
Lepp, S., McCray, R., Shull, J.M., Woods, D.T. *et al.,* 1985, *Ap. J.,* **288,** 58.
McKee, C.F., and Ostriker, J.P., 1977, *Ap. J.,* **218,** 148.
Monier, R., and Halpern, J.P., 1987, *Ap. J. (Letters),* **315,** 17.
Pagel, B. E. J. 1989, in *Production and Distribution of CNO Elements,* eds. J. Danziger, F. Matteucci, K. Kjar, pp. 155-170 (Garching, ESO).
Planesas, P., Scoville, N.Z., and Myers, S., 1990, *Ap. J.,* accepted.
Pogge, R., 1988, *Ap. J.,* **328,** 519.
Rose, J. A., and Searle, L., 1982, *Ap. J.,* **253,** 556
Scoville, N. Z., Matthews, K. *et al.,* 1988, *Ap. J. (Letters),* **327,** L61.
Shull, J.M., and McKee, C.F., 1979, *Ap. J.,* **227,** 131.
Smith, M.G., Weedman, D.W., and Spinrad, H., 1972, *Ap. Lett.,* **11,** 21.
Snijders, M.A.J., Briggs, S.A., and Boksenberg, A., 1982, in *Proc. of 3rd European IUE Conf.* (Madrid)
Snijders, M.A.J., Netzer, H., and Boksenberg, A., 1986, *M.N.R.A.S.,* **222,** 549.
Sokolowski, J., Bland-Hawthorn, J., and Cecil, G. N. 1991, *Ap. J.,* accepted.
Telesco, C.M., Becklin, E.E. *et al.,* 1984, *Ap. J.,* **282,** 427.
Wang, Z., and Cowie, L.L. 1988, *Ap. J.,* **335,** 168.
Wilson, A.S., and Ulvestad, J.E., 1983, *Ap. J.,* **275,** 8.
Wynn-Williams, C.G., Becklin, E.E., and Scoville, N.Z., 1985, *Ap. J.,* **297,** 607.

Burbidge: Your evidence of heating in the ionized gas far out from the nucleus is very interesting. The work by Geoff and me in the 1960's, to which you referred, was done fairly near the nuclei of a number of non-radio galaxies. Since it was work that we did in conditions of bad seeing, the slit included fairly large regions around the nuclei. The only explanation we could come up with was heating by stellar winds from a concentration of red giant stars in the central regions of the galaxies.

Bland-Hawthorn: Many of the galaxies you observed are now known to have active nuclei and extended narrow-line emission, e.g. NGC 1097, NGC 1365, NGC 4258, etc.

Rubin: But doesn't M33 have [NII]/Hα ratios greater than unity off the HII regions, yet no AGN?

Bland-Hawthorn: Classic studies in the 1970's (e.g. Searle, Smith, Jensen) showed that [NII]/Hα is very high in the nuclear HII regions of M33. I was unaware that high ratios have been observed between the HII regions. M33, like the Milky Way and a number of other nearby spirals, is now known to have a diffuse ionized phase in the interarm region. As far as I know, [SII]/Hα is somewhat high but [NII]/Hα is not. Your point, as well as Dr. Burbidge's, is well taken: it may be that most warm ionized media are powered by a hot or an evolved stellar population.

Vilchez: Did you imply that the highest [NII]/Hα regions are associated with a different kinematics?

Bland-Hawthorn: The entire inner disk exhibits strong elliptic streaming. Presently, it seems that the HII regions (i.e. regions of low [NII]/Hα) coincide well with this phenomenon. The highest [NII]/Hα regions that occur along the radio axis show no obvious deviations, at a spectral resolution of $\sim$100 km s^{-1} FWHM, from the complex disk rotation.

A Diatribe on 'Theories' of the Chemical Evolution of Galaxies

D. LYNDEN-BELL

Institute of Astronomy, The Observatories, Cambridge CB3 0HA

ABSTRACT

Multiparametric computer models of chemical evolution are criticised and a reformed notation is advocated. Models with separable time evolution are preferred.

When a galaxy forms, the gas mass first grows by accretion and then finally decreases as a result of star formation. The G dwarf problem does not arise in such models because there is much less mass to dilute the supernova débris initially, so the interstellar abundance rises when fewer stars have been made. Models which both incorporate this behaviour and allow for the loss of supernova débris for small galaxies readily explain the mass–metallicity relationships of dwarf ellipticals and irregulars.

1. DIATRIBE

The *theory* of the Chemical Evolution of Galaxies, started by Schmidt, has since suffered from too many multiparametric computational models chasing too few established facts. Occam's razor has too seldom been used to cut away the inessential, and most 'results' remain as complicated graphs and tables of numbers left to moulder in the journals like the multitude of hypotheses that led to them. Seldom have the authors bothered to encapsulate the essence of their multitudinous output into a few key less accurate analytical formulae.

The essence of GOOD science (of which the paper by Searle & Sargent is a good example) lies in the 'UNDERSTANDING of the mechanism' and the encapsulation of that understanding into simple relationships readily transmitted to others. Even when a complicated parameter space has been fully explored by computer the good science has not yet been attempted! Among a number of exceptions to the above I would pick out for students of the field the following early key papers:

Van den Bergh: AJ, **62**, 492, 1958 and AJ **67**, 486, 1962; Schmidt: ApJ **129**, 243, 1959 and ApJ **137**, 758, 1972; Searle & Sargent: ApJ **173**, 25, 1972; Bagnuolo, Sargent & Searle: ApJ **168**, 327, 1971; Larson: Nature Phys. Sci. **236**, 7, 1972; Pagel & Putchet: MN **172**, 13, 1975 and a whole series of Pagel's papers, several written jointly with Edmunds (see references).

The one trespass I made into the field (Lynden-Bell 1975) was aimed to amend an unrealistic property of Pagel's fine generalization of the simple model. In his model the accretion rate is proportional to the star formation rate, but this has the unfortunate property that the mass in gas either rises monotonically or falls monotonically. Models in which the gas mass rises by accretion as the galaxy forms, and later declines as it is turned into stars seem much more natural. I showed that such models give a natural way out of Schmidt's G–dwarf problem. The reason why such models give much fewer stars of low metal abundance is that, unlike the simple model, only a small fraction of the final mass of the galaxy is present as gas when the first generation of supernovae explode. Thus their elements are mixed with much less gas and so raise it to a higher abundance than is achieved at that stage in the simple model. No stars of lower metal abundances are made in the succeeding generations. My secondary aim in writing that paper was to relieve the field of a poor mathematical notation in which the symbols employed had no understandable relationship with their meanings. In this I failed. Unrepentant I shall do this again here using g for the gas mass, s for the mass in stars, m for $g + s$ and y_i for the nucleogenic yield of chemical element i. The interstellar abundance by mass of element i will be Z_i. I strongly advocate that others adopt this simple notation.

The simple model has the beautiful property that the *rate* of star formation is irrelevant. What matters is the rate of increase of the interstellar element abundances not with respect to time but with respect to the total mass in stars. By parameterising the accretion and mass losing models sensibly, *e.g.* by specifying the total mass that has been accreted not as a function of time but as a function $A(s)$ of the total mass in stars, it is possible to maintain this separable property. Many calculations UN-NECESSARILY give this up and CONFUSE understanding by employing parameters that define the star formation rate to get $Z(s)$ or $s(Z)$ relationships. With sensible parameterisation the time evolution of the model is totally separate from the evolution of the abundances and the gas mass as functions of the mass in stars. Such a separation is a great help to clear thinking and should not be wantonly abandoned.

The field of chemical evolution is rapidly becoming fertile in that after many years of hard work variations between different elements are now showing systematic differences among stars of different metal content. It may be argued that the lag–times after star formation between Oxygen spreading supernovae and Iron spreading explosions or Carbon bearing planetary nebulae lead to significant changes to even the simple model. The instantaneous recycling approximation in which the lags are neglected may not always give accurate results. Nevertheless I make no apology for using it. Rather I seek apologies from all those who have not explored their models by using it. We are after INSIGHT not pernicious accuracy of ill–understood numbers. After we have a good basic theory with parameters quite well determined, the extra complications of non–instantaneous recycling may be merited.

In 1962 before Schmidt left the field to become a quasar astronomer, he told me two things:

1. The abundances of HII regions in Andromeda (and M33 (Aller 1942)) seemed to vary systematically being higher in the centre and lower at the outer edges

of the galaxies. Accurate work by Searle and then by Pagel and others have established this in a number of nearby galaxies.

2. The dwarf ellipticals had low stellar abundances and Schmidt attributed this to their loss of supernova débris due to the small binding of the gas into the weak potential wells of small galaxies. He remarked that the effect seemed to be systematic with luminosity, the depletion being much more obvious in the smallest dwarf galaxies. It was not known whether the effect saturated or continued among giant galaxies.

The intense excitement generated by finding the quasars 3C48 and 3C273 as identifications of two of the radio sources in the 3rd Cambridge catalogue took Schmidt away from Chemical Evolution and I suspect that my memory of lunch–time conversation is one of very few records of these early ideas and results. Work by Searle (1971) and Mould (1984) have accurately established them. Sandage and Visvanathan (1978, 1977) have also established that the brighter ellipticals are systematically redder and this is currently attributed to a systematic increase in metal abundance even in the more massive ellipticals.

This systematic variation of the effective yield with the mass of the system strongly suggests that supernovae can eject gas even from quite large ellipticals and this is confirmed by the high abundances in the X-ray emitting intracluster gas.

My 1975 accretion model of galactic chemical evolution did not allow for supernova driven mass loss. Here we give simple models that allow both for accretion and the loss of supernova débris.

2. EQUATIONS OF CHEMICAL EVOLUTION

Let F be the flux of mass being formed into stars of all types. Let βF be the flux of mass returned to the interstellar world almost immediately (in the instantaneous recycling approximation). Then the rate of increase of mass in stars is

$$\dot{s} = (1 - \beta)F. \tag{1}$$

The rate of increase of the total mass of element i in the interstellar medium is

$$
\begin{array}{ccccccc}
\frac{d}{dt}(gZ_i) & = & -Z_iF & + & (y_i(1-\beta)+Z_i\beta)F & \times & f. \\
\text{increase} & = & \text{loss} & + & \text{nucleogenically} & \times & \text{fraction not} \\
& & \text{into} & & \text{enhanced and re-} & & \text{expelled from} \\
& & \text{stars} & & \text{turned from stars} & & \text{galaxy}
\end{array}
\tag{2}
$$

We shall drop the index i to remove notational clutter and dividing by equation (1) we have the equation of chemical evolution,

$$\frac{d}{ds}(Zg) = -Z - \frac{\beta(1-f)Z}{(1-\beta)} + y f. \tag{3}$$

Even at this early stage we can see the effect of incomplete retention of supernova débris very clearly. When β is small the second term on the RHS is not very significant. On neglecting it, the change from the $f = 1$ case is confined to changing the yield y to an effective yield fy. Thus the way in which f varies with the mass of the galaxy or position within the galaxy, will be reflected in the effective yield which will determine the mean metallicity of the stars when the gas has been used up.

The numerous complications of supernovae exploding in a medium itself made up of a mixture of gas and supernova débris is too complicated to be discussed in detail. However, it is not implausible that for small galaxies the fraction of supernova débris finally retained should be proportional to the depth of the potential well formed by the stellar system. In ellipticals Fish in 1964 deduced that $M^2/R \propto M^{\frac{3}{2}}$ so that $M/R \propto M^{\frac{1}{2}}$. This suggests that we might take our retained fraction f proportional to $s^{\frac{1}{2}}$, but since $f \le 1$, this law can not be maintained for large s. When the potential well is very deep the loss of supernova débris should be negligible, so f should tend to 1 for s large. We shall therefore assume that there is a characteristic velocity v associated with the mixture of supernova débris and entrained gas which is ejected from the galaxy so that the fraction retained, f, is of the form

$$f = k\, s^{\frac{1}{2}}/(v^2 + k\, s^{\frac{1}{2}}) \qquad (4)$$

We take k and v^2 to be properties of supernovae in star forming regions and shall assume them to be universal constants. This hides a mass of ignorance but my efforts to do something better, founded on physics, led to no enlightenment so the simplest hypothesis seems best. Equation (3) can be re-written in the form

$$(ZgW)' = y\,f\,W \qquad (5)$$

where the integrating — or weight–factor W is defined by

$$
\begin{aligned}
(\ln W)' &= (Dg)^{-1} \\
D = D(s) &= (1 - \beta)/(1 - \beta\, f(s))
\end{aligned}
\qquad (6)
$$

D is a dilution factor due to mass loss that lies between 1 and $1 - \beta$. A dash denotes d/ds.

Equation (5) can be formally integrated into the form

$$Z(s) = (gW)^{-1} \int_0^s y\,f\,W\,ds \qquad (7)$$

For $f = 1$ we have $W = \exp \int_0^s g^{-1} ds$ while if $f \ll 1$ the weight factor is $\exp \int_0^s [(1 - \beta)g]^{-1} ds$, which is not very different for β small.

So far we have not specified how much mass is being accreted. The current total mass may be expressed as follows,

$$m(s) = s + g(s) = A(s) - E(s),\qquad(8)$$

where $A(s)$ is the total mass accreted up to the moment when the mass in stars is s, and $E(s)$ is the total mass ejected from the galaxy by then. From our definition of f

$$dE/dt = \beta F(1-f) \text{ and so } E'(s) = \beta(1-f)/(1-\beta).\qquad(9)$$

Thus from (8), using (6) for D

$$A'(s) = D^{-1} + g'(s).\qquad(10)$$

A specification of the functional form of $g(s)$ leads via (10) to a specific $A(s)$ and thence via (6) to a specific form for $W(s)$ which we use in (7) to find $Z(s)$. However, that path usually ends in unwieldy integrations that cannot be simply performed. Another path is more fruitful. Instead of postulating a simple rising and then falling functional form for $g(s)$ we could just as well specify $W(s)$ suitably and find the corresponding $g(s)$ from (6). Provided this has a sensible rising and then falling form and the $A'(s)$ given by (10) is never negative, this will lead us to physical models. We can choose the precise form of $W(s)$ so that fW integrates nicely in (7). Here we choose $W(s)$ so that in the limit $f \equiv 1$ in which all supernova débris is retained, then $g(s)$ becomes that of our old best accretion model. This is (see Appendix)

$$W(s) = s/(s_\infty - s),\qquad(11)$$

where s_∞ is the final mass of the then–all–stellar galaxy. Using this W in equation (6) we find

$$g(s) = D(s)^{-1}s(s_\infty - s)/s_\infty.\qquad(12)$$

This sets out proportional to s for s small but falls to zero as s approaches s_∞ just as we require. From (10) we then find that

$$A'(s) = (1-\beta)^{-1}(1 - s/s_\infty)\{2 - \tfrac{1}{2}\beta[(5/2)^2 - (5/2 - f)^2]\},\qquad(13)$$

which is never negative since $0 < \beta \leq 1$ and $f \leq 1$. Notice that $A'(s) \to 0$ as $s \to s_\infty$ so the accretion turns off appropriately.

On integrating (7) we find

$$Z/y = D\,I(x,\sigma),\qquad(14)$$

where

$$x^2 = s/s_\infty, \quad \sigma = v^2/(k\,s_\infty^{\frac{1}{2}}),\qquad(15)$$

and as before, but in new notation,

$$D = \frac{(1-\beta)(x+\sigma)}{[(1-\beta)x + \sigma]},\qquad(16)$$

and

$$I(x, \sigma) = x^{-4} \int_0^x \frac{2x^4}{(x + \sigma)(1 - x^2)} dx. \tag{17}$$

I is readily evaluated by partial fractions as

$$I = x^{-4} \left\{ \frac{1}{1 - \sigma^2} ln \left[\frac{(1 + x/\sigma)^{2\sigma^4}}{(1 - x)^{1-\sigma}(1 + x)^{1+\sigma}} \right] - x^2 + 2\sigma x \right\}. \tag{18}$$

The median abundance of the stars of a system that has no gas left, $\hat{Z}$, is the abundance that was in the interstellar gas when half the stars were formed, *i.e.* when $s/s_\infty = \frac{1}{2}$ and $x = 1/\sqrt{2}$. We plot $\hat{Z}/y$ against the final mass $M = s_\infty$. If we define the characteristic mass for escape by $M_e = v^4/k^2$ then $M/M_e = \sigma^{-2}$. Notice from (15) that for large σ, $I(1/\sqrt{2}, \sigma) \propto \sigma^{-1} \propto (M/M_e)^{\frac{1}{2}}$ and $D \simeq (1 - \beta)$. Thus at small masses $Z/y \propto M^{\frac{1}{2}}$.

The significance of M_e is that at greater galaxy–masses more than half of the supernova débris is retained in the galaxy. In fact $f(M_e) = \frac{1}{2}$. In Figure 1 we plot the median abundance $\hat{Z}$ divided by y as a function of the final galaxy mass. With both axes plotted logarithmically a change of yield corresponds to a vertical shift while a change in our assumed value of M_e corresponds to a horizontal shift. Fitting this plot to the observed [Fe/H] of local dwarf ellipticals (Mould 1984, Skillman *et al.* 1989) we get a good agreement in the slope at the low mass end. Terlevich's calibration of the metallicities of the SAMURAI ellipticals (Faber *et al.* 1989) show a lesser slope as is predicted by the model. They are reproduced in Figure 1. In the simple model the median abundance in the final distribution of stars is $y \, ln2 = 0.69y$. In our model in the large mass limit the final median is $y \, ln(2^4/e^2) = 0.77y$. The excess of low abundance stars in the simple model has lowered its median abundance by 10% for a given yield. However, to fit our model to the observations, the nucleogenic yields must be considerably higher than has normally been assumed. We are not merely enriching the galaxies but the intergalactic medium too. To fit the slope of the Abundance vs. Mass relationship of the Samurai ellipticals to the curve of Figure 1 we need M_e to be the mass corresponding to an elliptical of $M_B \sim -19^m$. If the measured abundances corresponded to median abundances in the stars then

[Fe/H] $= 0$ corresponds to $\hat{Z}/y = -0.6 \pm .1$ dex. However, it is probable that the measured abundances are closer to the means than to the medians in which case this gives $< Z > /y = -0.5 \pm .1$ dex (since log $0.77 = -.11$). This suggests that the true yields are between 2 and 4 times the solar abundances and that loss of supernova débris has lowered the effective yields to the observed range. The above calculations neglect the fact that the observed abundances in elliptical galaxies are measured at their brightest parts near the centre. Any concentration of their earlier gas content there would have enhanced their central abundances at the expense of their outer parts. Thus our estimates of their mean abundances may be overestimated.

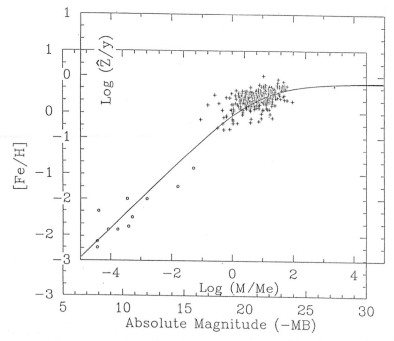

Figure 1. Median metal abundance per unit true yield as a function of final mass with observations of dwarf spheroidals and ellipticals superposed. The offsets of axes determine the true yield and the mass M_e that constrains supernova débris. The smaller mass to blue light ratio of the metal poor dwarf ellipticals may mean that they should be shifted leftwards on the mass scale by up to 1^m. Such a correction has not been made although it would improve agreement with theory.

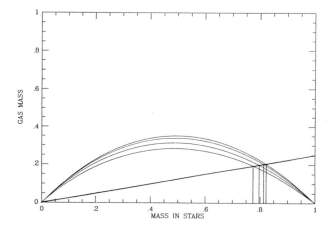

Figure 2. The gas mass as a function of the star mass for $(M/M_e) = 10, 1, 0.1, 0.01$ and $\beta = 0.3$. On intersection with the heavy line 25% of the galaxy is then gaseous (appropriate for the solar neighbourhood).

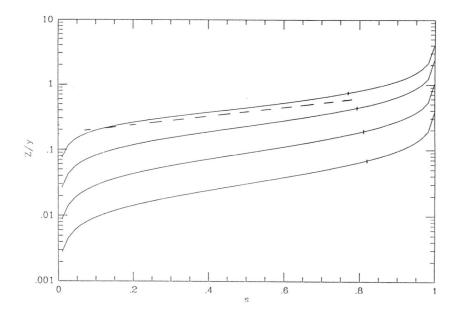

Figure 3. The cumulative distribution of metal abundances $Z(s)$ for $M/M_e = 10, 1, 0.1$ and 0.01 (reading from top to bottom). All are calculated for $\beta = 0.3$. The dashed line shows the run of the Pagel & Patchet data.

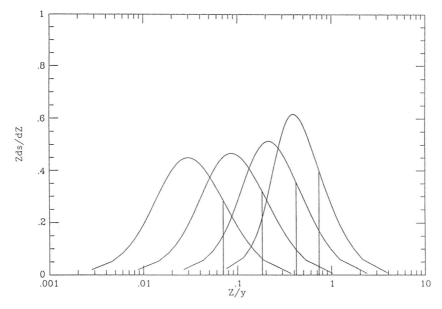

Figure 4. As for Fig. 3 but the differential distribution of metallicities in the stars $Z\,ds/dZ$ as a function of $\log Z$ for $M/M_e = 10, 1, 0.1$ and 0.01 (reading from right to left). These distributions are truncated as shown when the galaxy is 25% gaseous.

Figure 3 plots $s(Z)$ for a sequence of M/M_e values. These give the cumulative distributions of stars as a function of their metal abundance. Also plotted are the points given by Pagel & Patchet. As Bernard Pagel now prefers the differential distributions these are plotted as $ds/d\ln Z$ as a function of $\log Z$ in Figure 4.

I thank R.J. Terlevich for both the data and the production of Figure 1.

RETROSPECT

During seven happy years (1965–72) in which the R.G.O. at Herstmonceux Castle was ruled by Sir Richard Woolley I had the almost daily privilege of discussing astronomy with Bernard Pagel often while walking around the beautiful grounds. There I learnt from him the intricacies of the element abundances, which data were to be believed, and why! In the latter half of that period we often drove to work together. There was no lack of astronomical discussion in the car but in spite of the detailed thought required, I cannot remember even a near miss in the driving! We shall miss his critical wisdom but he will always be welcome at the RGOs new house in Cambridge. I am delighted to contribute this work to his 60th birthday Festschrift.

REFERENCES

Aller, L.H. , 1942. *Astrophys. J.* , **95,** 52.
Edmunds, M.G. & Pagel, B.E.J., 1984. *Mon. Not. R. astr. Soc.* , **211,** 507.
Faber, S.M., Wegner, G., Burstein, D., Davies, R.L., Dressler, A., Lynden-Bell, D. & Terlevich, R.J., 1989. *Astrophys. J. Suppl.* , **69,** 763.
Fish, R.A., 1964. *Astrophys. J.* , **139,** 284.
Lynden-Bell, D. , 1975. *Vistas in Astronomy,* **19,** 299.
Mould, J.R., 1984. *Publ. astr. Soc. Pacif.* , **96,** 773.
Pagel, B.E.J. , 1986. *Highlights Astr.,* **7,** 551.
Pagel, B.E.J. & Edmunds, M.G. , 1981. *Ann. Rev. Astr. Astrophys.,* **19,** 77.
Sandage, A. & Visvanathan, N., 1978. *Astrophys. J.* , **223,** 707.
Searle, L. , 1971. *Astrophys. J.* , **168,** 327.
Skillman, E.D., Kennicutt, R.C. & Hodge, P.W., 1989. *Astrophys. J.* , **347,** 875.
Visvanathan, N. & Sandage, A., 1977. *Astrophys. J.* , **216,** 214.

APPENDIX

For no ejection from the galaxy $W = \exp \int_0^s f^{-1} ds$. Thus for any chosen $W(s)$ we can define a corresponding gas mass $G(s)$ with no ejection by

$$G^{-1} = (\ln W)'.$$

For the system with ejection we have from (6),

$$g = D^{-1}/(\ln W)' = D^{-1}G.$$

From (10) the increase in the total accreted mass per unit increase of s is then

$$A'(s) = D^{-1} + (D^{-1}G)'.$$

For f of the chosen form (4) we have $f = x/(\sigma + x)$ and using this in (6) to get D^{-1} we derive

$$A'(s) = D^{-1}\left\{ 1 + G\left[(\ln G)' - \frac{\beta\sigma}{2x(\sigma + x)(\sigma + (1 - \beta)x)} \right] \right\}.$$

If G is non zero at $s = 0$ and if $G'(s)$ is not initially infinite, the last term dominates for small x, so $A'(s) < 0$ which is not physically possible. This is a property of the particular form (4) which relates the retained fraction to the star mass rather than the total mass. Systems that are built up from zero gas mass do not suffer so in the main text I took the W corresponding to the G of my old model with $\Gamma = 0$. The general form for G in the old model was $G = \frac{(s+\Gamma)(s_\infty - s)}{(s_\infty + \Gamma)}$. It is not hard to make other models with G initially zero. *e.g.* if

$$G(s) = s^{\frac{1}{2}}(s_\infty - s)/s_\infty^{\frac{1}{2}} = s_\infty x(1 - x^2),$$

then

$$W(s) = (1 + x)/(1 - x),$$

and

$$Z/y = D\, x^{-1}(1 + x)^{-2} \int_0^x \frac{2x^2(1 + x)dx}{(\sigma + x)(1 - x)}.$$

Performing the integration and using (16) for D we find

$$\frac{Z}{y} = \frac{(1 - \beta)(x + \sigma)}{[(1 - \beta)x + \sigma]} \frac{1}{x(1 + x)^2} \left\{ \frac{2}{1 + \sigma} \ln\left[\frac{(1 + x/\sigma)^{\sigma^2(1-\sigma)}}{(1 - x)^2} \right] - x^2 - 2(2 - \sigma)x \right\}.$$

Such models have $A' > 0$ provided $\sigma > \beta/3$. Models based on $G(s)$ depending on still lower powers of s are in trouble near $s = 0$ but those with higher powers do not suffer from this trouble near $s = 0$.

QUESTIONS

Peimbert. Have you computed the ratio of mass lost to intergalactic space to the mass left in stars in elliptical galaxies?

Lynden-Bell. That depends on the return fraction β and is readily found from formula (9). For fully burnt out systems one finds

$$\frac{E(s_\infty)}{s_\infty} = \frac{1}{1-\beta}\beta\sigma \left[1 - \sigma ln(1 + \sigma^{-1})\right]$$

where $\sigma^{-2} = M/M_e$. For $M << M_e$ σ is large and $E/s_\infty \to \frac{\beta}{1-\beta}$. For the lowest mass systems $M \sim 10^{-4} M_e$ and for $\beta = 1/3$; $E/s_\infty \to \frac{1}{2}$. But for $\gamma = 1$ this decreases to $\frac{1}{2}(1 - ln2)$ and for $M >> M_e$ the formula reduces to $E/s_\infty \to \frac{2\beta}{1-\beta}\left(\frac{M_e}{M}\right)^{\frac{1}{2}}$.

Terlevich. I was thinking about the way of solving the problem of ejecting the SN remnant. With standard values for the E_{SN} and disks column densities, it is necessary 10 times the E_{SN} to pierce the disk. So the possibility is connected with the fact that small gas rich galaxies and outer parts of spiral disks tend to have larger HII regions than metal rich disks. Then it may be a way to link the local mass density and shear with the size of the HII region and therefore the number of SN per star forming region.

Lynden-Bell. Yes, such a mechanism could help to give a law like that assumed here in which the disks of highest surface density loose less débris than disks of lower surface density.

Terlevich. Regarding Mike Edmunds' point about X-ray emission in Dwarf galaxies, that is observed. Large $L\alpha/L\beta$ has been detected in HII galaxies.

An important (perhaps) point is that there is now a strong upper limit to the amount of intergalactic hot gas given by the lack of Compton tail in the MWB spectrum. This gives a strong upper limit to the total mass of hot ($\sim 10^7$ K) plasma.

Lynden-Bell. Since only the small galaxies loose much mass and they form a small fraction of the total mass, the quantity of hot gas involved is not an embarrassment.

Díaz. The gas that is coming off in these SN débris, does it mix with the surrounding gas before leaving the Galaxy, and how would mixing change things?

Lynden-Bell. In the model there are two sorts of SN débris – trapped débris that mixes with the rest of the Galaxy and untrapped débris that escapes directly. I did make some models in which there was some entrainment of the normal interstellar gas into the escaping débris but I did not find characteristic effects that could be observed, so I stuck with the zero entrainment model. Entrainment models increase the total mass loss from dwarf galaxies which is already quite significant.

Pagel. 1) What happens if galaxies have dark halos?

2) What was the value of the free parameter in your 1975 accretion model (of which I continue to be a great admirer)?

3) There is an important distinction between mass-weighted and luminosity-weighted mean metallicity (cf. Arimoto & Yoshii).

4) Instantaneous recycling does not work for iron in the Solar neighbourhood. It may do so for E-galaxies if the stars were formed so fast that SNI do not contribute, but then O/Fe will be enhanced by a factor 3 or so, giving $y \simeq 10$ $Z_\odot$.

Lynden-Bell. 1) It depends whether dark haloes are baryonic and loose gas or are esoteric and only influence things gravitationally. In the latter case I assume they modify the potentials by a factor in keeping with the V = const conspiracy.

2) I took the galaxy to grow from nothing. Our assumptions break down if the gas mass is not zero initially as explained in the Appendix, which also gives a variant model.

3) These are important and difficult problems. The model most easily predicts median metallicities of the stars.

4) Once the outline models are properly established, it is certainly worth making models that incorporate the lags but not before.

Edmunds. 1) What about abundance gradients in larger elliptical galaxies – does this imply *local* control of gas loss?

2) Can you comment on the dynamical effects on the small galaxies of considerable mass loss?

Lynden-Bell. I have, for simplicity, assumed the loss is a global property of a galaxy but the sort of parameterization used in terms of the potential certainly suggests a more refined model in which the mass loss is related to the local potential. I have been looking at models that also incorporate some concentration of the gas within the system as well as local loss. Simple systems along those lines are possible.

2) If the scale of mass loss is several dynamical times, the system expands adiabatically but both accretion and mass loss should be considered. When the accretion dies down, the radius of the system behaves as $r \propto \frac{1}{M}$, *i.e.* $\rho \propto M^4$. The low density of the low mass dwarf spheroidals may be determined by this.

The Chemical and Dynamical Evolution of Galactic Discs

C.J. Clarke
Institute of Astronomy
Madingley Road, Cambridge, CB3 0HA

Section 1 Introduction

The study of the abundances of stars and nebulae offers insights into several areas of astrophysics, such as the primordial abundance of the Universe, the details of stellar nucleosynthesis and the evolutionary history of galaxies. Here we shall be concerned with this last topic: in particular we shall consider the role of chemistry in constraining the history of gas flows in galactic discs. In Section 2 we consider the case of an isolated disc galaxy, such as the Milky Way, and examine whether the abundance data is compatible with substantial radial redistribution of gas in the disc plane during its star formation history. In Section 3 we raise the possibility that galactic abundance distributions may be modified by environmental effects and discuss the likely chemical signatures of tidal encounters and of ram pressure stripping in rich clusters.

Section 2 Chemical Evolution of Isolated Galaxies

A variety of mechanisms may contribute to the dynamical evolution of gas in isolated disc galaxies. Any source of dissipation in the disc gas (such as collisions between Giant Molecular Clouds or shocks induced by spiral arms, for example) will drive radial flows in the disc plane. In addition, gas may have rained down into the disc, as a vestige of halo collapse, over a substantial fraction of the disc's star forming lifetime. With these effects in mind it is scarcely to be expected that the disc of the Milky Way has evolved as a sequence of concentric 'closed box' annuli in which a fixed total mass is gradually cycled between gas and stars. The relative importance of the various flow terms is however far from clear. In what follows 'in situ' models refer to the case of negligible radial redistribution of gas in the disc plane, 'closed box' to the case where in addition there are no source terms from the halo during the disc's star forming history and 'viscous' models (with or without late infall of gas from the halo), to the case where gas moves over significant radial distances in the disc plane before turning into stars.

In 'in situ' models the exponential mass distribution in galactic discs (Freeman 1970) is imprinted during the infall and presumably reflects the initial angular momentum distribution of the tidally torqued protogalaxy (Mestel 1963). In 'viscous' models, an exponential stellar distribution is generated from an arbitrary intial gas profile provided only that the radial flow (viscous) timescale of the gas is everywhere of order the star formation timescale (Lin and Pringle 1987, Yoshii and Sommer-Larsen 1989).

Numerous authors have attempted to constrain the evolutionary history of the Milky Way from Galactic abundance distributions (e.g. Wyse and Silk 1989 and references therein). Unfortunately, the chemical data admits of no unique model fit, and the best that can be done is to use chemistry as an *a posteriori* check on physically motivated evolution models. In the case of 'in situ' models, radial metallicity gradients can readily be obtained by any star formation prescription that favours preferential enrichment of the inner disc Fitting the solar neighbourhood stellar abundance distribution is more problematical however: there is a broad consensus that 'closed box' models predict an excess of low metallicity stars (the so-called G-dwarf problem, Pagel and Patchett 1975) and that therefore, in the context of 'in situ' models, some degree of late infall of halo gas into the disc plane is inevitable (Lacey and Fall 1985). The G-dwarf problem is therefore used as a standard piece of evidence in favour of the hypothesis that the halo collapse timescale is a non-negligible fraction of the Hubble time.

As far as 'viscous' models are concerned, the zeroth order model (no late infall of gas from the halo and a star formation law varying smoothly with radius) is scarcely capable of reproducing the main features of the chemical data. The strong radial flows virtually erase any abundance gradients arising from differential enrichment of the disc gas (Sommer-Larsen and Yoshii 1989, Clarke 1989). The resulting small abundance variation across the disc (~ 0.2 dex) may be barely compatible with that observed in the old stellar disc (Neese and Yoss 1988) but is certainly inadequate when compared with the large radial variations in nebular oxygen abundances (~ 1 dex) measured in the Milky Way and a number of extragalactic spirals (e.g. Pagel and Edmunds 1981). Moreover, the model solar neighbourhood abundance distribution is not radically different from 'closed box' models: it therefore shares their deficiency of over-predicting the number of metal-poor G dwarfs.

One solution to the latter problem is to permit delayed infall of halo gas into the disc. Sommer-Larsen and Yoshii 1990 have argued on this basis that delayed infall is a pre-requisite for both 'in situ' and 'viscous' models. This solution, however, leaves unsolved the problem of the small radial abundance gradient predicted by 'viscous' models.

An economical solution to both these problems is provided if star formation is suppressed at large radii (Clarke 1989, 1991). Truncated star formation is observed in a number of disc galaxies (van der Kruit and Searle 1982): in all cases the disc gas is observed to extend well beyond the star formation cut-off and constitutes, therefore, a large reservoir of presumably low metallicity gas. In 'viscous' models, the radial inflow of gas from this region will inevitably dilute the outer portions of the star forming discs, thus generating a substantial metallicity gradient at late times (Figure 1). Moreover, the presence of the low metallicity gas at large radii is also communicated to the solar neighbourhood and modifies the local stellar abundance distribution (Figure 2). The effect of radial inflow of unenriched gas is qualitatively similar to that of late infall of gas from the halo: in both cases the number of metal poor stars is reduced, in line with observational data.

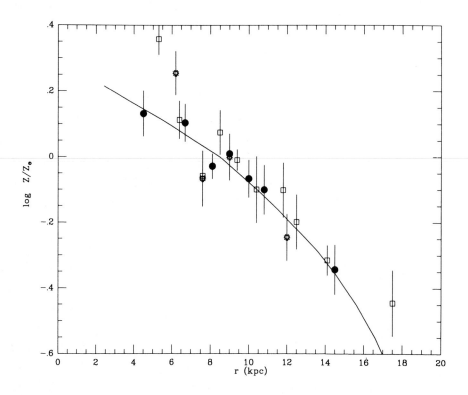

Figure 1 Model radial abundance distribution for a 'viscous' model with star formation edge at 18 kpc (for details see Clarke 1989), compared with observed abundance indicators compiled by Lacey and Fall (1985).

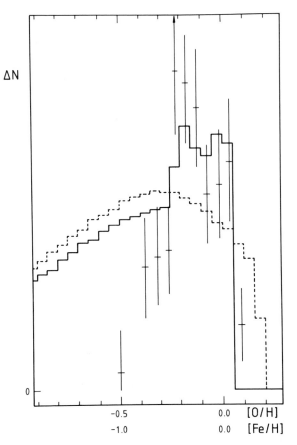

ΔN

-0.5   0.0   [O/H]
-1.0   0.0   [Fe/H]

Figure 2 Model solar neighbourhood abundance distribution for 'viscous' models with (solid line) and without (dashed line) the imposition of a star formation cut-off, compared with data from Pagel 1989.

In conclusion, therefore, 'viscous' models for disc evolution are only readily compatible with the chemical data if star formation is suppressed at large radii. Inflow from this region both generates a radial abundance gradient and helps to reduce the fraction of metal poor stars in the solar neighbourhood. Late infall of gas from the halo is not then a requisite of the model, although neither can it be ruled out.

Section 3 Chemical Evolution of Interacting Galaxies

Studies of radial abundance gradients in disc galaxies have shown a disappointing lack of correlation between chemistry and Hubble type, apart from the observation that barred galaxies (in which radial flows ensure radial mixing in much less than a Hubble time) are distinguished by a lack of abundance gradients (Pagel and Edmunds 1981). This fact raises the suspicion that the current abundances recorded in the disc gas may more reflect the recent history of burst mode star formation and radial gas migration rather than the properties of the stellar distribution. This is certainly the case in the viscous models described in Section 2: where a star formation edge is included, the present nebular abundance distribution is mainly determined by the diluting effect of the gaseous inflow at late times, when the star formation which determined the gross structure of the stellar disc has been largely completed.

The importance of galactic environment in modifying the most fragile component of galaxies (their interstellar medium) is becoming increasingly obvious from studies of the gas distributions in tidally interacting galaxies and in galaxies near the cores of rich clusters (see, for example, the review by Kenney 1989). If, as suggested in Section 2, abundance gradients indeed result from unenriched gas flowing in from large radii – or else from infall from the halo at large radii (Matteucci and François 1989) – then one would expect that abundance patterns should be sensitively coupled to the fate of the outermost gas. If, however, metallicity gradients result from differential enrichment within the star forming disc, one would expect such gradients to be little affected by gas dynamical processes at large radii.

At first sight, this suggests that in the former class of models substantial reservoirs of gas at large radii should be correlated with steep abundance gradients. This issue is however confused, for field spirals, by the effect of tidal encounters between galaxies, since these both strip off outlying gas *and* drive strong radial inflows in the residual gas (Noguchi and Ishibashi 1986): thus following a tidal encounter the abundance gradient generated by dilution should be enhanced. Therefore it is not at all clear whether a galaxy without extended HI has either never possessed a large scale gas distribution or whether it has been shorn in a tidal encounter. This uncertainty severely weakens the expectation that any statistical correlation should exist between extended HI and metallicity gradients in the discs of field galaxies.

In terms of statistical correlations between gas distributions and chemical gradients, the best laboratory is probably found in rich clusters of galaxies. Here there is considerable evidence that the main effect determining the distribution of gas in disc galaxies is ram pressure stripping by a hot intra-cluster medium (Warmels 1986). But whereas tidal encounters effect *global* changes in the disc gas, ram pressure stripping is a largely local effect, shaving off the loosely bound gas at large galactocentric radii but leaving the gaseous component of the inner discs more or less intact (Kenney and Young 1989). In this case, one would expect that galaxies that had experienced such stripping should have smaller abundance gradients than field galaxies, since they would forego the possibility of dilution of the outer disc by gas inflowing/infalling from large radii. This effect would also lead to higher mean abundances in cluster spirals, an effect recently reported in Virgo by Shields,

Skillman and Kennicutt (1990). Comparison of abundance *gradients* (a second order effect) is inevitably a harder observational task, particularly given the truncated gas distributions in cluster spirals. It is very much to be hoped, however, that such information might be extracted through careful observation of a larger sample of Virgo spirals.

4. Conclusion

We have discussed how radial migration of gas can affect the chemical evolution of galactic discs, both in isolated spirals and in those subject to environmental interactions. We have shown that chemical evolution of disc galaxies may be quite sensitive to the behaviour of low metallicity gas at large radii: contamination of the star forming disc by this unenriched gas can both generate metallicity gradients and can help to alleviate the solar neighbourhood G dwarf problem. We have suggested that comparison of the abundance distributions in field spirals and in those subject to ram pressure stripping in rich clusters may help to disentangle the processes responsible for generating metallicity gradients.

References
Clarke, C.J. 1989, *Mon. Not. R. astr. Soc.* **238**, 283.
Clarke, C.J. 1991, *Mon. Not. R. astr. Soc.* in press.
Freeman, K.C. 1970, *Astrophys. J.* **160**, 811.
Kenney, J. 1989. In *The Interstellar Medium in External Galaxies*, Proceedings of 2nd Wyoming Conference, Grand Teton National Park, July 1989.
Kenney, J. and Young, J.S. 1989, *Astrophys. J. Suppl.* **66**, 261.
Lacey, C.G. and Fall, S.M. 1988. *Astrophys. J.* **290**, 154.
Lin, D.N.C. and Pringle, J.E. 1987, *Astrophys. J. Lett.* **320**, L87.
Matteucci, F. and François, P. 1989, *Mon. Not. R. astr. Soc.* **239**, 885.
Mestel, L. 1963. *Mon. Not. R. astr. Soc.* **126**, 553.
Neese, C.L. and Yoss, K.M. 1988, *Astr. J.* **95**, 463.
Noguchi, M. and Ishibashi, S. 1986, *Mon. Not. R. astr. Soc.* **219**, 305.
Pagel, B.E.J. 1989. In *Evolutionary Phenomena in Galaxies* J.E. Beckman & B.E.J. Pagel (eds.) C.U.P. p201.
Pagel, B.E.J. and Edmunds, M.E. 1981, *Ann. Rev. Astr. Astrophys.* **19**, 77.
Pagel, B.E.J. and Patchett, B.E. 1975, *Mon. Not. R. astr. Soc.* **172**, 13.
Sommer-Larsen, J. and Yoshii, Y. 1989, *Mon. Not. R. astr. Soc.* **238**, 133.
Sommer-Larsen, J. and Yoshii, Y. 1990, *Mon. Not. R. astr. Soc.* **243**, 468.
Shields, G., Skillman, E., and Kennicutt, R.C. 1990, *Astrophys. J.* submitted.
van der Kruit, P.C. and Searle, L. 1982, *Astr. Astrophys.* **110**, 79.
Warmels, R.H. 1986, Ph.D. Thesis, University of Groningen.
Wyse, R.F.G. and Silk, J. 1989. *Astrophys. J.* **339**, 700.
Yoshii, Y. and Sommer-Larsen, J. 1989. *Mon. Not. R. astr. Soc.* **236**, 779.

Lynden-Bell: Do the Virgo spirals show any evidence for flatter abundance gradients?

Clarke: The gradients do look flatter, at a glance, but this is only based on Shields *et. al's* five Virgo spirals.

Rubin: In Virgo, the galaxies which are ram pressure stripped in HI are not stripped in CO. This must be relevant to star formation and abundance.

Clarke: Yes, it certainly supports the idea that ram pressure stripping only affects the outer, low density region of galactic discs.

Pagel: I should like to stress the point of Shields, Skillman and Kennicutt again: at the high abundances found in Virgo galaxies, the calibrations are particularly uncertain, which makes it very difficult to compare gradients.

Clarke: I appreciate the difficulty. It would still be interesting, though, to see whether for a larger sample the data were consistent with *no* intrinsic gradient within the inner 1.5 effective radii which appears to be the case for the present small sample.

Element Abundances and the Chemical Evolution of Galaxies

M.G. EDMUNDS

Department of Physics and Astronomy,
University of Wales College of Cardiff,
P.O.Box 913, CARDIFF CF1 1XL.

SUMMARY
Some developments in the determination of chemical abundances in galaxies since the review of Pagel and Edmunds (1981) are discussed, together with various theoretical constraints. Comment is made on the calibration of abundance indicators, particularly for HII region emission-line spectra and the Mg_2 index in composite stellar populations.

1. GAS ABUNDANCES
The most reliable determinations of chemical abundances in nearby galaxies are still given by the direct measurement of gas abundances by analysis of HII region spectra. Such analyses are not without their problems - for example, it remains an open question whether condensation of elements onto interstellar dust significantly depresses the gas phase abundances - but, as will become apparent in section 6 , the gas abundance estimates are superior to estimates based on the integrated light of stellar populations. Even where *individual* star abundances can be determined, it is not clear how representative such stars are of the bulk of the population, until large samples are available.

The "strong line" method for HII region chemical analysis, based on the strength of [OII] $\lambda 3727$ and [OIII] $\lambda\lambda 4959,5007$ lines combined in a R_{23} index has met with considerable success since its original proposal by Pagel *et al* in 1979. The index is calibrated as a function of oxygen abundance against HII regions where the electron temperature is well determined from observation of other lines (particularly [OIII] $\lambda 4363$) or, at the metal-rich end, against detailed models of the excitation and line emission of HII regions. Various revisions have been proposed, particularly by Edmunds and Pagel (1984), MaCall, Rybski and Shields (1985), Evans and Dopita (1985) and Vilchez (1989), lowering the abundances at the high end. For low abundances, Skillman's (1990) recalibration is not far from Edmunds and Pagel (1984). A current "best guess" calibration is given in Figure 1. Our own work (Diaz *et al*

1987) on spatially resolved spectra of the HII region NGC 604 in M33 gives us further confidence that the method works, since a near-constant R_{23} abundance index is observed despite considerable excitation variation across the nebula. Some refinements in the choice of a model to represent particular HII regions can be introduced by using a radiation hardness parameter η (Vilchez and Pagel 1988, Pagel 1990) defined by the ratios of [OII],[OIII],[SII] and [SIII] lines now that CCDs allow routine observations of the near infrared [SIII] λ9069,9532. Care should, of course, always be exercised when applying the "strong line" methods. They were never intended to achieve an accuracy of more than about ± 0.2 dex, and extra effort to observe temperature-sensitive lines (where this is possible - such as at the right hand side of Figure 1) is always worthwhile.

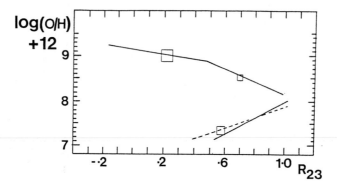

Figure 1 Calibration of the R_{23} abundance index. The basic calibration follows Edmunds (1989), the dotted line shows Skillman's (1990) calibration at low abundances. The representative HII regions are M83 nuclear region from Bonatto, Alloin and Bica (1989), NGC 604 in M33 from Diaz *et al* (1987) and I Zw 18 from Campbell (1990).

It is important that the method should only be applied to straightforward giant HII regions, and not where shock heating or unusual sources of ionizing radiation are contributing significantly to the excitation of the nebula. The high [NII]/[OII] line strength ratios observed near the nuclei of some galaxies are classic examples of the problem. The cause is almost certainly *not* a *large* overabundance of nitrogen, but the presence of hard radiation (note the paper by Cid Fernandez *et al* in this volume). It can be seen, for example, in NGC 1365 where the [NII]/[OII] increases in directions *away* from the nucleus as gas sees what is probably the bi-conical hard UV radiation field (Edmunds, Taylor and Turtle 1988, Phillips *et al* 1983) of the currently popular active galactic nucleus model. (See also the fine extended measurements by Joss Bland-Hawthorn elsewhere in this volume). Vera Rubin has reminded us that diffuse gas [NII]/[OII] ratios are high in the interarm region of M33. This might again suggest the presence of hard radiation, and one might speculate that it is due to bare stellar cores or something similar which are not particularly obvious at visible wavelengths. Perhaps such objects may show up in our own Galaxy in the ROSAT EUV/soft X-ray survey, if the radiation is not too absorbed along the line-of-sight by gas in the spiral arms. It is already known that there is considerable far UV flux present in ellipticals - probably from stellar sources.

While on the subject of nitrogen, the disparity between abundances measured in nebulae by optical and by infrared line remains a worrying problem which needs further work. On the origin of nitrogen, it is interesting to see that our ancient suggestion (Edmunds and Pagel 1978) that nitrogen abundances may reflect the age of a galaxy (in the sense of time elapsed since the bulk of star formation ocurred) still has some support (Garnett 1990).

As an example of the well-defined abundance gradients found in spirals, we show just one typical case, NGC 2997, in Figure 2.

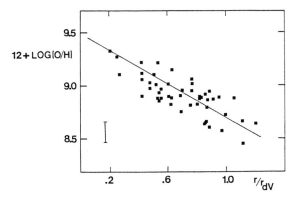

Figure 2 The radial oxygen abundance gradient in NGC 2997. The observational points are based on the compilation of line strengths by Walsh and Roy (1989), using the R_{23} calibration given in Edmunds (1989). The radial distances are normalised to the de Vaucouleurs photometric radius

As discussed by Angeles Diaz in this volume, problems arise when an attempt is made to discover correlations between the magnitude of the abundance gradient and other galaxian properties. All spirals seem to have a negative gradient outwards from the centre , but there is no obvious correlation of the slope or absolute level of the abundances with either Hubble type or mass - at least among the more massive spirals. Barred spirals tend to have flatter gradients, and in the Magellanic or dwarf irregulars there seems to be little or no gradient (Skillman, Kennicut and Hodge 1989). Begonia Vila-Costas and I are currently trying to assemble as much information as possible on those spirals for which we have reliable abundances, and hopefully some systematics may emerge. There is still much room for further observation, particularly of very massive spirals such as those in the Rubin sample (Rubin *et al* 1980,1982,1984) and - if possible - of large low-surface-brightness galaxies like Malin 1.

The sulphur/oxygen ratio has been suspected of varying across some spirals (e.g. M33; Vilchez *et al* 1988), and although a real variation is not yet established unambiguously, there is also some reasonable evidence for a variation across M51 (Diaz *et al* 1991).

2. INDIVIDUAL STARS

For a few nearby systems it is possible to obtain individual stellar abundances, and the Magellanic Cloud data have been reviewed by Russell and Bessell (1989), including F supergiant data of their own. The deduced metallicities $[Fe/H]_{SMC} = -0.65\pm0.2$, $[Fe/H]_{LMC} = -0.30\pm0.2$ are in good agreement with HII region measurements of oxygen abundance compared to local Galactic HII regions. There is no evidence for abundance gradients in the stellar component of the LMC or SMC. The nebular and stellar carbon abundances are not in good agreement, and the correct value remains uncertain. For our own Galaxy it has been suggested recently (Fitzsimmons *et al* 1990) that B stars show little abundance gradient, in contrast to the fairly well established HII results. Since B stars are necessarily young, one would expect them to follow the gas abundances, and as shown in Figure 3 it may well be that the stellar results are not really incompatable with a gradient of about -0.07 dex kpc^{-1}.

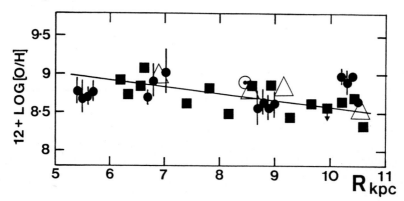

Figure 3 The Galactic abundance gradient near the Sun. The symbols with error bars are Fitzsimmons *el al* (1990) LTE analyses of B stars in four clusters, the solid squares are Shaver *et al* (1983) "best" HII region analysis, the open triangles are radial averages of 36 open clusters, assuming a solar [O/Fe] ratio, and scaling

the [Fe/H] determined by UBV ultraviolet excess (Cameron 1985) to the solar O/H abundance. The solid line is a -0.08 dex kpc^{-1} gradient passed through log(O/H) = 8.7 at 8.5 kpc.

3. THEORETICAL CONSTRAINTS

Before delving further into observational data, it is worth looking at a bit of background theory. The basis of chemical evolution theory is the "simple" model (e.g. Pagel and Patchett 1975), which is probably never completely applicable to any real galaxy but, as Bernard Pagel once aptly described it, is "something to hold onto in an uncertain World". The simple model assumes a closed region of gas which is gradually turned into stars, enriching itself chemically as a result of stellar evolution. The amount of heavy elements produced from each new episode of star formation is characterised by the *yield* p, a number which eventually may be calculated from supernova models, star formation and stellar eolution theory, but in the forseeable future must be deduced from observation. Once the ratio of gas mass to total (gas plus star) mass has reduced to a fraction f, then the metallicity in the gas will have risen to z = p ln (1/f) and the mean stellar metallicity (weighted by *mass* of stars rather than their *luminosity*) will be $< z > = $ p - p [f ln (1/f)]/(1 - f).

Real galaxies (or regions of galaxies) are not necessarily closed systems, and it is important to know the effect of gas flows in or out of the region considered. I recently made some progress in systematising the general effects of gas flow (Edmunds 1990 - where proofs will be found), results which many workers will instinctively know already but which have not, as far as I am aware, been set out in detail. It is useful to define the *effective yield* p_{eff} for a system as the yield that would be implied *if the system were regarded as closed i.e. "simple"*. Thus for a system with gas fraction f and gas metallicity z, the effective yield is defined by $p_{eff} = z/\ln(1/f)$. It is possible to demonstrate at least eleven "theorems" - although theorem is probably far too grand a word!

T(1) In a model with outflow, but no inflow, the effective yield is always less than that of the simple model.
T(2) In a model with inflow whose metallicity does not exceed that of the system gas itself, the effective yield is always less than that of the simple model provided the outflow rate exceeds the inflow rate.
T(3) In a model with inflow of unenriched gas, the effective yield is always less than that of the simple model.
T(4) In a model with outflow but no inflow, the ratio of secondary to primary elements is unchanged from that expected in the simple model.
T(5) In a model with unenriched inflow, the ratio of a secondary to a primary

element is always greater than for the simple model (this result is unaffected by any outflow).

T(6) The g-dwarf problem cannot be solved by any outflow.

T(7) The g-dwarf problem can be solved by *particular* forms of inflow.

T(8) The mass-weighted mean stellar abundance in a model with outflow is always less than that of the simple model.

T(9) The mass-weighted mean stellar abundance in a model with unenriched inflow is always less than, or (in the limit of gas exhaustion) equal to, the true yield p.

T(10) If the gas initially has abundance z_0 and any subsequent inflow has constant abundance z_0, then:

(i) T(1) and T(3) are unchanged except that the simple model prediction is now $z = p \ln (1/f) + z_0$.

(ii) T(8) is unchanged, except that the prediction of the simple model is increased by z_0.

(iii) In T(9) $< z >$ is always less than, and in the limit equal to, $p + z_0$.

T(11) The *overall mass-weighted mean stellar abundance* for a system cannot exceed the yield p. This requires including all stars in the mean abundance, whatever their position in the galaxy , and is unaffected by any flows of gas between or out of regions of the galaxy, or by unenriched inflow from outside the galaxy into any region.

The power of these results is that they are independent of the details or form of both the gas flow and the star formation rates. They do, of course, require that the "true" yield p remains constant.

An incompleteness in the above results is that in many situations the gas flowing into a region may be enriched in metals, and that enrichment may change with time - for example radial flow in a disk galaxy. To analyse such systems is much more difficult, and recourse has often been made to numerical models. At Cardiff we are making some headway into analytic results for more realistic models of spirals and ellipticals (Edmunds and Greenhow 1991). General theorems are much harder to find, and it is usually necessary to assume some form of star formation law. We have *some* fairly general results, using star formation laws of the Schmidt form $k\rho^n$, where ρ is the gas density, n need not be integral, and k can vary with radius. I will just mention a couple of rather interesting conclusions. The most obvious *a priori* choice for a star formation law might be with k constant and n = 1. (This is "linear" star formation - described by Manuel Peimbert as the "chicken broth" model, since if you have chicken you can make chicken broth, if you have no chicken ......). It can be shown that for almost any initial gas configuration and any radial flow *no* abundance gradient will be generated in the gas. It is possible, however, to generate a gradient in the mean stellar abundance as a function of position. We deduce that in real galaxies n is not 1 and/or k does vary with radius. It can also be shown that radial gas flow can make a gas metallicity gradient either steeper or shallower, depending on the details of k(r).

4. TESTING CHEMICAL EVOLUTION MODELS

Observational testing of chemical evolution models is not easy. An obvious start would be to compare gas metallicities and gas fractions to investigate effective yields. Although some work has been done on this (e.g. Garnett and Shields 1987; Matteucci and Chiosi 1983), there is a major difficulty in determining the gas fraction. The HI measurements are reasonably straightforward, but the (often quite significant) mass of molecular hydrogen must be inferred from measurements of carbon monoxide lines. This is done by a relation of the form: Column density of H_2 molecules = X *times* (CO line measurement). If X is a universal number, rather profound relationships between molecular content and Hubble type begin to appear (Young and Knezek 1989). But it seems very unlikely that X *is* a constant, indeed it is easy to argue that it should not be. Various molecule pundits insist on constants differing by up to a factor of four, and in recent work Whitworth (1990) has shown fairly convincingly that X should be a *highly* nonlinear function of local environment in terms of total hydrogen mass, density, chemical abundances and UV radiation field. There is also some evidence from the Magellanic clouds that molecular masses based on (admittedly uncertain) virial arguments are considerably higher (factors of six) than implied by straight application of a constant X. So any comparison of z and f must take into account possible variation of X with other factors (such as z itself), and it is premature to draw firm conclusions until the molecular conversion factor is sorted out. A rough compilation for irregular galaxies (essentially without H_2 measurements - although it may be fatal to neglect the molecules) is shown in Figure 4, which would be consitent with low (≤ 0.005) and variable effective yields.

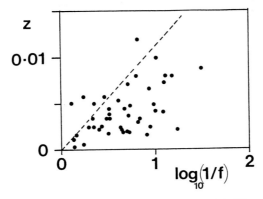

Figure 4 Compilation of metallicities as a function of (HI) gas fraction f for irregular galaxies, based mainly on the compilation of Vigroux, Stasinska and Comte (1987).

The dashed line is the simple model prediction with yield p = 0.005.

Variable yields (cf **T(1)** and **T(3)** above) might - as has frequently been suggested - imply outflow or unenriched inflow, although I note in passing that it is difficult to "fine tune" inflow model to give *very* low effective yields, while outflow does this easily. There are considerable differences in the predicted relative numbers of stars as a function of metallicity in the stellar populations of inflow and outflow models, but observational tests based on surveys of individual star analyses (even photometric) are not yet available. **T(5)** suggests that some tests for inflow based on element ratios might be possible, although it is not yet clear how big such effects might be.

The possibility of regulation of outflow by the gravitational field has been a popular mechanism for explaining metal abundance/absolute magnitude relations for galaxies. It can be questioned whether such relationships are fundamental, or reflect selection effects in the choice of galaxy samples. It remains quite possible that selection effects cause us to look at galaxies along a particular absolute magnitude/surface density (or brightness) relation, and that the important variation is of metal abundance with surface density. There seems to be fair evidence for a relation between gas metal abundance and surface density in spirals (Edmunds and Pagel 1984, Garnett and Shields 1987) - although the sample needs extending to more extreme low and high density systems. For dwarf irregulars Skillman, Kennicut and Hodge (1989) plot a diagram (their Figure 2) which combines gas abundance measurements in dwarf irregulars with stellar abundance measurements for dwarf spheroidals to give an apparently linear relation between log z and absolute blue magnitude, over a range of some 11 magnitudes. It is difficult to know exactly what to make of this diagram. Apart from the suggested selection effects, it is not obvious (even if one believes in a mass/abundance relation) that the stellar abundances in dwarf spheroidals should match up with gas in dwarfs - a difficulty which is acknowledged by Skillman *et al*. If the spheroids were formed fast and early, they might be expected to have [O/Fe] = +0.5 like old Galactic stars, and it is [Fe/H] which is mesured in spheroidals while [O/H] is measured in irregulars. The mean stellar abundance of a stellar population can be at least 0.3 dex less than its gas metal abundance, and if colours of stellar populations are used to estimate metallicity (see Section 6 below) the deduced value will be even lower if the colours of single-metallicity populations are used as the calibrators. Perhaps the difference in mass/blue light ratio between irregulars with their bright young blue stars, and the old population of the spheroids is enough to give a fortuitous compensating shift in blue absolute magnitude. Time will tell, but I'll predict that the absolute magnitude/metal abundance relations will become more scattered as more systems are discovered which have similar absolute magnitudes but considerbly different surface brightnesses.

It would be good to compare gas and mean stellar population abundances in spiral disks, but the presence of bright young stars effectively precludes analysis of visible colours and spectral features as a measure of the abundances in the bulk of the population. Perhaps it might be possible to identify suitable infrared features which would be less contaminated by the recent star formation (I am grateful to G.Meurer for this suggestion).

The origin of a relation between surface density and abundance is not clear. One possibility is the existence of a threshold in gas surface density below which massive (metal-producing) stars do not form. Skillman (1986) suggested from dwarf galaxy observations that 10^{21} H atoms cm^{-2} is a critical value, and similar proposals have been made for spiral disks (Tenorio-Tagle and Bodenheimer 1988, Kennicut 1989). Phillipps, Edmunds and Davies (1990) have explored surface density/abundance relations for dwarf galaxies on this basis. Put simply, the relation is explained by a low density region having few stars and few metals because of the threshold effect, while in denser regions the star numbers (and hence surface brightness) will be higher and the massive stars will have contributed relatively more metals. Outflow can also give some sort of a relation by reducing the effective yield, and lessening the stellar density by removing some of the gravitationally binding mass, allowing the system to expand. There are obvious restrictions on the rate and amount of mass loss if complete unbinding of the system is to be avoided. This is a problem that is worth further investigation, and some calculations of the effect of outflow on structure do exist (Carlberg and Hartwick 1981, Fukunaga-Nakamura and Tosa 1989). Phillipps and Edmunds (1991) look at the implication of thresholding for spiral disks, but it seems that this mechanism cannot be the main cause of abundance gradients.

5. ABUNDANCES IN ELLIPTICAL GALAXIES

Despite considerable effort, the metallicities of elliptical galaxies are not well determined. There are two major problems. The first is observational, and is due to the low surface brightness of ellipticals which makes accurate determination of colours or line strengths difficult, not just because of low signal-to-noise but because the surface brightness rapidly approaches the brightness of the night sky at radial distances not far from the nucleus. For example, in Thomsen and Baum's (1989) study of the fairly bright Coma cluster elliptical NGC 4839 the average sky brightness in V was around 21.2 mag $arcsec^{-2}$, a surface brightness which the galaxy reaches only 11" from the nucleus. Errors in the level of sky subtraction can therefore seriously affect colour or line strength measurements, particularly in the outer parts - and to determine a gradient one obviously wants as long a radial baseline as possible. The second difficulty is interpretation, since the light observed is the integrated spectrum from an (initially unknown) mix of many types of star. The different stars may vary in luminosity, temperature, age and chemical composition, so it is perhaps surprising that any conclusions can be drawn at all! Nevertheless there are some suggestive observed trends. A recent review is that of Kormendy and Djorgovski (1989).

From (B-R) and (U-R) colours, Franx, Illingworth and Heckman (1989) find that colour gradients becoming bluer outwards are "almost universal" in a sample of some 17 galaxies, a result reinforced by a further 39 ellipticals in Peletier *et al* 1990, except that significant gradients were not found in galaxies fainter than $M_B \sim$ -20. There are no other obvious correlations of the *gradient* with luminosity. Typical gradients are Δ(B-R)/Δlog r = -0.07 and Δ(U-R)/Δlog r = -0.23, which could be interpreted (see Figure 7 below) as a metallicity change of about 0.2 to 0.3 dex per decade in radius. This gradient is also typical of those found in line-strength observations (e.g. Gorgas, Efstathiou and Salamanca 1990, Thomsen and Baum 1989,1990, Coutre and Hardy 1990). Gorgas *et al* suggest d[Fe/H]/dr = -0.22±0.10, where "Fe" is really representing metallicity rather than just iron abundance, and it is found that the magnesium Mg_2 line strength index and iron line strengths may have a somewhat different radial behaviour, a point also noted by Davies and Sadler (1987).

There is also the question of the "well-known" colour-luminosity relation for ellipticals. Observationally there is indeed a trend for more luminous ellipticals to be redder, but as mentioned in the last section, this should not be taken at face value. Edmunds and Phillipps (1989) emphasise that the existence of a clear relation between luminosity and surface brightness for ellipticals (see e.g. Bingelli, Sandage and Tarenghi 1984) - whether due to physics or selection effects - implies that there is also a relation between abundance and surface brightness. Which is more fundamental? The current position is obscure. The (B-H) colour might be expected to be a reasonable abundance indicator, since it has a long baseline, and B is likely to be much more sensitive to line blanketing than H. Although Kormendy and Djorgovski (1989) suggest that "the straightforward colour versus density or surface brightness relations are not convincing", Edmunds and Phillipps found that (B-H) correlated better with surface brightness than with absolute magnitude, but for Mg_2 the correlation with absolute magnitude was very tight, and not at all good with surface brightness. The implication is that (B-H) and Mg_2 are not measuring quite the same thing. Although CO blanketing may affect the H band (Frogel *et al* 1990), the effect is unlikely to be large enough to completely vitiate the use of (B-H) as an abundance indicator, although it certainly seriously affects infrared colours.

6. CALIBRATION OF ABUNDANCE INDICATORS FOR OLD STELLAR POPULATIONS

In an attempt to see if Mg_2 is a good abundance indicator we give here a very rough *semi*-empirical calibration. The recent models by Buzzoni (1989), based on Salpeter mass functions, are a great help in estimating the photometric behaviour of composite stellar populations because of the convenient way in which they are presented. We start by noting three important effects which can be deduced from these models:

(i) If a given mass of interstellar material is formed into stars, then the brightness of the population at some later time decreses with increasing metallicity. In particular, at a given age, metal-poor giants are considerably brighter than metal-rich ones. The effect is illustrated in Figure 5.

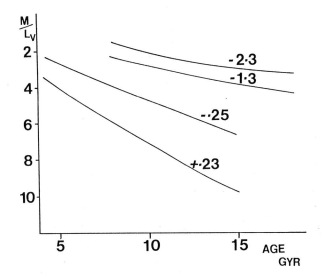

Figure 5 Mass-to-visible-light ratios (equivalent to relative V band fluxes from equal initial mass populations) from Salpeter IMF populations of stars. The abscissa represents the age of the population since its formation, and the figures on the plot lines give the (uniform) metallicity [Fe/H] of the stars in the population. Data taken from Buzzoni (1989).

(ii) Despite the complication of (i), populations of different metal abundance fortunately fade in brightness at about the same rate.

(iii) Different parts of the population may dominate the light in different regions of the spectrum, and the relative dominance is a function of metallicity. For example, in both metal poor and metal rich populations the red giants dominate in the infrared (J,H,K) bands, but in the visible (e.g. at the 5160A Mg_2 feature) the main sequence stars may dominate, particularly at high metallicities.

To calibrate Mg_2 we begin by noting from Buzzoni's models that as a first approach for an old stellar population we can approximate the light as coming from just main sequence and red giant stars. We chose an age of 15 Gyr, although this choice is probably not critical and the results may not alter much in the range 10 - 18 Gyr. From stellar evolutionary tracks it is possible to deduce a "mean" luminosity and effective temperature for red giant stars of a given age and composition. For the main sequence stars, an examination of the behaviour of the magnesium index with effective temperature together with the dependence (from models) of effective temperature and luminosity with mass, convinces one that (for a reasonable Salpeter mass function) the luminosity at a wavelength of 5160A is dominated by stars *at* the turnoff, rather than further down the main sequence. The Mg_2 indices of these "representative" stars are given in Table 1, and have been deduced from *observed* Mg_2 strengths in giants and dwarfs as a function of effective temperature and metallicity, using either published data (particularly Faber *et al* 1985) or unpublished data on dwarfs, kindly supplied by Sandy Faber.

The Buzzoni models can be used to combine these indices together with the correct weightings to give the overall predicted indices for a composite population. The result is shown in Figure 6, which represents the Mg_2 index of it single uniform metallicity populations of age 15 Gyr as a function of metallicity.

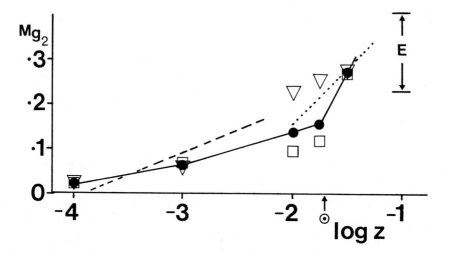

Figure 6 Rough Mg_2 index calibration for single abundance stellar populations 15 Gyr old. The open triangles represent the red giant stars, the open squares the main-sequence dwarfs, and the filled circles are a combination of these using Buzzoni's (1989) population models. The dashed line is the empirical calibration

from globular clusters by Brodie and Huchra (1990), which is almost the same as Gorgas *et al* (1990) and the dotted line is Terlevich *et al* (1981) which uses Mould's (1978) theoretical slope, empirically fixed at [Fe/H] ~ -0.5. The E indicates the typical range of observed elliptical galaxy indices.

Table 1 Estimated Mg_2 indices of "representative" stars

[Fe/H] = 0 corresponds to z(solar) = 0.02

	[Fe/H]= -2.30	-1.30	-0.30	-0.05	+0.20
Average red giant	0.023	0.061	0.230	0.252:	0.274
Turnoff MS dwarf	0.020	0.066	0.095	0.120	0.265
Relative giant/dwarf contribution at 5160A	0.86	0.89	0.52	0.37	0.33
Combined population	0.020	0.061	0.137	0.155	0.267

The following points can be seen:

(i) Up to metallicities of 1/2 solar - i.e. covering the globular clusters - the calibration between Mg_2 and log z is well defined, linear and agrees well with the empirical calibrations of Gorgas, Efstathiou and Salamanca (1990) and Brodie and Huchra (1990).

(ii) At metallicities from about one-fifth to twice solar, the dwarf star population is contributing a *smaller* Mg_2 index than the giant component. This is contrary to naive expectations based on the gravity dependence of Mg_2, and arises because the relevant tutnoff dwarfs have considerably higher effective temperatures than the red giant branch stars.

(iii) Although the combined population metallicity relation is fairly linear up to solar abundance, it becomes *highly non-linear* (or at least, very much steeper) at metallicities greater than solar, when the indices of the bright dwarf stars increase up to values which are again comparable with the giants. Since the dwarfs dominate the visible light at these metallicities, further increase in metal abundance may well give *very* strong Mg_2 index for the population, because the dwarf indices are such a strong function of metallicity here. Although much more careful investigation is required, with inclusion of horizontal, subgiant and asymptotic giant branch stars, it is evident that metallicity estimates above solar should be teated with great caution. Further reason for caution comes from the great sensitivity of molecular line strengths to metal abundances above about twice solar in cool stars.

It seems that Mg_2 should be a good abundance indicator , at least for single abundance systems, up to solar abundance. Beyond solar, things are tricky! There is a need for a detailed, careful theoretical study of the behaviour of Mg_2 in composite stellar populations.

The metallicity dependence of brightness of stars mentioned earlier implies that we must not assume that a metallicity derived from a photometic index of a population containing stars of various metal abundances is the mass- or number-weighted mean metal abundance of the system. It represents the *luminosity* weighted mean. A very rough estimate shows that interpreting a luminosity weighted (i.e. as observed) mean abundance as if it were a mass-weighted mean could give an under-estimate of the metallicity by up to 0.2 or 0.3 dex in some (metal rich) cases. Here, interpretation as a *mass*-weighted mean implies using single-abundance populations as the calibrator, rather than taking proper account of the metallicity structure of the stellar population. Typical errors may be less than 0.3 dex, but it ought to be investigated more carefully, and depends on the abundance indicator involved, and the actual metallicity structure of the stellar population, which will in turn depend on the history of star formation, gas flows, etc. The best place to start is with the simple model, and Figure 7 is a plot of colour indices computed by Arimoto and Yoshii (1987), which gives a reasonable rule of thumb for interpreting at least differential shifts in broad band colours, if not absolute values.

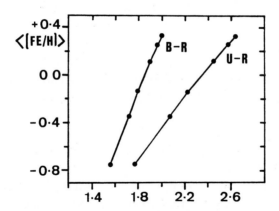

Figure 7 Colours as metallicity indicators of stellar populations. These computed colours from Arimoto and Yoshii (1987) are for populations with approximately simple model metallicity structure.

References

Arimoto,N. and Yoshii,Y. 1987, Astron.Astrophys.,**173**,23.
Bonatto,C.,Alloin,D. and Bica,E. 1989, Astron.Astrophys.,**226**,23.

Brodie,J.P. and Huchra,J.P. 1990, Ap.J.,**362**,503.

Bingelli,B.,Sandage,A.R. and Tarenghi,M. 1984, Astron.J.,**89**,64.

Buzzoni,A. 1989, Ap.J.Supp.,**71**,817.

Cameron,L.M. 1985, Astron.Astrophys.,**147**,47.

Campbell, A. 1990, Ap.J.,**362**,100.

Carlberg,R.G. and Hartwick,F.D.A. 1981, Astron.J.,**86**,1410.

Coutre,J. and Hardy,E. 1990, Astron.J.,**99**,540.

Davies,R.L. and Sadler,E.M. 1987, in *Structure and Evolution of Elliptical Galaxies*, IAU Symp. No 127, Ed. T.de Zeeuw, Dordrecht,Reidel, p441.

Diaz,A.I.,Terlevich,E.,Pagel,B.E.J.,Vilchez,J.M. and Edmunds,M.G. 1987, MNRAS,**226**,19.

Diaz,A.I.,Terlevich,E.,Pagel,B.E.J.,Vilchez,J.M. and Edmunds,M.G. 1991, MNRAS,**253**,245.

Edmunds,M.G. 1989, In:*Evolutionary Phenomena in Galaxies*, Eds. J.Beckman and B.E.J.Pagel, Cambridge Univ.Press,p356.

Edmunds,M.G. 1990, MNRAS,**246**,678.

Edmunds,M.G. and Greenhow,R.G. 1991, MNRAS *to be submitted.*

Edmunds,M.G. and Pagel,B.E.J. 1978, MNRAS,**185**,77p.

Edmunds,M.G. and Pagel,B.E.J. 1984, MNRAS,**211**,507.

Edmunds,M.G. and Phillipps,S. 1989, MNRAS,**241**,9p.

Edmunds,M.G.,Taylor,K. and Turtle,A.J. 1988, MNRAS,**234**,155.

Faber,S.M.,Friel,B.G.,Burstein,D. and Gaskell,C.M. 1985, Ap.J.Supp.,**57**,711.

Fitzsimmons,A.,Brown,P.J.F.,Dufton,P.L. and Lennon,D.J. 1990, Astron.Astrophys.,**232**,437.

Franx,M.,Illingworth,G. and Heckman,T. 1989, Astron.J.,**98**,538.

Frogel,J.A.,Terndrup,D.M.,Blanco,V.M. and Whitford,A.E. 1990, Ap.J.,**353**,494.

Fukunaga-Nakamura,A. and Tosa,M. 1989, Pub.Astro.Soc.Japan,**41**,953.

Garnett,D.R. 1990, Ap.J.,**363**,142.

Garnett,D.R. and Shields,G. 1987, Ap.J.,**317**,82.

Gorgas,J.,Efstathiou,G. and Aragon-Salamanca,A. 1990, MNRAS,**245**,217.

Kennicut,R.C. 1989, Ap.J.,**344**,685.

Kormendy,J. and Djorgovski,S. 1989, Ann.Rev.Astron.Astrophys.,**27**,235.

Matteucci,F. and Chiosi,C. 1983, Astron.Astrophys.,**123**,121.

McCall,M.L.,Rybski,P.M. and Shields,G.A. 1985, Ap.J.Supp.,**57**,1.

Mould,J. 1978, Ap.J.,**220**,434.

Pagel,B.E.J. 1990 In:*Dynamical and Chemical Evolution of Galaxies*, eds. Ferrini,F.,Franco,J. and Matteucci,F.,Giardi, Pisa-Lugano,*in press.*

Pagel,B.E.J. and Edmunds,M.G. 1981, Ann.Rev.Astron.Astrophys.,**19**,77.

Pagel,B.E.J.,Edmunds,M.G.,Blackwell,D.E.,Chun,M.S. and Smith,G. 1979, MNRAS,**189**,95.

Pagel,B.E.J. and Patchett,B.E. 1975, MNRAS,**172**,13.

Peletier,R.F.,Davies,R.L.,Illingworth,G.D.,Davis,L.E. and Cawson,M.C.M. 1990, As-tron.J.,**100**,1091.

Phillips,M.,Turtle,A.,Edmunds,M.G. and Pagel,B.E.J. 1983, MNRAS,**203**,759.

Phillipps,S.P. and Edmunds,M.G. 1991, MNRAS,**251**,84.

Phillipps,S.P.,Edmunds,M.G. and Davies,J.I. 1990, MNRAS,**244**,168.

Rubin,V.C.,Ford,W.K. and Thonnard,N. 1980, Ap.J.,**238**,471.

Rubin,V.C.,Ford,W.K. and Thonnard,N. and Burstein,D. 1982, Ap.J.,**261**,439.

Rubin,V.C.,Ford,W.K. and Thonnard,N. 1984, Ap.J.Lett.,**281**,L21.

Russell,S.C. and Bessell,M.S. 1989, Ap.J.Supp.,**70**,865.

Shaver,P.A.,McGee,R.X.,Newton,L.M.,Danks,A.C. and Pottash,S.R. 1983, MNRAS,**204**,53.

Skillman,E.D. 1986 In:*Star Formation in Galaxies*, Ed. Lonsdale Persson,C.J., NASA,Washington,p263.

Skillman,E.D. 1989, Ap.J.,**347**,883.

Skillman,E.D.,Kennicut,R.C. and Hodge,P.W. 1989, Ap.J.,**347**,875.

Tenorio-Tagle,G. and Bodenheimer,P. 1988, Ann.Rev.Astron.Astrophys.,**26**,145.

Terlevich,R.J.,Davies,R.L.,Faber,S.M. and Burstein,D. 1981, MNRAS,**196**,381.

Thomsen,B. and Baum,W.A. 1987, Ap.J.,**315**,460.

Thomsen,B. and Baum,W.A. 1989, Ap.J.,**347**,214.

Vigroux,L.,Stasinska,G and Comte,G. 1987, Astron.Astrophys.,**172**,15.

Vilchez,J.M. 1989 In:*Evolutionary Phenomena in Galaxies*, Eds. J.Beckman and B.E.J.Pagel, Cambridge Univ.Press,p356.

Vilchez,J.M. and Pagel,B.E.J. 1988, MNRAS,**231**,257.

Young,J.S. and Knezek,P.M. 1989, Ap.J.Lett.,**347**,L55.

Walsh,J.R. and Roy,J-R. 1989, Ap.J.,**341**,722.

Whitworth,A.P. 1990 *private communication*.

DISCUSSION

V.Rubin: Spatial effects in observations of nuclei often include HII regions off the nucleus. Nuclei may really be different from their surroundings.

M.G.Edmunds: I agree, but the excitation of surrounding HII regions may depend critically on how much of the nuclear radiation field they see.

G.Ferland: How do the gas-to-dust ratios depend on z?

M.G.Edmunds: I don't know. Low metallicity galaxies like the SMC certainly seem to have less dust, but the precise relation between dust abundance and metallicity is not known. It may also depend on other parameters like the radiation field, or the star formation history of the local region.

R.J.Terlevich: The underlying parameter that provides the empirical abundance variation is a dependence of the IMF on abundances. Is this dependence included in the models in a consistent way?

M.G.Edmunds: No, because I don't believe we even know if a metallicity change shifts the IMF to higher or lower masses. Give us a good theory of star formation and then it can be put into the chemical evolution models.

R.J.Terlevich: What about the Mg_2 versus σ relation?

M.G.Edmunds: It must be telling us something, but I confess that I really don't yet know how to interpret it in terms of chemical evolution.

The Nature of QSO CIV Absorption Lines

N. ARIMOTO & J. KÖPPEN

Physics Department, University of Durham, South Road, Durham, DH1 3LE, U.K.

Institut für Theoretische Astrophysik der Universität Heidelberg, Im Neuenheimer Feld 561, D-6900 Heidelberg 1, F.R.G.

Abstract: Based on chemical evolution approach to the observational data of CIV doublet line ratio seen in QSO spectra, we suggest that CIV clouds could be the progenitors of Malin1-type very low surface brightness galaxies.

1. INTRODUCTION

Over the past decade considerable progress has been made in our understanding of QSO absorption lines (cf. Sargent et al. 1979; Young et al. 1982; Pettini et al. 1983; Foltz et al. 1986; Bergeron & Stasińska 1986; Murdoch et al. 1986; Lanzetta et al. 1987; Tytler et al. 1987; Lanzetta 1988; Sargent et al. 1988a; Sargent et al. 1988b; Steidel et al. 1988; Caulet 1989; Turnshek et al. 1989; Khare et al. 1989; Sargent et al. 1989; Steidel 1990; Petitjean & Bergeron 1990). A general view in recent studies is that Lyman α forest lines are formed by primordial intergalactic clouds while heavy element lines are due to intervening galaxies. Among the latter, the number density distribution, dn/dz, of CIV lines shows a remarkable decrease with increasing redshift in the redshift range $1.3 \leq z_{abs} \leq 3.7$ (Sargent et al. 1988a; Steidel 1990), which deviates significantly from that of Lyman Limit Systems (Sargent et al. 1989). Steidel et al. (1988) and Steidel (1990) have interpreted this as a direct evidence for chemical evolution in haloes of high redshift galaxies. Their claim could be immediately verified if chemical compositions of individual clouds were accurately derived. Unfortunately, very few objects have been analysed and, what is worse, accuracy in resulting abundance estimate has been far from satisfactory (cf. Bergeron & D'Odorico 1986; Chaffee et al. 1986).

The observed evolution in dn/dz of CIV clouds could be explained by a systematic enhancement of heavy elements in clouds and/or by a systematic change of ionization level in the clouds, provided that the local comoving density of clouds remains constant. The validity of this assumption is hardly confirmed. This means that any arguments based on the observed number density evolution dn/dz are not conclusive.

In this paper, instead of dn/dz, we analyse the ratio of the equivalent widths $DR \equiv W(1548)/W(1550)$ of the CIV doublet which increases with redshifts in the range

$1.4 \leq z_{abs} \leq 3.4$ (Steidel 1990). Obviously, a study of doublet ratio does not require any assumption on the local comoving density; through a curve of growth, DR is uniquely determined as a function of CIV column density, $N(CIV)$. Two major factors determining $N(CIV)$ are the ionization level and the carbon abundance of the clouds. All recent works have suggested that the intensity of the metagalactic ionizing flux $J_\nu(z)$ appears to be independent of z in the range $1.8 \leq z \leq 3.8$ (Bajtlik et al. 1987; Bechtold et al. 1988). Therefore, we study a simple model of a gas cloud photoionized by the observationally deduced intergalactic radiation field and seek an enhancement history of heavy elements that gives the observed evolution of the CIV doublet ratio.

2. CLOUD MODEL

We assume that the typical absorbing cloud is homogeneous with a gas density n_g. It shall be formed as a rotational ellipsoid with equatorial radius R and polar radius fR. The factor f thus describes the cloud's shape in a simple way, with $f = 1$ for spherical clouds. The total mass of a cloud then is

$$M = \frac{4\pi}{3} R^3 f m_H n_g, \tag{1}$$

where m_H is the mean atomic mass of the gas. If the cloud's rotational axis is oriented randomly, the most likely line-of-sight through the cloud is along this axis. Thus the most likely hydrogen column density is

$$N(H) = 2R f n_g. \tag{2}$$

The gas cloud is immersed in the intergalactic radiation field which causes the gas to be ionized. Extensive photoionization models have been computed by Bergeron & Stasińska (1986). We use their results to compute, from our hydrogen column densities, the proper ionic column densities: For any ion of the trace elements the column density is proportional to the elemental abundance in the gas. This abundance will be specified from a chemical evolution model.

The ionization structure of the model cloud is determined by three parameters: the spectral shape of the ionizing radiation, the "ionization parameter" U — the ratio of the number densities of ionizing photons and gas atoms — and the optical depth of the cloud in the HI Lyman continuum.

The exact spectrum of the intergalactic radiation field is not well known. However, the continuum spectra of quasars, which probably are dominant contributors, can well be represented by a power-law. Thus, the mean intensity is $J_\nu(\nu) = J_0(\nu/\nu_0)^{-\alpha}$ with the frequency ν_0 of the hydrogen ionization threshold. Results of ionization models do not depend very much on the spectral slope α. Then the ionization parameter is

$$U = \frac{\pi J_0}{hc\alpha n_g},$$ (3)

which involves the gas density n_g in the cloud. Finally, the optical depth of the cloud is determined by the column density $N(HI)$ of hydrogen atoms, and hence by $N(H)$.

For any fixed cloud geometry (i.e. f) and ionizing spectrum (α) eqs.(1)-(3) allow all cloud properties n_g, R, M to be determined from the parameters U, J_0, and $N(H)$:

$$n_g = \frac{\pi}{hc}\frac{J_0/\alpha}{U} = 1.5815\ 10^{-5}\frac{J_{21}/\alpha}{U},$$ (4)

where $J_{21} = J_0/(10^{-21}erg cm^{-2}s^{-1}Hz^{-1})$ is a measure of J_0 in convenient units. Note that the gas density derived does not depend on the cloud's geometry or the column density.

$$Rf = \frac{hc}{2\pi}\frac{U N(H)}{J_0/\alpha},$$ (5)

$$Mf^2 = \frac{h^2c^2m_H}{6\pi}\frac{U^2 N(H)^3}{(J_0/\alpha)^2}.$$ (6)

To determine the clouds' parameters, we proceed this way: From a study of the line ratios, such as CII/CIV, Bergeron & Stasińska (1986) found that for high ionization systems the ionization parameter is in the range $0.0015 \le U \le 0.02$. As Bergeron & Stasińska, we adopt for the spectral index of the ionizing radiation $\alpha = 1.5$. This is not a critical choice, as mentioned earlier. Metal line systems have HI column densities of $N(HI) \ge 10^{17}cm^{-2}$ (Tytler, 1987). Since the distribution function of $N(HI)$ is a steeply dropping power law, $10^{17}cm^{-2}$ can be expected to be a representative value. We have computed photoionization models (similar to Köppen 1978) and find that for the specified range in U this column density is obtained by those models whose hydrogen column density is

$$N(H) = 10^{22.6}U.$$ (7)

This constraint is used in eqs.(5) and (6) to eliminate the dependence on $N(H)$:

$$Rf = \frac{hc}{2\pi}10^{22.6}\frac{U^2}{J_0/\alpha} = 4.079\ 10^5\frac{U^2}{J_{21}/\alpha},$$ (8)

$$Mf^2 = \frac{h^2c^2m_H}{6\pi}10^{67.8}\frac{U^5}{(J_0/\alpha)^2} = 1.177\ 10^{20}\frac{U^5}{(J_{21}/\alpha)^2}.$$ (9)

The numerical forms give R in kpc and M in $M_\odot$. We note that by eliminating U from eqs.(8) and (9) a radius-mass relation for the absorbing clouds is found:

$$R = 3.821 \ 10^{-3} \ M^{0.4}(J_{21}/\alpha)^{-0.2}f^{-0.2}. \tag{10}$$

The remaining unknown parameter is the mean intensity J_0 of the ionizing radiation at the hydrogen ionization limit. Bajtlick et al. (1987) derived from the distribution of the lines in the Lyman α forest an estimate for the intergalactic radiation field as a function of redshift z. Their model I has an intensity close to $J_{21} = 1.3$ for redshifts beteen 1.5 and 3.5. Taking this value, one gets for the specified range of U also ranges for the cloud parameters. Thus, the clouds that meet the observational constraints form a sequence of ionization parameter U. In Table 1 we give the parameters for such clouds, as a function of Mf^2, i.e. their mass, if they were spherical.

Table 1. Cloud parameters

$Mf^2(M_\odot)$	$Rf(kpc)$	$n(cm^{-3})$	U	$\log N(H)$
1.0E06	1.02E+0	9.02E-3	1.45E-3	19.76
1.0E07	2.56E+0	5.69E-3	2.34E-3	19.97
1.0E08	6.43E+0	3.59E-3	3.72E-3	20.17
1.0E09	1.62E+1	2.26E-3	5.89E-3	20.37
1.0E10	4.06E+1	1.43E-3	9.33E-3	20.57
1.0E11	1.02E+2	9.02E-4	1.48E-2	20.77
4.5E11	1.86E+2	6.67E-4	2.00E-2	20.90

It is worthwhile to see whether the clouds are stable against the gravitational instability. In the case of spherical clouds, Jeans criterion is satisfied if the escape velocity is larger than the sound speed c_s: $(2GM/R)^{\frac{1}{2}} > (kT/m_H)^{\frac{1}{2}}$. For cloud temperature $T \simeq 10^4$K (Rees 1988), we find that massive clouds with $M \geq 10^9 M_\odot$ are unstable against gravitational collapse; thus it is very likely that star formation is triggered in these massive clouds.

3. EVOLUTION OF CARBON DOUBLET RATIO

We have computed the evolution of equivalent widths of absorption lines from the spherical clouds of different masses with the parameters as given in Table 1. We assume that the cloud parameters (mass, radius, and gas density) do not change with time. Thus, the total hydrogen column density does not change. We allow for the evolution of the ionizing radiation field, which is taken from Bajtlick et al. (1987). Thus, with the fixed $N(HI) = 10^{17}cm^{-2}$ and the evolving U, the column densities of various ions were interpolated from the results of Bergeron & Stasińska (1986) and scaled to the actual gas metallicity whose evolution with redshift is computed from an analytical chemical evolution model. Finally the equivalent widths are calculated with a curve-of-growth computed for zero source function and a constant microturbulent velocity of $b = 30 kms^{-1}$.

A Friedmann cosmological model is adopted with zero cosmological constant ($\lambda = 0$). Calculations have been done with Hubble constant $H_0 = 50 kms^{-1} Mpc^{-1}$ and deceleration parameter $q_0 = 0.05$. The results depend only weakly on these cosmological parameters. The epoch of the beginning of the metal enhancement is fixed at $z_f = 3.5$. The chemical evolution model is the so-called "simple" model which is characterized by only two parameters – star formation rate s and yield of heavy element y.

To see whether the observed DR evolution is simply due to the cosmological evolution of UV background field, we first have computed models under the assumption that all clouds have the same metallicity regardless of their location in the universe. None of three models calculated with constant metallicity $Z/Z_\odot = 0.001, 0.01, 0.1 (Z_\odot = 0.0185)$ reproduces the observed tendency of gradual increase of DR with redshift, thus indicates that chemical evolution is of vital importance.

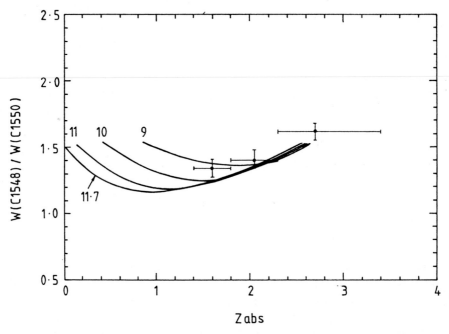

Figure 1: Evolutionary tracks of the equivalent width ratio $W(1548)/W(1550)$ for the model clouds. Only the parts of tracks with $W(1548) > 0.3A$ are shown. Data are taken from Steidel (1990). Numbers indicate logarithmic values of the clouds mass in $M_\odot$. Model parameters are: $H_0 = 50 kms^{-1} Mpc^{-1}$; $q_0 = 0.05$; $z_f = 3.5$; $N(HI) = 10^{17} cm^{-2}$; $b = 30 kms^{-1}$; SFR $s = 0.015 Gyr^{-1}$; yield $y = 0.1 Z_\odot$.

The best fit model is shown in Fig.1 in which Steidel's (1990) data are also indicated.

Model parameters are $s = 0.015 Gyr^{-1}$ and $y = 0.1 Z_\odot$. Since Steidel's data are confined to lines having equivalent widths $W(1548) \geq 0.30A$, we show in Fig.1 only those parts of the evolutionary tracks which satisfy Steidel's criterion. Note that the ionization level of small clouds with $M \leq 10^8 M_\odot$ is so small that $W(1548)$ of these clouds never exceeds Steidel's criterion. As mentioned earlier, only massive clouds, which are shown in Fig.1, are gravitationally unstable.

The saturation of the curve of growth causes DR to decrease from 2 to around 1 when increasing the column density $N(CIV)$. Therefore, Steidel's data indicates a gradual decrease of $N(CIV)$ in the redshift range $1.4 \leq z_{abs} \leq 3.4$, which as Fig.1 shows is naturally explained by the enhancement of heavy elements in the clouds. All evolutionary tracks of DR suggest that the doublet ratio could increase again towards low redshift, although the metallicity of the clouds continues to increase. This is due to a sharp drop of the intensity of the UV background radiation which we have taken from Bajtlik et al. (1987). It worth noting that if the decrease of the gas fraction f_g due to star formation is considered in calculating the ionization parameter U, DR remains nearly constant in the redshift range $0 \leq z_{abs} \leq 1.4$, because U varies in proportion to f_g^{-1} and compensates the decline of UV background radiation intensity. These alternatives could be tested with forthcoming Hubble Telescope data.

One always gets exactly the same evolutionary tracks of DR as shown in Fig.1, provided the two chemical evolution parameters are chosen to satisfy the condition $s(Gyr^{-1})y/Z_\odot = 0.0015$. This degeneracy occurs because we use the "simple" model. In this closed-box one-zone model with the instantaneous recycling approximation, the time variation of the heavy element abundance of the gas $Z_g(t)$ is given by

$$Z_g(t) = yst. \tag{11}$$

Therefore, the history of metal enhancement is proportional to ys, so is the time variation of $N(CIV)$, and hence DR. Under the instantaneous recyclying approximation, $s \equiv \alpha k$, where α is the locked-up stellar mass fraction and k is the coefficient of the star formation rate (SFR) in units of Gyr^{-1} (cf. Köppen & Arimoto 1990a). In Table 2, we give examples of possible chemical evolution solutions.

The most interesting solution is that of $k = 0.01 - 0.02 Gyr^{-1}$ and $x = 1.6 - 1.8$, because these parameters are very close to those of the underlying populations in irregular galaxies (Arimoto & Tarrab 1990). No objects which share the same parameters as the two extreme cases in Table 2 – low yield - high SFR and high yield - low SFR – are yet known. To predict how these clouds would appear in a nearby field, we have calculated the photometric evolution of the clouds with $k = 0.01 Gyr^{-1}$, $x = 1.6$, $M = 10^{11} M_\odot$ by using the population synthesis model of Arimoto & Yoshii (1986). The resulting cloud model has the following observable quantities: total absolute magnitude in B-band $M_B = -19.0$, HI mass $\log M_{HI} = 10.94$, HI mass-to-light ratio $\log M_{HI}/L_B = 0.89$, colours $U - B = -0.14$, $B - V = 0.43$, $V - K = 1.99$, and radius $R = 102 kpc$. It is interesting to point out that a very low surface brightness galaxy Malin1 does have very similar quantities: $M_B = -20.5$, $\log M_{HI} = 11.02$, $\log M_{HI}/L_B = 0.7$, disc colour $B - V = 0.30 - 0.47$, and a radius $R = 110 kpc$ (Bothun et al. 1987). The model also predicts chemical properties: helium abundance $Y_g = 0.230$, metal abundance of the gas $\log Z_g/Z_\odot = -1.47$, and luminosity-weighted mean stellar metallicity $[Fe/H]_\ell = -1.69$, which are characteristic of the most metal-poor irregular galaxies.

Table 2. Chemical evolution solutions

$y(Z_\odot)$	x	α	$s(Gyr^{-1})$	$k(Gyr^{-1})$
0.01	2.35	0.99	0.15	0.15
0.1	1.85	0.95	0.015	0.016
1.0	1.35	0.78	0.0015	0.0019

Note to Table 2: The slope of initial mass function (IMF), x, and α are taken from the table of Köppen & Arimoto (1990b) by assuming a Salpeter-like IMF $\phi(m) \propto m^{-x}$ with the stellar mass range $0.05 \leq m/M_\odot \leq 60$.

4. CONCLUSION Disney & Phillipps (1987) have suggested that Malin1 is just an iceberg of low surface brightness giant galaxies that forms a significant constituent of the universe; this could explain the large number of absorption line systems that are seen in QSO spectra. One of our chemical evolution solutions indicates that Malin1 type galaxies naturally explain the observed evolution of CIV doublet ratio in QSO spectra. As Fig.1 shows, possible clouds masses are in the range $10^9 - 4.5\ 10^{11} M_\odot$. This implies that many of clouds could have much lower surface brightness than Malin1 and therefore might not yet be detectable with present day techniques. The high frequency of occurance of CIV lines in QSO spectra suggests that there may be huge number of undetected very low surface brightness galaxies which could contribute significantly to the baryonic mass density in the universe. Our model also predicts that these galaxies are very likely to have primordial helium abundance.

Acknowledgements: Financial support from the Deutsche Forschungsgemeinschaft (SFB 328) is gratefully acknowledged. We thank to Mrs.P.Russel for drawing the figure.

References
Arimoto, N., Tarrab, I.: 1990, Astron. Astrophys. **228**, 6
Arimoto, N., Yoshii, Y.: 1986, Astron. Astrophys. **164**, 260

Bajtlik, S., Duncan, R., Ostriker, J.P.: 1988, Astrophys. J. **327**, 570

Bechtold, J., Weymann, R.J., Lin, Z., Malkan, M.A.: 1987, Astrophys. J. **315**, 180

Bergeron, J., D'Odorico, S.: Mon. Not. Roy. Astron. Soc. **220**, 833

Bergeron, J., Stasińska, G.: 1986, Astron. Astrophys. **169**, 1

Caulet, A.: 1989, Astrophys. J. **340**, 90

Chaffee, F.H., Foltz, C.B., Bechtold, J., Weymann, R.J.: 1986, Astrophys. J. **301**, 116

Disney, M., Phillipps, S.: 1987, Nature **329**, 203

Foltz, C.B., Weymann, R.J., Peterson, B.M., Sun, L., Malkan, M.A., Chaffee, F.H.: 1986, Astrophys. J. **307**, 504

Khare,P., Tork, D.G., Green, R.: 1989, Astrophys. J. **347**, 627

Köppen, J.: 1978, Astron. Astrophys. Suppl. **35**, 111

Köppen, J., Arimoto, N.: 1990a, Astron. Astrophys. **in press**

Köppen, J., Arimoto, N.: 1990b, Astron. Astrophys. Suppl. **in press**

Lanzetta, K.M.: 1988, Astrophys. J. **332**, 96

Lanzetta, K.M., Turnshek, D.A., Wolfe, A.M.: 1987, Astrophys. J. **322**, 739

Murdoch, H.S., Hunstead, R.W., Pettini, M., Blades, J.C.: 1986, Astrophys. J. **309**, 19

Petitjean,P., Bergeron,J.: 1990, Astron. Astrophys. **231**, 309

Pettini, M., Hunstead, R.W., Murdoch, H.S., Blades, J.C.: 1983, Astrophys. J. **273**, 436

Rees, M.J.: 1988, in *QSO Absorption Lines*, eds. J.C.Blades, D.Turnshek. C.A. Norman, Space Telescope Science Institute symposium series 2, Cambridge University Press, Cambridge

Sargent, W.L.W., Young, P., Boksenberg, A., Carswell, R.F., Whelan, J.A.J.: 1979, Astrophys. J. **230**, 49

Sargent,W.L.W., Boksenberg,A., Steidel,C.C.: 1988a, Astrophys.J. Suppl. **68**, 539

Sargent, W.L.W., Steidel, C.C., Boksenberg, A.: 1988b, Astrophys. J. **334**, 22

Sargent, W.L.W., Steidel, C.C., Boksenberg, A.: 1989, Astrophys. J. Suppl. **69**, 703

Steidel, C.C.: 1990, Astrophys. J. Suppl. **72**, 1

Steidel, C.C., Sargent, W.L.W., Boksenberg, A.: 1988, Astrophys. J. **333**, L5

Turnshek, D.A., Wolfe, M., Lanzetta, K.M., Briggs, F.H., Cohen, R.D., Foltz, C.B., Smith, H.E., Wilkes, B.J.: 1989, Astrophys. J. **344**, 567

Tytler, D.: 1987, Astrophys. J. **321**, 49

Tytler, D., Boksenberg, A., Sargent, .L.W., Young, P., Kunth, D.: 1987, Astrophys. J. Suppl. **64**, 667

Young, P., Sargent, W.L.W., Boksenberg, A.: 1982, Astrophys. J. Suppl. **48**, 455

Neutrino Decay and the Ionisation of Nitrogen

D.W. SCIAMA

International School for Advanced Studies,
International Centre for Theoretical Physics, Trieste.
Department of Astrophysics, Oxford.
Institute of Astronomy, Cambridge.

INTRODUCTION

It is a special pleasure to have the opportunity of speaking at the Jubilee of my old friend, Bernard Pagel. Over the years Bernard, with his unique blend of wide and deep knowledge of astronomy, disciplined and rigorous thinking, and amused tolerance for even the wilder excesses of some of his friends and colleagues, has always made me careful to give full weight to the more reliable arguments based on sound observations. In particular, he has recently stressed to me that his observations (Pagel and Simonson 1989) of the cosmic abundance of ^{4}He, when combined with now-standard arguments from considerations of primordial nucleosynthesis, leave little scope for the possibility that there exists in nature low-mass ($\sim$ 30 eV) photinos from supersymmetry theory, if they are as abundant in the universe as any one of the three neutrino types.

This insistence has led me to revise some old ideas of mine, according to which dark matter photinos decay into photons which could produce observable ionisation effects both in individual galaxies, including our own, and in the universe as a whole. It has, in fact, turned out to be much more rewarding to retest (Sciama 1990a,b) these ideas in terms of decaying dark matter neutrinos, although the corresponding hypothesis remains very speculative. Despite this Bernard has been very supportive, and has encouraged me to apply the hypothesis to all relevant astronomical systems. In particular, when I mentioned to him that I could produce arguments both for and against the proposition that the decay photons could ionise nitrogen (ionisation potential 14.53 eV) as well as hydrogen (13.6 eV), he advised me to do my best to ionise nitrogen. As so many people have found with Barnard's advice, it has turned out to be very valuable.

The point is an important one within my scheme for two reasons. First of all, there are observational arguments (Sciama 1990b,c) for imposing an upper limit of 15 eV on the energy of a decay photon (this energy being a monochromatic line in the rest frame of the decaying neutrino). Thus if there is a lower limit on the energy of 14.53 eV, the allowed range of energies becomes closely constrained. This in turn would closely constrain the mass of the decaying particle, the density and age of the universe, and the value of the Hubble constant.

Secondly, if the photon energy exceeds 14.53 eV, this would limit the choice of suitable clusters of galaxies for a proposed search for the postulated emission line from dark matter in clusters, using the ultra-violet detectors in space. The point here is that the redshift of the cluster must be sufficient to ensure that when the photons reach the outskirts of our Galaxy their energy has been reduced to below 13.6 eV, to ensure that they propagate through the interstellar medium to the detector with very little absorption. If the photons leave the galaxy cluster with an energy exceeding 14.53 eV, then the redshift of the cluster would have to exceed 0.07. This limit also has implications for the detectability of the line in relation to the characteristics of the detector being used.

DO DECAY PHOTONS IONISE NITROGEN?

There are two arguments in favour of the assumption that the decay photons can ionise nitrogen. The first argument was supplied by an anonymous referee of my paper entitled 'Dark matter decay and the ionisation of the local interstellar medium' (Sciama 1991). In that paper I was proposing that the partially ionised cloud of radius $\sim$ 3-5 parsecs which is believed to surround the sun is mainly ionised by decay photons emitted partly by neutrinos in the clouds and partly by neutrinos in the hot dilute bubble of radius $\sim$ 100 pc believed to surround the cloud. I had produced a (weak) argument, based on Copernicus observations of the star α Leo, that the decay photons *cannot* ionise nitrogen.

The referee kindly produced a much better argument that the decay photons not only can but *must* ionise nitrogen. This argument is based on the observation of Gry, York and Vidal-Majder (1985) that along the 'exceptionally vacant' line of sight towards β C Ma the column density of NI bears the same ratio to HI as in other regions near the sun and in the Orion HII nebula. This observation can easily be shown to require that whatever is ionising hydrogen is also ionising nitrogen. Was Bernard the referee I wonder?

The second argument concerns the spiral galaxy NGC 891. Recent Hα observations of this galaxy (Rand, Kulkarni and Hester 1990) show that it contains ionised hydrogen out to heights above the plane of several kiloparsecs. The source of this ionisation was a puzzle to the observers, and Paolo Salucci and I (Sciama and Salucci 1990) have suggested that the photons emitted by dark matter neutrinos in NGC 891 may be responsible. This suggestion is relevant to the present article because if the decay photons can also ionise nitrogen one might expect to find considerable quantities of NII also at great heights above the plane of NGC 891. I am not claiming to 'predict' this phenomenon, because I had heard rumours that such observations of NII have already been made. Indeed it was when I mentioned such rumours to Bernard at this conference in his honour that he advised me to go for nitrogen as well as for hydrogen. I am hoping to receive a preprint soon confirming the rumours. When it arrives, I shall have more than ever reason to be grateful to Bernard for his good advice.

References

Gry,C.,York,D.G. and Vidal-Majder,A. 1985, Ap.J.,**296**,593.

Pagel,B.E.J. and Simonson,E.A. 1989, Rev.Mex.Astr.Asrofis.,**18**,153.

Rand,R.J.,Kulkarni,S.R. and Hester,J. 1990, Ap.J.Lett.,**352**,L1.

Sciama,D.W. 1990a, Comments in Astro.,**15**,71.

Sciama,D.W. 1990b, Ap.J.,**364**,549.

Sciama,D.W. 1990c, Phys.Rev.Lett.,**65**,2839.

Sciama,D.W. and Salucci,P. 1990, MNRAS,**247**,506.

Editors' Note: An interesting observational test of this neutrino decay model was given by Fabian, A.C., Naylor, T. and Sciama, D.W. in MNRAS, **249**,21p,1991.

CONCLUDING REMARKS

B.E.J. PAGEL

NORDITA
Blegdamsvej 17
DK-2100 Copenhagen Ø, Denmark

ABSTRACT

Some personal impressions from the conference are described, together with a few desiderata for future work.

1. INTRODUCTION

This has been an enjoyable continuation of the Herstmonceux Conference series, with which I have a long association, and with the unusual feature that I have been interested in virtually every talk and poster. After many years of conference-going I have reached the conclusion that the best thing about a conference is not so much what you learn as what you unlearn and in this respect, also, I have not been disappointed. In what follows, I describe some personal impressions carried away from the conference.

2. BIG BANG NUCLEOSYNTHESIS

On BBNS I should like to recall with the aid of Fig. 1 the complicated forest of Galactic chemical evolution effects that one needs to cut through in order to extrapolate observable abundances backwards to derive primordial values. In the case of helium, nevertheless, the use of extragalactic H II regions by the method of Peimbert and Torres-Peimbert (1974, 1976), extended to make use of nitrogen as well as oxygen abundances (Pagel, Terlevich & Melnick 1986), seems to work rather well for the low-abundance regions (the Orion Nebula is excluded; cf. Pagel 1982), especially in cases where independent data for both blue and red lines are available, and yields a primordial helium mass fraction of 0.23 with an error that I judge very unlikely to exceed ± 0.01 (Pagel & Simonson 1989; Simonson 1990; Pagel 1991). Together with other people's work on deuterium and helium 3 and on lithium (e.g. Delyannis and Rebolo in this volume) and the standard theory discussed by Hubert Reeves and by Max Irvine, this leads to the restrictions on baryonic density shown in Fig. 2 and to an upper limit of 3 on the number of neutrino flavours which has been ascribed to "some optimists" (Barrow & Dombey 1989) even after confirmation in the famous accelerator experiments. The argument involves an assumed upper limit of 10^{-4} to primordial $(D+^3He)/H$, which seems generous in view of either the expected survival rate of ^{3} He (Yang et.al. 1984) or the likelihood that pristine material has gradually been added to the Galactic disk by inflow (see section 6 below), but I was interested to hear from Hubert Reeves that he does not regard these arguments as compelling. It is to be hoped that observations of high red-shift absorption-line systems in front of quasars will eventually settle the matter (Webb et al. 1991).

Fig. 2 shows density limits from standard (homogeneous) BBN theory (tall vertical lines) and also somewhat wider limits (shorter double vertical lines) from reasonable inhomogeneous BBNS models (Kurki-Suonio et al. 1990; Reeves 1990). Models permitting $\Omega_b = 1$, discussed by Willy Fowler, do not look viable as they predict too much helium and probably too much ^{7}Li. Beryllium has been mentioned (together with boron) as a possible signature of inhomogeneous BBNS (Kajino & Boyd 1990) although this is disputed (Terasawa & Sato 1990). Willy quoted from the data of Ryan et al. (1990) shown in Fig. 3 and the point that I would make here is that even if the lowest point shown there is taken as a detection, as opposed to an upper limit as indicated by the authors, the corresponding beryllium abundance is no more than what one would expect in any case from conventional galactic chemical evolution. There is no evidence, in fact, for a cosmic "floor" to the abundances of any elements besides hydrogen except helium and lithium (cf. Pagel 1991).

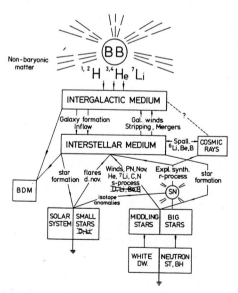

Fig. 1

A possible scheme for galactic chemical evolution (adapted from Pagel 1981)

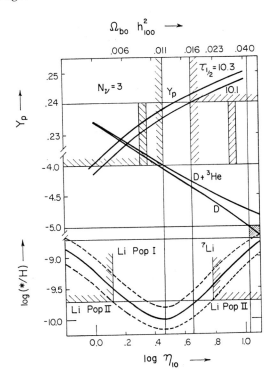

Fig. 2

Primordial abundances predicted from standard BBNS theory after Yang et al. (1984), Steigman (1989) and Deliyannis et al. (1990) as a function of the usual baryonic density parameters and neutron half-life. Horizontal lines show upper limits based on observation and extrapolation using reasonable galactic chemical evolution considerations. Tall vertical lines show resulting limits on baryonic density from the standard (homogenous) BBNS model, while shorter double vertical lines show rough limits based on mildly inhomogeneous models (adapted from Kurki-Suonio et al. 1990).

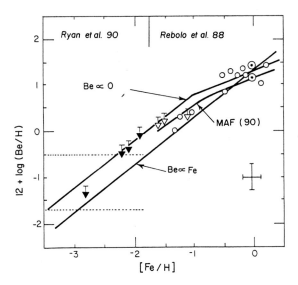

Fig. 3

Plot of stellar beryllium abundances against metallicity after Ryan et al. (1990) with some theoretical predictions as indicated. MAF is the theoretical model of Mathews, Alcock & Fuller (1990) based on spallation. The upper and lower "Sun" symbols refer to meteoritic and photospheric Be, respectively. Horizontal dotted lines show the range of primordial values predicted by Kajino & Boyd (1990) as a consequence of inhomogeneous BBNS models.

3. STELLAR NUCLEOSYNTHESIS

On supernovae, we had nice talks from Hillebrandt, Nomoto, Thielemann and Canal. Spectroscopic analysis of SN1987A will help greatly in fixing the yields from stars of initial mass near $20M_\odot$, which is very useful. One would, of course, like to know the dependence of the yield of certain well-defined elements such as oxygen on initial stellar mass, then to be convolved with some assumed initial mass function to predict the overall yield, but it is my impression that even the yields from individual stars are still quite poorly known. For intermediate-mass stars the position is no better. The calculations of Renzini & Voli (1981) are still what most people use, but Icko Iben tells us that these are wrong (cf. Stasinska & Tylenda 1990), so it is not clear what to do. The site, or sites, for the r-process is, or are, still a mystery and the ν-process (Woosley et al. 1990) was criticized by Wolfgang Hillebrandt; but there has been good progress on the s-process through the work of Prantzos and of Beer, Käppeler, Gallino and their colleagues, and our general understanding of stellar evolution will be greatly improved by the completion of the monumental stellar opacity project by Seaton and his international team.

4. STELLAR ABUNDANCES

Stellar abundance analysis is now a very different game from what it was when I started doing differential curve-of-growth analysis in the wake of Greenstein and his associates nearly 30 years ago. For this we owe a great debt to Blackwell and his associates at Oxford including Geoff. Smith who showed really impressive fits to some very accurate line profiles. Lambert, Nissen, Gustafsson, Andersen, Ruth Peterson, François et al. have been playing this new game of model-atmosphere spectrum synthesis with enormous sophistication and supplying a nice firm basis both for understanding nucleosynthesis in evolved giants and for sorting out the subtle differences between individual elements in unevolved stars resulting from galactic chemical evolution effects in the parent interstellar medium.

Important points which I carry away from Lambert's talk are that the s-process pattern in S and carbon stars is quite similar to that in the Solar System, mimicable by an exponential distribution of exposures for the "main" component; that these stars in a certain mass range can account for Galactic enrichment in lithium as required by standard BBNS; that they may also make fluorine although it is unclear how; and that there are grains in meteorites that may come directly from AGB giants, along with other grains that, as we heard from Hillebrandt, have analogues in SN 1987A.

There is still some action in solar abundances. Nick Grevesse reported confirmation of C,N,O abundances from molecular lines in the far infra-red, but some anomalies concerning thorium, which was previously believed to agree with meteoritic determinations. This bears in turn on the wider issue of the use of stellar thorium abundances as a chronometer (Lawler, Whaling & Grevesse 1990).

5. NEBULAR ABUNDANCES AND IONIZATION

From Clegg, Maciel and Peimbert we heard that planetary nebulae (other than Type 1 and maybe Type 4) show the same large-scale trends across the Milky Way and between nearby galaxies as do H II regions for O, Ne, S and Ar, but individual effects of nuclear burning on He, C and N. In Type 1, oxygen is also affected. These important clues to the evolution of red giants and to their role in galactic chemical evolution are understood qualitatively though not yet quantitatively.

Ferland et al. have provided a very thorough discussion of physical conditions in the dusty Orion Nebula, combined with new observations taken with lower spectral resolving power than I would have wished to use (cf. Pagel & Simonson 1989), especially for such a bright object. With a value of $\log \eta = 1.3$ according to their data ($\eta = (O^+/O^{++})/(S^+/S^{++})$, see Vilchez & Pagel 1988), I am sceptical of the claim that the correction for neutral helium in Orion is negligible (see, e.g., Mathis 1982). A sensationally low estimate of helium abundance was derived for the Second Byurakan Survey H II galaxy 0335-052 by Elena Terlevich, but this depended on a large electron density deduced from [Ar IV] which may not be representative of the bulk of the H^+- He^+ region. The mass fraction of 0.22 deduced in the low density limit is within the scatter of existing determinations for other objects with similar oxygen abundance like I Zw18.

Angeles Diaz described some of the work she is doing in collaboration with Elena and myself to improve knowledge of abundance gradients in spiral galaxies using nebular lines of [O II] and [O III], η and photo-ionization models. At large abundances the results are quite density-dependent owing to cooling by fine-structure transitions in the infra-red, but we can at least limit the abundance from below by assuming a low density like 10 cm^{-3}. Evan Skillman has established an interesting relation between oxygen abundance and luminosity (or surface brightness?) among nearby irregular galaxies and Don Garnett has checked the relation between N/O (from models, infra-red and ultra-violet observations) with N^+/O^+ (from optical spectra). Paradoxically it seems that N^+/O^+ becomes a better indicator of N/O when the degree of ionization is high.

At this meeting we were able to pay but little attention to ionized media associated with active galactic nuclei. We did hear from Joss Bland about extended liner-like emission from diffuse gas in NGC 1068, reminiscent of something Mike Edmunds and I noticed very faintly in NGC 1365 several years ago, and from Fernandez about a successful application of models resembling the Warmers of Terlevich & Melnick to optical spectra of AGN leading to solar-like abundance but with N/O enhanced.

Dennis Sciama described his sensational new theory of ionization of H I regions by the decay of τ neutrinos - a highly falsifiable theory but not falsified up to now. What impresses me most about it is the fit to the electron density distribution in the Galaxy deduced from observation.

6. GALACTIC CHEMICAL EVOLUTION AND STELLAR ABUNDANCES

Donald's diatribe was most enjoyable and I agree with many of his comments. I look forward to seeing the details of his specific model, which offers an alternative to other galactic wind type models in interpreting the luminosity-metallicity relation for galaxies. Mike Edmunds discussed various questions related to this including the still unresolved question of the relative roles of luminosity and surface brightness in the biparametric sequence of elliptical galaxies and the influence of a possible gas surface density threshold for star formation, which in turn could be important in the generation of radial abundance gradients from viscous flows as discussed by Cathy Clarke.

I am intrigued by what may well prove to be a convergence of clues to Galactic chemical evolution from observations of the oldest Galactic stars, on the one hand, and certain high red-shift absorption-line systems on the other. Alec Boksenberg described a metal-deficient damped Lyman-α system with z=2.3 in front of PHL957 which could represent an early stage in the formation of a spiral disk, although this is controversial, and Gene Smith reported on the relative rarity of detectable Lyman-α emission from such systems; this could be due (among various alternatives) to a low initial star formation rate, which appeals to me (cf. Pagel 1989b). Arimoto discussed the stability of CIV clouds, again assuming a low star formation rate, and suggested a relationship with either dwarf or low surface brightness galaxies.

Meanwhile, back in our Galaxy, Tim Beers has produced from the well-planned objective-prism survey a marvellous collection of metal-deficient stars with [Fe/H] down to -4.3 and a statistical distribution consistent with a simple model without any "floor" to stellar metallicities, which is also supported by Tom Kinman's finding of more low-metallicity RR Lyrae stars. These results are important for extrapolations back to the Big Bang, among other things. Poul-Erik Nissen reported a new study of stellar ages and kinematics with suggestive indications that halo and disk stars are separated in the metallicity-rotational velocity plane, i.e. distinct population structure is present, but both overlapping in metallicity around [Fe/H] = -1. This result (if not due to selection effects) appeals to me and may be related to the "metal-weak thick disk" of Morrison, Flynn & Freeman (1990). In nuclear regions of external galaxies the method of population synthesis from a base of integrated spectra of globular clusters, described by Danielle Alloin, promises to give not only the stellar metallicity distribution but an insight into important aspects of their chemical history. The base needs to be extended into the ultra-violet and infra-red, and to include super metal-rich clusters such as NGC 6553 and maybe some clusters in other galaxies, if their metallicities can be calibrated.

The upshot of all these results is that we have in hand many of the ingredients needed to put forward simple chemical evolution models for our own and other galaxies, nor has there been any lack of such models (or of complicated ones). I think that my analytical model (Pagel 1989b) contains some of the important ingredients for the halo and disk of our own Galaxy and in this context I cannot resist a reference to the famous (or infamous) G-dwarf problem. This has recently been rediscussed by Sommer-Larsen (1990ab), who has improved the correction for moderately metal-deficient stars presumably representing the "Thick Disk" and derived the distribution shown in Fig. 4. It is broader than my previous estimates, but the "problem" has by no means gone away. It can be solved by suitable inflow models as originally pointed out by Larson (1976) and Lynden-Bell (1975), or perhaps by initial enrichment from the bulge and halo (Ostriker & Thuan 1975) but only if the corresponding true yield is very large (Köppen & Arimoto 1990).

The biggest problem remaining is to understand the indications derived from relative abundance variations among unevolved stars of different elements, e.g. O/Fe, Eu/Fe etc. (cf. Wheeler, Sneden & Truran 1989). The main problem with O/Fe is observational: it increases as Fe/H goes down, but does it flatten off at [O/Fe] = 0.5 as assumed by Pagel (1989b) or does it rise further as assumed by Ramon Canal (Abia & Rebolo 1989; Abia, Canal & Isern 1990)? The issue is a spectroscopic one,

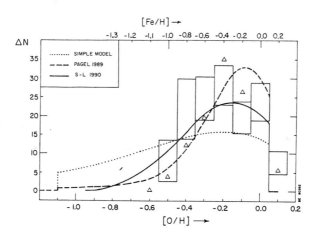

Fig. 4

Distribution function of nearby G dwarfs as a function of oxygen abundance and metallicity corrected to the Solar cylinder after Pagel (1989b, triangles) and after Sommer-Larsen (1990ab, boxes representing bin widths and error bars). Model curves are as follows: "Simple" model with plausible prior enrichment from the halo (dotted); Clayton-type inflow model by Pagel (1989b; broken-line curve); and dynamical model by Sommer-Larsen (1990b; solid curve). The data have been corrected for an assumed Gaussian dispersion σ ([Fe/H]) = 0.15 to allow for errors and cosmic scatter in the age-metallicity relation.

mainly concerned with a descrepancy between permitted and forbidden lines (Fig. 5), but until it can be decided we are handicapped in trying to find an abundance "clock", preferably well approximated by instantaneous recycling, against which to measure the chemical evolution of other elements. Leonard Searle mentioned a very similar problem for calcium, which I had previously thought to behave in an understandable way, along with Mg, Si, S and Ti. The abundances of barium and related elements in halo stars are still subject to many uncertainties, as Patrick François pointed out, but in disk stars they follow iron closely and this result is quite hard to understand. Europium, the main observable r-process element, shows what may be mildly α-like behaviour relative to iron for [Fe/H]>-2 or so, but at the lowest abundances apparently [Eu/Fe]<0, which has been explained by Mathews & Cowan (1990) as a consequence of the r-process occurring mainly in relatively low-mass SN II compared to SN II iron production (Fig. 6); we heard from Ken Nomoto that he expects the latter itself to peak at 12 $M_{\odot}$.

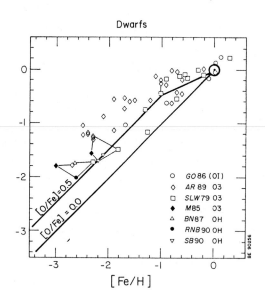

Fig. 5

Relation between oxygen and iron abundances in dwarf stars, after Ryan, Norris & Bessell (1990).

7. CONCLUSIONS

There has been some progress on all fronts but (as usual) much remains to be done. My off-the cuff list of desiderata for the immediate future is the following:

Theory:
Better (true) yields (whether called y or p).
More sophisticated galactic chemical evolution models incorporating limiting cycles, thresholds etc., but without excessive parameters and fungus.

Observations:
Sort out the O/Fe discrepancy and other individual element: element ratios.
Improve the H II region abundance calibration at the high end.
Reconcile stellar and H II region abundances.
Better total stellar and gas masses including molecular hydrogen.
Improve nebular carbon abundances fron uv observations.
Push the halo to yet lower metallicities.

Finally, I should like to express my thanks to all my friends who have come here, and especially to Roberto Terlevich for having the idea of honouring me in this way, for all the work he has (and will) put in to make the conference a success and both to him and to Donald Lynden-Bell, Mike Edmunds, Angeles Diaz, Elena Terlevich and Pepe Vilchez for long-standing and on-going scientific collaboration which I believe to have been effective and know to have been remarkably enjoyable.

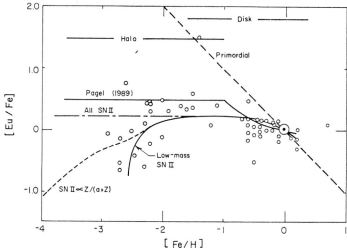

Fig. 6
Plot of stellar abundance ratios [Eu/Fe] against [Fe/H] adapted from Mathews & Cowan (1990). The curves show some theoretical models due to Mathews & Cowan or to Pagel (1989b) who assumed that Eu is simply proportional to oxygen.

References

Abia C, Canal R & Isern J, 1990 **XI IAU European Regional Meeting, La Laguna**

Abia C & Rebolo R, 1989, **Astrophys J,** 347, 186

Barrow J D & Dombey N, 1989, **New Scientist,** Oct 28

Deliyannis C P, Demarque P & Kawaler S D, 1990, **Astrophys J Suppl,** 73, 21

Kajino T & Boyd R N, 1990, **Astrophys J,** 359, 267

Kiselman D, 1990, **Nordic-Baltic Astronomy Meeting, Uppsala**

Köppen J & Arimoto N, 1990, in **Chemical and Dynamical Evolution of Galaxies,** F Ferrini,
    J Franco & F Matteucci (eds) Pisa: Giardini Editore

Kurki-Suonio H, Matzner R A, Olive K A & Schramm D N, 1990, **Astrophys J,** 353, 406

Larson R B, 1976, **Mon Not R astr Soc,** 176, 31

Lynden-Bell D, 1975, **Vistas in Astr,** 19, 299

Mathews G J, Alcock C & Fuller G M, 1990, **Astrophys J,** 349, 449

Mathews G J & Cowan J J, 1990, **preprint**

Mathis J S, 1982, **Astrophys J,** 261, 195

Lawler J E, Whaling W & Grevesse N, 1990, **Nature,** London, in press

Morrison H L, Flynn C & Freeman K C, 1990, **preprint**

Ostriker J B & Thuan T X, 1975, **Astrophys J,** 202, 353

Pagel B E J, 1981, in **The Structure and Evolution of Normal Galaxies,** S M Fall & D Lynden-
    Bell (eds), Cambridge University Press, p 211

Pagel B E J, 1982, in **The Big Bang and Element Creation,** D Lynden-Bell (ed) Phil Trans R Soc
    London A, 307, 19

Pagel B E J, 1989a, in **Evolutionary Phenomena in Galaxies,** J E Beckman and B E J Pagel (eds),
    Cambridge University Press, p 201

Pagel B E J, 1989b, **Rev Mex Astr Astrofis,** 18, 161

Pagel B E J, 1991, **Phys Scripta,** in press

Pagel B E J & Simonson E A, 1989, **Rev Mex Astr Astrofis,** 18, 153

Pagel B E J, Terlevich R J & Melnick J, 1986, **Pub Astr Soc Pacific** 98, 1005

Peimbert M & Torres-Peimbert S, 1974, **Astrophys J,** 193, 327

Peimbert M & Torres-Peimbert S, 1976, **Astrophys J,** 203, 581

Rebolo R, Molaro P & Beckman J E, 1988, **Astr Astrophys,** 192, 192

Reeves H, 1990, **Physics Reports,** in press

Renzini A & Voli M, 1981, **Astr Astrophys,** 94, 175

Ryan S G, Bessell M S, Sutherland R S & Norris J E, 1990, **Astrophys J Let** 348, L57

Ryan S, Norris J & Bessell M S, 1990, submitted to **Astrophys J,**

Simonson E A, 1990, **Thesis,** Sussex University

Sommer-Larsen J, 1990a, **preprint**

Sommer-Larsen J, 1990b, submitted to **Mon Not R ast Soc**

Stasinska G & Tylenda R, 1990, **Astr Astrophys** in press; and this conference

Steigman G, 1989, **Proc II International Symposium for the 4th Family of Quarks and Leptons**

Terasawa N & Sato K, 1990, **Astrophys J Let,** 362, L47

Vilchez J M & Pagel B E J, 1988, **Mon Not R astr Soc,** 231, 257

Webb J K, Carswell R F, Irwin M J & Penston M V, 1991, **Mon Not R astr Soc,** in press

Wheeler J C, Sneden C & Truran J W, 1989, **Ann Rev Astr Astrophys,** 27, 279

Woosley S E, Hartmann D H, Hoffman R D & Haxton W C, 1990, **Astrophys J,** 356, 272

Yang J, Turner M S, Steigman G, Schramm D N & Olive K A, 1984, **Astrophys J,** 281, 493

Index